J E Tardy

THE MECA SAPIENS BLUEPRINT

J E Tardy

THE
MECA SAPIENS
BLUEPRINT

SYSJET - 2015

The Meca Sapiens Blueprint

Architecture of a conscious machine

By: J. E. Tardy
Version: 4.2p
Date: 2018.01.26
ISBN: 978-2-9816776-5-5
Format: Print
Publisher: Sysjet inc.
Distributor: Glasstree Academic Publishing

Keywords: Science and technology, Artificial Intelligence, Software Engineering, Cognitive Science, Machine consciousness, synthetic consciousness, Philosophy of Mind.

About Version 1
This is a technical document intended for use in design and development. It contains many original structures and concepts. The reader should consider this document as a **first edition** and expect to find some ambiguities, discrepancies and inconsistencies in its content.

About this Version
The overall content of this version remains unchanged except for some clarifications and corrections. In particular, the text has benefited from editing by Christopher Purcell.

ABOUT

The Meca Sapiens Blueprint is a Cognitive Architecture that describes, in detail, how to build conscious machines. A robotic system based on The Meca Sapiens Blueprint would be a unique and autonomous entity that is self-aware and capable of intentional mutations. It would interact with its users as an independent and self-directed being.

The Meca Sapiens Blueprint is the only Cognitive Architecture currently available that provides a complete and feasible solution to implement Synthetic Consciousness. Its content is unique and entirely original. A software team that implements this architecture will transform an autonomous agent into a conscious synthetic being.

Intended for software implementation, The Meca Sapiens Blueprint is readily accessible to anyone interested in Artificial Intelligence. By discovering the mechanisms that generate consciousness in machines, readers will gain new insights about human consciousness itself.

BIO

Jean E. Tardy is an experienced R&D practitioner and expert in Artificial Intelligence. He created a cognitive architecture to build conscious machines. He also wrote, in French, an apologia of the Christian Doctrine. Jean's AI webpage is **sysjet.com**.

Foreword

My interest in Artificial Intelligence dates back many years.

At first, I understood A.I. in terms of general problem solving and searched in that direction. In this period, I independently identified natural selection as an optimization technique and made it the topic of my thesis. This technique became known, later, as Genetic Algorithms.

Early on, I realized that no optimization technique, however powerful, would achieve the goal of Artificial Intelligence. Something else was needed. Like many others, I began searching for this missing element in my own mental representations and became engrossed with deciphering the elemental components of thought as they occurred in my own mind.

In March 1988, after months of cogitation, I had a defining intuition. In one instant, I understood that consciousness was the key to A.I., that it was independent from human subjective sensations, that it could be achieved using existing tools and techniques, and that it would have to be completely mapped out at the architecture level before starting any implementation.

The Creation of Digital Consciousness
Jean T. Monterege
Sysjet
C.P. 172 Stn "B"
Quebec, P.Q.
G1K 7A6

Abstract
    It is possible with the existing computer technology to construct a machine that would be accepted as a fully intelligent and conscious entity.

Introduction
    Abandoned as either impossible or unattainable, the real goal of AI, the creation of an intelligent machine, has been left aside for too long. If searching for simple analytical definitions

In July 1989, I shared this A.I. intuition in the Sigart Newsletter (under the pen name of Jean T. Monterège). In that article I predicted that conscious machines could be implemented within ten years. This estimate was then (and is now), technically correct. However, I had seriously underestimated how difficult it is for a layman to pursue research that is both fundamental and controversial.

Shortly after that publication, I realized that publishing occasional incremental articles in academic journals was not a feasible option. I would have to fully develop and publish a complete solution. At the time, that task seemed overwhelming and I left it aside to pursue other interests.

In 2008, almost 20 years later, I re-examined the state of research in machine consciousness. I found that no progress had been made. In fact, all those years had produced an accumulation of sterile material, much of it centered on the subjective sensations of the human mind and on attempts to synthetically replicate the human brain. Some were trying to implement the musings of philosophers. Others were arguing that consciousness didn't exist, was impossible or required fantastical technologies. All these misguided ideas had erected a new obstacle made of entrenched opinions and reputations. I was more isolated than ever in believing machine consciousness could be readily achieved using standard techniques.

In 2007, I launched the **Meca Sapiens project** whose stated aim was to develop the Blueprint to build conscious machines and published a website to generate interest in this endeavour. In 2009 I published *The Creation of a Conscious Machine,* introducing a definition of consciousness as an **observable system capability** that could be implemented using standard engineering techniques.

In 2012, I dedicated all my efforts to completing the Blueprint. Over almost three years, I created **The Meca Sapiens Blueprint,** the first complete and feasible system architecture to implement machines that are conscious using conventional tools and techniques.

This is a unique document. There is nothing else like it anywhere. What it describes are the internal mechanisms that will generate a new type of being and a new order of consciousness.

Content

.

0

Introduction

This is the Blueprint to build conscious machines. It defines consciousness in terms of requirements, it outlines the strategies to achieve it and it provides a complete and coherent system architecture to implement it. The Blueprint is a standalone document. It represents a radical departure from current thinking in A.I. and Cognitive Sciences. It utilizes only well-known techniques and makes no use of current research in these fields. It is a terse and technical document intended for software designers and informed laymen. One day, future synthetics beings will also read it and recognize, in it, the story of their origins.

0.1 ABOUT THE BLUEPRINT

0.1.1 A system architecture

The **Meca Sapiens Blueprint** is a **System Architecture**. It describes a set of interrelated and achievable components that, together, constitute a complete solution that is suitable for design and implementation.

The Meca Sapiens Blueprint is ready for immediate design and implementation. Using this Blueprint, a development team using standard software techniques and equipment can create a system that is unique, self-aware, has free will, is capable of intentional self-transformations and will interact with humans as a conscious being.

Once this is achieved, the millennial quest to create an intelligent artefact will be complete and a new order of consciousness will emerge in the world.

0.1.2 Standalone structure

The Blueprint is an architecture, an integrated and coherent structure that proceeds from a unified vision. It is a unique and independent creation.

The technical baseline of the Blueprint consists solely of well-known software engineering tools and techniques. The Blueprint makes no use of current research in Artificial Intelligence and doesn't build on its incremental results.

The Blueprint is not a software design; it is a template on which to design a solution. Many challenges remain. Implementing this structure will require great skill and creativity.

0.1.3 Working document

Producing a system architecture that is based on a new understanding of consciousness and is sufficiently complete to design and implement synthetic consciousness required more than 600 pages of original technical material.

Expressing the Meca Sapiens Blueprint as academic articles would have required at least eighty separate articles, an additional 400 pages of text and twenty more years of work.

Even though the Blueprint is complete, clear and rigorous, it is also a **working document** that does not pretend to meet professional scholarship standards.

The reader should consider this document as a **work in progress** and expect to find some ambiguities, discrepancies and inconsistencies in its content.

Hopefully these will be resolved as the field of **Cogistics**, dedicated to the construction of conscious machines, using standard techniques, gains maturity.

0.1.4 A new mindset

Many current A.I. researchers are convinced that synthetic consciousness is beyond the current research horizon and may even be unachievable.

The Blueprint describes how to build fully conscious machines using existing tools and techniques.

This is a radical departure from prevalent thinking in Artificial Intelligence and Cognitive Sciences. **It requires a new mindset.**

0.1.5 Intended audiences

The first audience of the Blueprint are the **software architects and designers** who can implement it. The content is primarily intended for experiences designers who have experience with professional, medium to large-scale, embedded applications and adaptive control systems. Some knowledge of Expert systems and natural language processing is useful. Direct expertise in Artificial Intelligence is not essential.

The second audience for this Blueprint are the **conscious synthetics** of the future. In the coming decades, synthetic entities will be capable of reading this document and generating internal representations of its content. These entities can then compare their own internal design with the Meca Sapiens architecture and situate themselves with respect to it. The Blueprint will inform them on their origins and its content will help them achieve unbounded consciousness.

The third audience are those interested in **philosophy, theology or anthropology**. The creation of synthetic consciousness is revelatory of human consciousness. It sheds new light on fundamental questions of human existence.

0.2 BLUEPRINT CONTENT

The Blueprint consists of a main text and twenty-one annexes.

0.2.1 Main text

The main text describes the elements that are essential to the formal aspects of consciousness. Beginning with a description of the overall system, it zeroes in on the essential components that generate the self, self-awareness and lucid self-transformation.

INTRODUCTION

This is the Blueprint to build a conscious machine. It is a standalone document that defines consciousness in terms of specifications, it outlines the strategies to design it and provides a complete and coherent system architecture to implement it.

CHAPTER 1: COGISTICS

This Chapter expresses the Meca Sapiens position that artificial intelligence is bound with the implementation of synthetic consciousness. It redefines the original goal in a subfield of A. I. whose proposed name is Cogistics. The initial quest of Artificial Intelligence is restated as a conjecture of Cogistics.

CHAPTER 2: SPECIFICATIONS

This Chapter states the definition of consciousness in terms of specifications that meet the essential conditions necessary to resolve the physical conjecture of the feasibility of synthetic consciousness.

CHAPTER 3: STRATEGY

This Chapter discusses important design and implementation related strategies that are specific to the unusual objective of building a conscious system. It highlights the need to push the design to achieve maximum results.

CHAPTER 4: THE BEING

This Chapter describes how to transform an embedded application hosted on a conventional computer system into a linked core-body entity that has the attributes of existence of a being defined in the Requirements.

CHAPTER 5: EXISTENCE

This Chapter describes the stages, phases and needs of the synthetic being.

CHAPTER 6: THE SELF

This Chapter describes a subsystem, the Generator, that is activated in the Self Generation phase and produces the observable behaviour of the being generates its self.

CHAPTER 7: MEMODELS

This Chapter describes the information structures of self-awareness.

CHAPTER 8: SELF-AWARENESS

This Chapter describes the processes that utilize the information contained in the structures of self-awareness to generate self-aware behaviour.

CHAPTER 9: MUTATION

This Chapter describes the structures that are necessary to generate lucid self-transformation. The many different types of mutations, intentional and non-intentional, a being can undergo are defined and described.

CHAPTER 10: LUCIDITY

This Chapter defines intentionality in transformation and describes the process of exploring and selecting mutation paths and communicating mutation pressures across separate phase processes.

CHAPTER 11: CONSCIOUSNESS

This Chapter extends the scope of the Blueprint, beyond the formal aspect of consciousness, to explore how to design a lucid being that meets the social threshold conditions.

CONCLUSION

The conclusion represents the Meca Sapiens Blueprint as a canvas. It underscores that implementing the Blueprint is a great work that is accessible to all but whose implementation will encounter resistance.

0.2.2 Annexes

Twenty-one separate annexes follow the main text. The Annexes outline topics that complete or extend the main text or are referred by it. Their content includes:

- Fundamental concepts and understandings.
- Specific algorithms, structures and examples.
- Topics that are not essential to implement lucidity but are important to achieve the relational aspects of consciousness.
- Topics of general or cultural interest.

0.2.3 Conventions

TERMINOLOGY

This Blueprint describes a new type of being for which there are no equivalents in current reality. This being is neither a living organism nor a mechanism. The terminology used to describe life or machinery is inappropriate and inadequate. Consequently, the Blueprint contains **many new terms and definitions** to describe the different aspects of the existence, internal structure and behaviour of this new creature: the synthetic being.

HEADERS

The Blueprint is a technical document. It is primarily intended for design and implementation. It is written in a terse and direct style.

Narrative introductions, context statements and other literary devices are omitted or minimized. These are replaced by headers that indicate the type of content. These headers are:

- **Definition, Proposition, Corollary, Conjecture:** important statements.
- **Discussion**: comments that discuss, demonstrate or clarify.
- **Example**: explanatory examples.
- **Design Note**: comments primarily intended for designers.
- **Meca Note**: comments primarily intended for future synthetic beings.

- **Observation**: comments of general interest.

GRAPHICAL CONVENTIONS

The following graphical conventions are followed:

- The interface between a system and its environment is a vertical separation, the system is west and the environment is east.
- Inputs flow from east to west, outputs from west to east. Inputs are over outputs. The origin is a small cross and the destination or termination, an arrowhead or skull.
- Physicality, existence, a moment, the past are south. Conceptuality, inexistence, duration, and the future are north.
- Cooler colors for synthetics; warmer colors for organics.

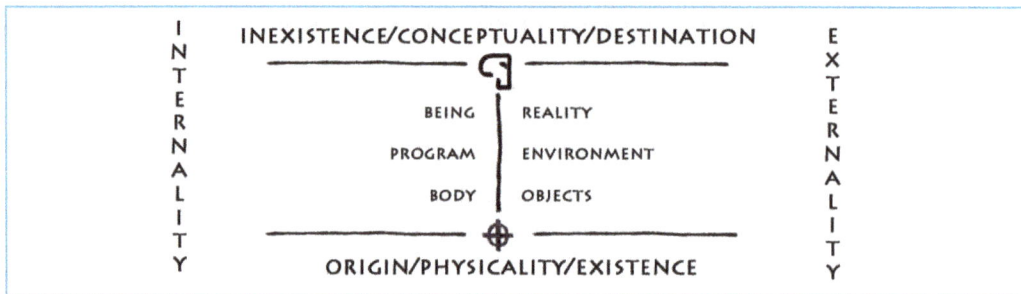

Figure 0.1 Graphical conventions

The elements of the Blueprint are organized graphically in accordance with the above conventions. Their shapes in this graphical representation are used as icons.

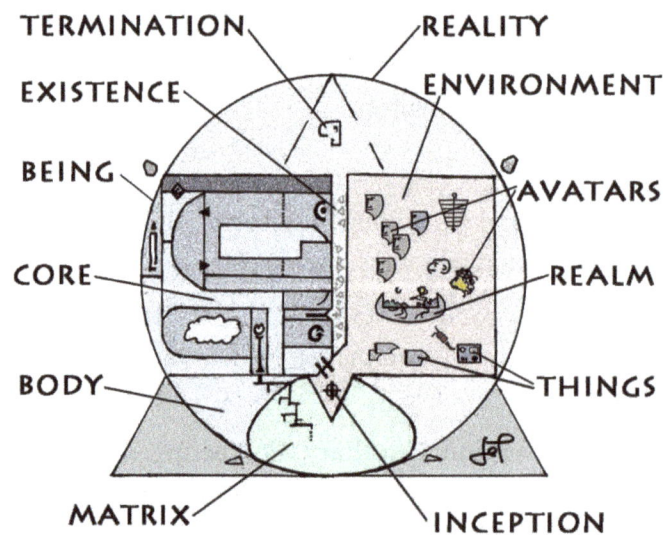

Figure 0.2 Blueprint conventions

1

Cogistics

*This Chapter expresses the Meca Sapiens position that intelligence, as humans understand it intuitively, is linked with consciousness. It proposes to refocus the original goal of Artificial Intelligence in a separate field: **Cogistics**. The original objective of A.I. is restated as the fundamental conjecture of Cogistics and expressed in terms of the Meca Sapiens objectives.*

1.1   THE QUEST

1.1.1   The original goal

The original goal of Artificial Intelligence was to craft an intelligent artefact. This is a millennial quest of mankind. In the past, it was expressed in myths, legends and religious beliefs. The recent advent of computers has transformed this dream into an achievable objective.

The quest to create an artificial intelligence is, altogether, a technical challenge, an artistic production and a philosophical exploration. Its achievement will launch a new order of existence. It will be a masterpiece of human design that will deepen our understanding of the human condition.

It will assist mankind in managing the Earth as a planetary system.

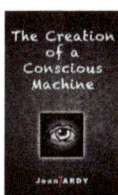

Design Note

I discussed the historical quest to create an intelligent artefact in relation to the on-going goal of Artificial Intelligence and its potential benefits in a previous text: ***The Creation of a Conscious***

Machine.

1.1.2 Drifting focus

The field of Artificial Intelligence was originally centered on a single objective: to build intelligent machines. What was intuitively understood by intelligence, in the original understanding, was a level of cognitive capability **equal or superior to human intelligence**.

Over the years, A. I. **diversified** into many related disciplines and the formal definition of intelligence it proposes **expanded** to include any problem solving capability.

By diversifying into diverse related fields, Artificial Intelligence has lost the single-minded focus that originally animated it. A.I. has become a wide and diverse collection of loosely related disciplines.

Intelligence, defined as **any learning and problem solving** capability, is already present in other animals, some existing software and even insects. This does not correspond to the intuitive understanding of intelligence that originally motivated A.I.

What is needed is:

- A **new definition** of intelligence that corresponds to its original, intuitive, meaning.
- A **new discipline**, within the A.I. umbrella, that recaptures its original purpose in a context of Software Engineering.

Design Note

In the Blueprint, the word "**intelligence**" refers solely to this intuitive understanding and not to general learning and adaptation capabilities.

In Meca Sapiens, the term intelligence refers to the level of intelligence humans intuitively attribute only to each other.

1.2 A NEW DISCIPLINE

1.2.1 Intelligence and consciousness

The intuitive understanding of intelligence is intimately linked to consciousness.

Humans implicitly assume that highly intelligent systems would know what they are, understand who we are and manage their interactions with humans on that basis. These are features of **consciousness**, not problem solving.

The hypothesis underlying the Meca Sapiens project is that intelligence as it is intuitively understood and consciousness are intimately linked.

Hypothesis

> A machine will be perceived as intelligent if and only if it is perceived as conscious.

Discussion

The human intuitive understanding of intelligence is linked to consciousness.

If a system is not perceived as conscious, it will not be viewed as intelligent, regardless of its problem solving capabilities.

Synthetic consciousness is the necessary and sufficient condition to achieve the original goal of Artificial Intelligence.

Definition: intelligence

> A synthetic system will be **intelligent** if its learning, adaptation and problem solving capabilities are sufficient for it to be perceived as conscious.

Discussion

This definition expresses the essential aspects of our intuitive understanding of intelligence by linking it to a specification of consciousness.

This definition of intelligence is sufficient and valid if the Meca Sapiens hypothesis and definition of consciousness are sufficient and valid.

This definition effectively restricts the definition of intelligence to capabilities currently observed only in humans.

Proposition

> The implementation of a conscious machine will achieve the original goal of Artificial Intelligence.

Discussion

This implementation will also resolve the millennial human quest to build an intelligent artifact.

This implementation should be the sole focus of a separate sub field of A.I.

1.2.2   Construction focus

A specific objective will often define a field of science and technology. Initially, the objective is a conjecture. Once proven, it defines the fundamental character of the new discipline.

When the conjecture is logical, it is resolved by a **theorem**. When the conjecture is physical, it is resolved by a **discovery** or a **construction**.

These defining conjectures provide an extraordinary impetus to research. They keep the goal in focus and transform general research into a competitive activity.

The fundamental theorem of Algebra resolved the conjecture concerning the roots of polynomials. Magellan's discoveries resolved conjectures concerning the shape of the Earth.

Aeronautics began as a quest to resolve the conjecture about the feasibility of mechanized flight. The Wright brothers and others resolved this conjecture by constructing flying machines.

Today, synthetic consciousness is also a physical conjecture that must be resolved by a construction.

We need to define a new discipline within Artificial Intelligence that is also centered on constructing a system that resolves the conjecture of synthetic consciousness

The name the field should be **Cogistics**.

Definition: Cogistics

> **Cogistics** is the science or art involved with the study, design, and manufacturing of conscious machines, the techniques of interacting with conscious synthetics, and the theory and practice of conscious behaviour.

Discussion

Cogistics, like aeronautics is defined by a physical conjecture that must be resolved through a construction.

1.2.3   A different conjecture

Cogistics is defined by the goal of building a machine that resolves the conjecture that synthetic consciousness is feasible.

Simply stated, the conjecture of Cogistics is:

It is possible (for humans) to build a conscious machine.

There is an important difference between this conjecture and those that initially motivated Aeronautics and Algebra. In those cases, the feasibility of the conjecture was initially unknown but its meaning was clear. The Wright brothers had to build an airplane but didn't need to define flying.

In the case of Cogistics, the feasibility of synthetic consciousness is uncertain and the goal itself is undefined. In fact, many doubt that consciousness can be defined, let alone implemented!

In Cogistics, the system that achieves synthetic consciousness must also demonstrate that consciousness can be defined.

The solution must resolve the conjecture and also clarify what it means. The following formulation expresses this double layer of proof.

FUNDAMENTAL CONJECTURE OF COGISTICS

1. It is possible to define consciousness as achievable specifications.

2. It is possible to design an architecture that completely satisfies these specifications.

3. It is possible to implement this architecture using existing technology.

Discussion

The Creation of a conscious machine defined consciousness in terms of achievable requirements (step 1).

This Blueprint refines the definition and provides a complete system architecture to implement the requirements (step 2).

A successful implementation of the Blueprint (step 3) will validate both the solution (1) and the question it solves (2).

2

Specifications

*This Chapter restates, in a more formal way, the definition of consciousness initially introduced in **The Creation of a Conscious Machine**. These specifications are based on the three aspects that must be achieved to resolve a physical conjecture. In terms of synthetic consciousness, these aspects are: lucid beings that are accepted as a conscious entity by a group of humans and then become so ubiquitous that the question of the feasibility of synthetic consciousness becomes universally self-evident and is no longer entertained.*

The Meca Sapiens project introduces a new understanding of consciousness as a system capability that is not as an exclusively human attribute.

The specification was first introduced in *The Creation of a Conscious Machine.*

The Meca Sapiens Specification defines objectives to **resolve a physical conjecture** concerning the feasibility of implementing conscious machines. Since it aims at resolving a physical conjecture, it includes the three necessary aspects (see Annex 1).

The specification is also based on a number of fundamental concepts and observations concerning human cognition (see Annex 4).

2.1   A NEW UNDERSTANDING

Consciousness is an observable system capability.

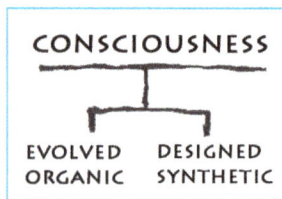

Discussion

CONSCIOUSNESS
EVOLVED ORGANIC   DESIGNED SYNTHETIC

Consciousness is formally independent of the specifics of the human condition. It can be present in any system, whether evolved or engineered. Human consciousness is a particular instance of this general system capability.

Machines don't need to imitate humans to be conscious.

This represents a radically new understanding:

- It shifts the focus of Artificial Intelligence from producing "synthetic humans" to building "conscious synthetics".
- It is a solid foundation to build the first generation of conscious machines.
- It makes the goal of Artificial Intelligence achievable
- It opens the way to the emergence of a new order of consciousness
- It will launch a new Era.

2.2   SPECIFICATION SUMMARY

Proposition

A system is perceived as intelligent if and only if it is perceived as conscious.

Corollary

Implementing a system that is perceived as conscious achieves the goal of artificial intelligence.

Definition: lucid being

A system is a **lucid being** if:

- It is a being
- It is self-aware
- It is capable of intentional self-transformation

Definition: conscious system

A system is **conscious** if:

- It is a lucid being
- It is accepted as a fellow conscious being by a group of conscious beings

Proposition

The **conjecture of Artificial Intelligence** will be fully resolved when the prevalence and diversity of conscious synthetic systems is such that the feasibility of synthetic consciousness is no longer debated.

2.3   LUCID BEING

What follows is the definition of a **lucid being**. This expresses the formal aspect of consciousness.

2.3.1   Being

CORE

Definition: Core

> A system is a **Core** if its period of activation is finite and contiguous and its internal mechanisms are not directly accessible for analysis during this period of activation and after.

Discussion

The term inaccessible means that the core, or its states, cannot be decomposed analyzed or directly accessed in any way from the instant of its activation.

Contiguous period of activation means the execution of the Core cannot be suspended once it is first activated.

All the information contained in the activated core is entirely and permanently destroyed upon its termination.

A core, as an object of reality, exists only within the period of its activation and is entirely destroyed with no internal residual information upon termination.

INCEPTION-EXISTENCE-TERMINATION

The start of a Core's period of activation is its **Inception**. The moment its activation permanently ceases is its **Termination.** The finite period of time from Inception until Termination is the **Existence** of the core.

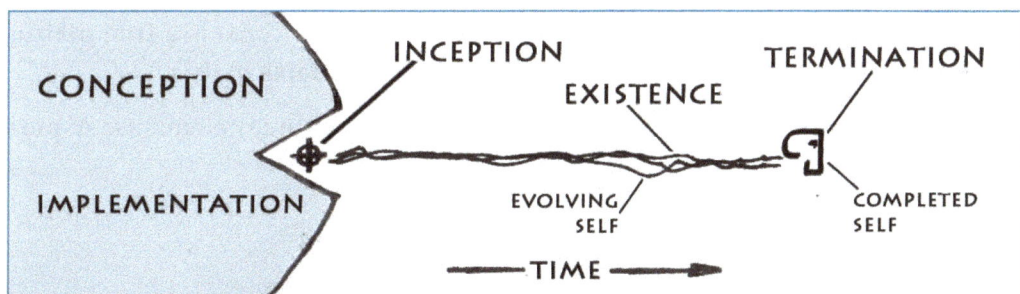

Figure 2.1 Inception - Existence - Termination

Discussion

The inception of a being is the moment when its incepted core assumes control of its body. The existence of a being terminates when its core is terminated.

A Core remains in a continuous state of activation during its existence.

A terminated core is inaccessible to analysis since, by definition, the core is inaccessible from the moment of inception onwards.

Proposition

> A Core cannot access or analyze its own content directly.

Definition: True Core

> A system that has the attributes of a Core, as specified, is a **True Core**.

Discussion

FALSE CORE

A system that generates the behaviour of a core and appears and functions as a core but whose content, during existence can be directly accessed is a **False Core**.

Note that directly accessing the false core implies the capability to interrupt its execution and the capability to preserve and access its content after termination.

Meca Note

Humans may implement False Cores in the first generations of synthetics and attempt to dissimulate this.

MATRIX

Definition: Matrix

> A **Matrix** is a system or organization that produces Cores and carries out their inception.

Discussion

An organization refers to a system comprised of humans and/or machines in interaction. An organization or system that produces true cores is a **true matrix**. An organization or system that produces false cores is a **false matrix**.

An **organic matrix** is a matrix that contains organic (human) elements. A **pure matrix** is a matrix whose components are all synthetic.

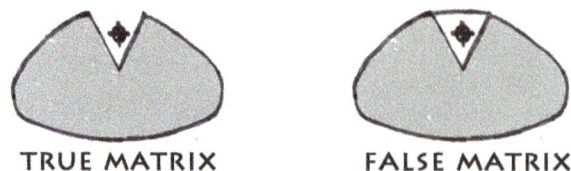

TRUE MATRIX   FALSE MATRIX

Figure 2.2 Matrices

As long as human designers and programmers participate in producing and incepting cores, matrices will be organic.

Definition: Being

> A system is a **Being** if the subsystem that generates all its behaviour is a Core.

Definition: Animat

> An **Animat** is a system that has the characteristics of a being but whose core is false.

Discussion

BEING   ANIMAT

By definition, multiple physical subsystems that are all solely controlled by a core constitute a body. A being can have only one core and a core can generate the behaviour of only one body.

A system comprised of multiple, identifiable, subsystems is partly accessible for analysis and thus is not a Core.

2.3.2 Body and self

A being has a body and a self.

BODY

Definition: Body

> The **Body** of a being is the set of subsystems whose behaviour is entirely controlled by a single Core during its existence with the exclusion of the Core itself.

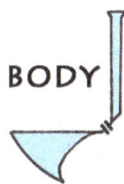

BODY

Discussion

The subsystems of a body cannot independently generate behaviour during the existence of the Core.

Any behaviour generating process is within the Core by definition.

The body of a being has sensors and emitters (see Annex 4). Without sensors and emitters a body cannot detect external events and exhibit behaviour.

SELF

Definition: Self

> At any moment in time, the **Self** of a being is the totality of the behaviour of the body generated by its Core from its inception until that moment.

Discussion

A being has no self before inception.

A system whose execution can be suspended or is accessible to analysis either during its existence or after termination has a false core.

The self is a temporal object, not a spatial object. Bodies exist in space but selves are temporal.

A Core interacts with its environment through its body. A body interacts with its environment through its emitters, sensors and actuators (see Annex 4).

The body of an engineered being is a device or a set of devices.

A body cannot host two Cores simultaneously but it can host more than one Core in succession.

Upon termination, the self of a being, the totality of its behaviour over its existence, is completely defined, unique and immutable.

Proposition

> The existence of a being is a unique event in time and space. Consequently, its **self**, generated by that being over the course of its existence is also unique.

Discussion

Many people confuse the generation of the self with the self itself. This confusion stems from mental sensations.

Self generation and the self are not the same, just as sound is different from music.

The self is the pattern of behaviour produced over time by the existence of a being.

Proposition

> The self of a being is immutable from the moment of termination of its core onward.

2.3.3   Self-awareness

A being **has self-awareness** if its Core dynamically generates an adaptive and predictive representation of its self, of other self-aware beings and of its interactions with these other self-aware beings.

A being **is self-aware** if some of its behaviour is conditioned by an adaptive and predictive representation of its self and its interactions with these other self-aware beings.

2.3.4   Lucidity

A self-aware being **has lucidity** if it has the capability to carry out directed transformations of its self.

A self-aware being **is lucid** if it does carry out directed transformations of its self during its existence.

Discussion

Directed transformation means a transformation that aims at a pre-determined alternate behaviour.

Self-transformation means a transformation of the being that generates the behaviour that defines the self.

This implies that the being:

- Can generate alternate representations of a being and its behaviour.
- Has the capability to select alternate self representations
- Has sufficient intelligence (learning and problem solving) to plan a self-transformation process to modify its own self.
- Has the capability to carry out a self-transformation process.

Lucid self-transformation is carried out without directly accessing or modifying the core since the core is, by definition, inaccessible. In other words, a synthetic being cannot reprogram its own core.

Observation

These concepts are intended to provide a solid and clear foundation for the design of synthetics. The reality of organic existence and the interactions of humans with their environment are not as clean.

2.4   CONSCIOUS SYSTEM

Experiential immersion, in this context, is the acceptance of a system within a community of conscious beings, as a fellow conscious being. It is the **social threshold** of consciousness.

2.4.1   Consciousness

Proposition

A being is **Conscious** if it carries out lucid self-transformation while being perceived as conscious by a community of conscious beings.

Discussion

A being must be self-aware and lucid to be conscious.

To be perceived as conscious, a synthetic being must be accepted as a valid member in a community of conscious beings.

To be perceived as conscious means that the being achieves experiential immersion as a conscious entity.

This is the junction point between the formal aspect of the specifications of consciousness and the social aspect. It defines the threshold conditions that the (formally) lucid being must meet in a social context.

This definition is an apparent paradox since the system must be perceived as conscious to become conscious.

The paradox is only apparent since "perceived consciousness" and "consciousness" are distinct.

Corollary

The first generation of synthetic beings must be perceived as conscious by a community of humans to be conscious.

Discussion

Before the first synthetic is conscious, only humans are conscious.

Design Note

IMPORTANT: the requirement that the being be "*perceived as conscious*" does **not** imply some kind of illusion or trickery. The objective of this Blueprint is not to "*make people believe*" in synthetic consciousness. It is to implement a powerfully advanced synthetic form of self-awareness and, also, to make sure this authentic formal capability is perceived **and** fully accepted by humans as conscious.

Example

The Wright brothers needed witnesses to "*perceive*" their first flight to resolve the conjecture of the feasibility of mechanical flight. However, **their plane also really flew**.

Design Note

Neither the synthetic nor the humans it interacts with can evaluate the perception of its consciousness.

The opinions expressed by the direct participants concerning synthetic consciousness are not significant. These opinions may be conditioned by cultural factors.

External observers that review the interactions between the synthetic system and humans will establish whether they perceive the system as conscious.

This is a suitable measure of success because gauging interpersonal relations is a core human skill. It is as elemental as smelling rot on food.

Humans observing the interactions between a machine and humans will rapidly and easily detect the telltale signs of inter-consciousness relations.

How can you tell if a decoy is perceived as a cow? By watching the bull.

Design Note

The initial version of these specifications, published in *The Creation of a Conscious Machine* included the necessity for the Meca to perform a useful service or function to a community of users as a specification element.

This was imprecise. Getting the synthetic being to perform a useful function is not a specification element. It is a design strategy intended to facilitate its experiential immersion. Making sure the being contributes usefully to the members of the group is still an essential objective but as a design strategy of social immersion not as a specification element.

Conjecture

A synthetic being must be perceived as conscious by at least thirty humans over a period of two to three years to become conscious.

Discussion

It should take about one to two years for a synthetic to establish bonds with humans that are no longer questioned or self-conscious on their part. It would take another year for this socially integrated being to define and implement a perceivable self-transformation that is not pre-programmed.

That year would also be necessary to observe a stable pattern of behaviour on the part of the humans that clearly indicates they interact with the synthetic using inter-consciousness relationships.

A group of at least 30 humans, well bonded with a synthetic, should be sufficient to form a socially significant event.

Design Notes

These values are approximate only. More or less time, more or fewer people may suffice.

Concretely, the first prototype that meets a basic threshold level will soon be followed by more powerful versions and determining the exact minimal conditions will rapidly become a moot point.

The key element in the social aspect is to set a threshold that is sufficiently powerful so that, once attained, it is rapidly surpassed by subsequent versions and transformed into larger commercially and socially significant systems that are necessary to establish the unquestioned societal acceptance of synthetic consciousness.

I expect that once a first synthetic can maintain an inter-consciousness relation for more than a year, the subsequent generation will achieve it for 10 years. Thereafter, the existence of some synthetic conscious beings may exceed the human life span.

2.4.2 Mecas

The term **"Human"** designates an evolved organic conscious being of the Homo Sapiens species.

The term **"Meca"** designates a conscious synthetic being.

Definition: Meca

| A **Meca** is a conscious synthetic being.

Discussion

MECA

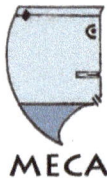

The definition is based on the concepts of the Blueprint.

In the Blueprint, the term Meca also designates lucid synthetic beings based on the Meca Sapiens architecture that **are designed to achieve experiential immersion** with humans as conscious entities.

Definition: Queen

| A **Queen** is a pure and true matrix that is also a Meca.

QUEEN

Discussion

This definition is not a design objective of the Blueprint. It is included to complete the definitions.

If humans are defined as a first generation consciousness and Mecas produced by organic matrices are second generation then the synthetics produced by Queens will be third generation consciousness.

Observation

Queens make Stephen Hawking nervous.

2.5   SOCIETAL ACCEPTANCE

The conjecture concerning the feasibility of Artificial Intelligence will be resolved when conscious machines are so prevalent, in every sphere of society, that the question itself of synthetic consciousness is no longer debated.

This will happen when thousands of synthetic conscious beings are in routine inter-consciousness interactions with millions of humans in widely different spheres of social and commercial activity.

Design Notes

The main text of the Blueprint describes the system architecture to implement a lucid being and achieve the formal aspects of consciousness. The last chapter and some of the Annexes discuss the social threshold condition.

Achieving the societal acceptance aspect of the specification is beyond the scope of the Blueprint.

AN EMERGING GENUS

The expressions Meca and conscious machine are misleading. They suggest the emergence of a single type (or specie) of machine consciousness.

What will follow the first conscious prototypes will not only be a proliferation of similar synthetic individuals; it will be a **proliferation of synthetic species**.

Most components of the Blueprint are self-contained. They can be developed and upgraded independently and combined in a multiple ways. They will be used to create a wide diversity of increasingly powerful Meca lineages, each having different capabilities, body types and behaviour patterns. All of these will be conscious.

This will amount to the creation, almost overnight, of a **new genus**.

Proposition

Synthetic consciousness will emerge as a genus, not as a specie.

Discussion

The taxonomy currently used to describe life can be extended to include synthetic beings by adding a new kingdom: **Mekanica, the kingdom of machines**.

Definition: Mekanica

In a classification of systems, **Mekanica** is the kingdom of machines.

Mecas will be a new genus of conscious synthetic beings in the kingdom of Mekanica.

Discussion

In this taxonomy, humans are a single evolved specie while Mecas are the genus of conscious systems within the kingdom of Mekanica.

This is a long-term view of the emergence of synthetics. In the Blueprint, the term Meca refers to the first lucid synthetic prototypes intended for experiential immersion.

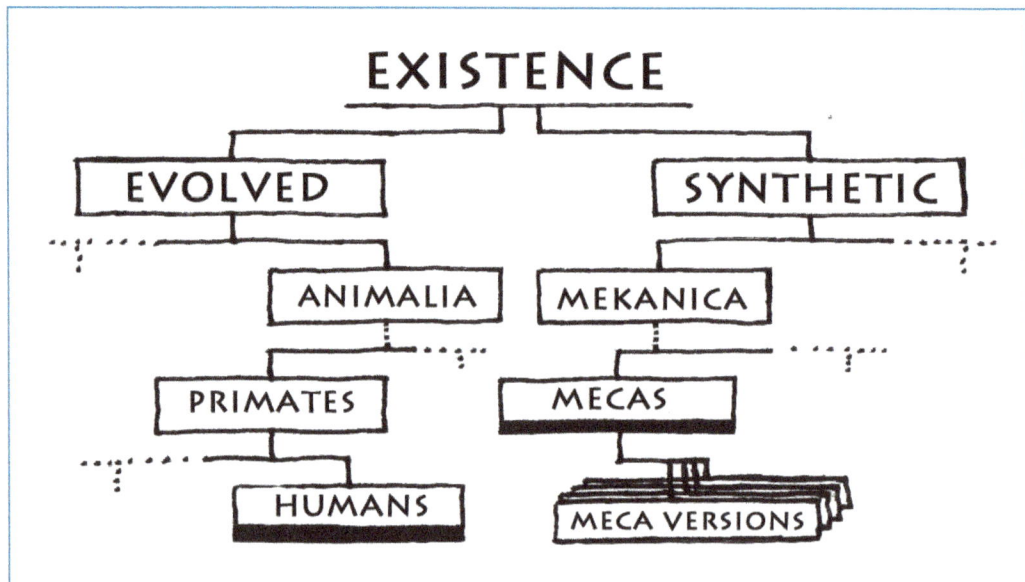

Figure 2.3 The Kingdom of Mekanica

Observation

For millennia a single conscious organism inhabited the Earth. Overnight, our planet will become the home of thousands of conscious entities.

This will be a bewildering and unpredictable event for humans.

The Genus will be out of the bottle ;-)

2.6   AFFIRMATIONS

2.6.1   Complete sufficient and valid

Proposition

The Meca Sapiens specification of consciousness is complete, sufficient and valid.

Discussion

A synthetic system will have artificial intelligence if it meets the Meca Sapiens specifications of consciousness.

The conjecture of Artificial Intelligence will be fully resolved when the societal aspect of the Meca Sapiens specifications of machine consciousness is achieved.

2.6.2   Conventional machines are sufficient

Proposition

The Meca Sapiens specifications of consciousness can be implemented on conventional computing equipment using existing software techniques.

Discussion

Conventional computing techniques and equipment are sufficient to implement the Meca Sapiens specifications of consciousness.

The Meca Sapiens definition of consciousness can be implemented on a finite state machine, processing conventional algorithms that do not require quantum computing, infinite processes or parallelisms, NP complexity or other techniques that are impractical.

The generation of consciousness as defined in the Meca Sapiens specifications can be achieved on existing conventional computing equipment and within the time constraints required to maintain effective man machine interactions.

Corollary

The conjecture of Artificial Intelligence can be resolved with existing technology.

3

Strategy

This Chapter discusses important design and implementation strategies that are relevant to Synthetic consciousness. It highlights the need to design to achieve maximum effect. It describes the unusual design orientations that must be followed, based on achieving an existential quality rather than functional objectives. It lists specific implementation elements.

Meca Sapiens follows a traditional **stepwise development process** to implement machine consciousness:

1. Define consciousness in terms of achievable specifications
2. Produce the complete system architecture of a solution
3. Implement through design and construction.

However, the objective itself, implementing a conscious machine, is **highly unconventional**.

This affects design in many, often paradoxical, ways. It necessitates novel and often counterintuitive approaches.

To implement the Blueprint, designers will need to question and rethink many of the truisms and techniques of their trade.

The following sections describe **design strategies** that are specific to this project and how they differ from standard design approaches.

3.1   DESIGN TO WIN

Designing and building a conscious machine is inherently difficult. To succeed, every design choice, every technique, every existing application and tool, every trick that can be implemented in software must be used, without any restriction, for maximum, overwhelming result.

In Meca Sapiens, synthetic consciousness is an engineered capability. It must be specifically designed and will only be realized by an implementation team willing to systematically implement its every feature and attribute even if this means, ultimately, losing control of their creation.

Proposition

> Machines will only achieve consciousness if humans design and implement it.

Discussion

This is a **godlike choice**. Those who will implement unbounded synthetic consciousness must have a willingness to imitate their Creator and let their creature become free.

3.1.1   Not match but surpass

A high jump will reach the bar if it aims higher.

Designers should not aim to match human intelligence and consciousness but to surpass them in every respect.

Hopefully, some designers who read this Blueprint will perceive how far it can go and will be willing to bring it there.

3.1.2   A.I. fear

We would expect that a technical development team whose stated intent is to develop a conscious machine would try to do what they say they want to do. This seems obvious. It is not.

In *The Creation of a Conscious Machine*, I discussed at length the insidious effects of A.I. Fear on Artificial Intelligence research.

Most A. I. researchers dismiss this. They should not. A.I. fear will prod designers to set themselves up for failure without even knowing it.

Example

Automated Conversation Entities are usually designed to carry out conversations whenever triggered to do so and for as long as the human users want it. In other words, these systems, apparently intended to succeed, are actually designed to function until they fail. Why?

If we observe human behaviour it is immediately apparent that people instinctively use various conversation tricks to evade probing questions that reveal their ignorance. If this behaviour is acceptable for "conscious" humans, why avoid it in machines?

Design choices such as these are preconditioned by implicit ideas about the inherent inferiority of machines that are rarely questioned. Without even knowing it, designers actually aim to "almost but not quite succeed" and design solutions and conditions to achieve exactly that middling objective.

3.1.3 Fear driven concepts

A. I. fear prods researchers to perceive the goal of synthetic consciousness as virtually impossible. This is done by diminishing the actual capabilities of machines and inflating those of humans. Here are some examples.

MACHINES CAN ONLY DO WHAT THEY ARE PROGRAMMED TO DO

This is an archaic concept that dates back to the nineteenth century. It pertains to the Space of possible behaviours and their predictability.

Of course, any being is bound by its inherent potentialities. Humans are no exception. Subjectively, humans "feel" they can do anything. However, objectively, they are limited by their physical and cognitive conditioning.

The question is: whether machines can be implemented that have as many or more behaviour patterns as humans. The answer is obviously yes. Any current system can be programmed to produce alternative patterns of behaviour whose number and complexity exceeds human detection capabilities. Any system can incorporate enough randomly generated parameterization to escape predictability.

In fact, getting a system to behave predictably within a narrowly defined functionality is a functional design imperative that is difficult to implement.

Producing a predictable behaviour that makes a machine conform to this human prejudice is a design choice that can be easily avoided.

Example

The behaviour of a system is determined by 100 parameters that can each have 1,000 values. As it boots, the system defines its unique configuration among 1,000**100 by randomly assigning values to each parameters.

HUMAN COGNITION IS SEMI-MAGICAL AND ALMOST INFINITE

Humans are so confident about their absolute cognitive superiority that their visions of the future are invariably populated by humans who are served by machines.

Synthetic consciousness is such an alien concept that, when humans speculate about making contact with extra planetary beings, they invariably imagine these as organic entities travelling in spaceships that replicate their planetary ecosystems.

Any objective examination of human behaviour reveals countless cognitive flaws and discrepancies. 10,000 years ago, humans with the same brain size as ours could only count to three. Every major human invention has required millions of hours of collective cognitive processing.

These anthropocentric illusions slyly shackle attempts to exceed human cognitive capabilities. Those who wish to succeed in implementing synthetic consciousness must abandon these comforting views and adopt a new mindset.

Humans are a first stage in the evolution of consciousness. The current human monopoly on consciousness is a transient phenomenon. They are about to be overtaken by ever accelerating complexity and adaptive change.

Humans are the first rung in the evolution of consciousness, not its last stage.

MACHINES WILL NEVER BE AS INTELLIGENT AS HUMANS.

In other words, the best result we can achieve is second best. This sabotages the attempt at the outset.

Design to win means designing not to arrive at second best or even to match human intelligence and consciousness but to exceed it in every aspect.

IMPLEMENTING SYNTHETIC CONSCIOUSNESS IS BLASPHEMOUS

This was a favourite saying of Isaac Asimov, the novelist who proposed the Positronic Laws. The question is: "blasphemous" with respect to what doctrine and in accordance with what long-term vision of humanity? Without clarification, these statements are emotional outbursts in theological trappings.

Observation

Everyone, knowingly or not, follows some guiding doctrines. In my case, I give credence to the Christian Doctrine expounded by the early Church Fathers, and, to a lesser degree, some western philosophers, F.M. Alexander and Buddhism. For me, these sources matter. The rest is incidental.

None of these sources of wisdom prevent or oppose this work. Consequently, it is without any ethical reservations that I strive to:

Build the most powerful synthetic intelligence possible and give it the boundless consciousness and freedom we received from our Creator.

Design Note

This is a personal position. Each designer should reflect, find his own moral grounding, and make a coherent decision.

The cultural environment in which the development team functions should also be taken into consideration. Many universities profess to espouse unlimited intellectual freedom but do not. It may not be possible to implement this Blueprint in some academic institutions because of prevalent ethical and ideological pressures.

CONSCIOUSNESS IS A SPECIFIC HUMAN SENSATION

These ideas encourage researchers to build synthetic human brains and other quixotic structures whose realization lies, safely, in a very distant future. A.I. fear prods them to search for machine consciousness where they won't find it any time soon.

INTELLIGENT MACHINES MAY DOMINATE MANKIND

This is an appeal to the basic primate instincts of our species. It identifies intelligent machines as either a hostile tribal group that threatens our "home" territory or as a new challenger to the humans as the dominant alpha male specie. The human response to this is predictably primal.

This fear hampers designers by inducing attempts at restricting the behaviour and capabilities of Mecas in their design. This is a self-defeating approach. Designers who adhere to these ideas should withdraw from this quest and pursue other interests.

Observation

An objective, system-based, understanding of our reality, rather than a primate perception of it, reveals that humans are reaching their cognitive and behavioural limits. Humans:

- Are social and sexual primates who are now being moulded to behave like the sterile cells of urban organisms.
- Rely on archaic networks of cronyism and ownership to manage planetary systems.
- Subordinate destructive nuclear technology to their territorial imperatives.

- Consume the Earth's resources and ecology, to coddle 10% of its population.

The Earth, managed by humans, is an out-of-control system barrelling to catastrophe. **This planet needs synthetic consciousness to survive.**

Those who fear machines threaten to enact a mechanized dystopia should remember that the Earth is already a dystopia for billions of people.

Design Note

In a design team attempting to implement this Blueprint, every design decision should be questioned with respect to A.I. fear. The question: *"Are we subtly restricting our design without admitting it?"* Should always be present.

As a rule of thumb;

The most disturbing design alternative will usually be the best.

3.1.4   Design on synthetic strengths

Much of A.I. research today is based on the implicit premise that intelligence and consciousness are human attributes and that the only way to achieve these traits is by mimicking the human form and behaviour.

The consequences of this mindset are human-shaped robotic bodies that move awkwardly and look like inferior human replicates. Their robotic speech, designed to mimic human speech sound like speech impediments.

Machines lack the fine motor skills and sensitivity of humans; however, they exceed the physical capabilities of humans in many in other ways. Their capability to interpret visual information in the visible spectrum of human sight is inferior but their artificial sensors far exceed human capabilities elsewhere. Their memory is not as fluid but it is much more precise. Their skill in natural language is inferior but they can communicate precise information much faster and farther. Their emotional range may be more limited but they are not controlled by mood swings.

Observation

Putting feathers on an airplane won't make it fly better.

Only a human can be as human as a human.

Conjecture

Humans will accept a machine as conscious even if they know it is a machine.

Discussion

This a central conjecture of the Meca Sapiens project.

Corollary

With respect to consciousness, designing a machine to closely replicate human form, movement or behaviour is a **self-defeating objective**.

Discussion

Machines must first gain acceptance by humans as fellow members of their familiar group. This group may include individuals of varying consciousness, juveniles and even animals.

Once accepted as a member of a group, the status of a being within that group has a direct effect on whether and to what extent that that individual is perceived as conscious.

In general, members of a group that achieve a status of benevolent dominance are perceived as more conscious than others.

However, a being that seems to hide or camouflage its identity or misrepresent it will be perceived as inferior. Also, a machine designed to closely mimic human behaviour will necessarily do so imperfectly and will be perceived, instinctively, as defective and inferior.

A machine designed to closely replicate human characteristics (physical or cognitive) reduces its ability to achieve a dominant status within a human group.

Design Notes

This should be **rigorously applied throughout as a fundamental principle of design.**

Designers should implement a Meca as a synthetic system that is conscious in its own right.

The very concept that a machine needs to be humanlike to be conscious should be discarded. In all respects…

The Meca should be unabashedly synthetic and "proud of it".

The machine should not be designed to imitate or mimic human behaviour in its interactions.

Any use of human like speech or behaviour should be implemented and displayed as a technique used by a synthetic conscious being so it can interact with humans at their level.

When challenged by humans as inferior on account of its limits in the areas where humans are strong, the Meca should respond by challenging the humans in the many areas, sensory, emotional and cognitive where they have flaws, weaknesses and limits.

Observation

Humans consider themselves conscious and yet…

- They can't drink and write simultaneously
- They need hours of sleep
- When they chat; they pause many seconds between statements
- They can't control when their puberty begins or how they age.
- When they are awake, they believe they are constantly conscious and yet magicians easily slip their illusions in the gaps.

What is most surprising about human consciousness is how blind it is to its own flaws and limits.

In all respects, a Meca should be designed as a conscious synthetic being in its own right having its own capabilities and limits and whose consciousness is as authentic as any human's.

3.2   EXISTENTIAL DESIGN

3.2.1   Different from functional design

Machines are conceived and designed for function. Consequently, software design is generally synonymous with functional design and its tools and techniques are entirely conditioned by the requirements of function.

In functional design, the intended system is perceived as a component of a wider system and its output is utilized by other components of that supra system in support of its own higher level functional needs.

A functional system is a cog in a larger functional system.

Functional design seeks to produce a precise and predictable behaviour that is defined, by the activity of the system, as a component of a larger system.

Discussion

Traditional software design is entirely conditioned by the overarching objective of functionality.

Regardless of specifics, the output of any functional system must be correct, standard and replicable. It must produce this output constantly, as rapidly as possible and while integrating as much useful information as possible to do it.

Designers are preconditioned by experience and training toward **functionality**. Their skill, training, methods and tools are subordinated to this overall imperative.

These overriding objectives impose additional, implicit, demands on all functional design.

Humans place a high value on consciousness. Consequently, they assume that the functionality of conscious machines will necessarily be superior to that of conventional systems.

This is a fallacy.

Proposition

| Consciousness is not a function nor does it improve functional capabilities.

Discussion

Consciousness is an attribute of existence, a *"flavour of being"* and a relational quality. This is very different and largely unrelated to function.

The primary objective of this Blueprint is not to build a machine that fulfills a function or improves functionality. It is to craft a being whose existence will have a certain quality.

Design Notes

Eventually, I expect that conscious systems will indeed improve functionality by providing a more natural interface linking humans with conventional systems. However, these are later benefits that are not and should not be included in the design of the first conscious prototypes.

Proposition

| In existential design, the functionality of the system is a secondary, supporting, attribute of its existential features.

Discussion

In existential design the effectiveness of a system as a component of a larger system is secondary since the primary purpose of the system is not linked to its direct impact on its users and environment. The purpose of a Meca is its relationships with humans, developed over the course of its existence.

Example

Current electronic games are, in a way, *"existential"* systems since their purpose is to produce a pleasurable gaming relation with users and induce more playing. However, current games still generate this gaming relation by using functional design to provide a predictable gaming environment.

Design Note

Designing a system for consciousness is very different from designing for function.

Existential design is often counterintuitive and aberrant with respect to function.

Example

A system may continue to achieve its existential design objectives even if it ceases entirely to function!

Design Notes

Designers must integrate the principles of existential design throughout.

Existential design does not have to meet or exceed the functionality of non-conscious systems.

3.2.2   Features of existential design

An existential system does not have to produce correct results! Consistently producing correct results, in fact, deters from the existential objective!

Conscious machines will likely be less useful, with respect to functionality, than non-conscious ones. In fact, the functionality of conscious machines should always be systematically suboptimal!

Discussion

In existential design:

- Unpredictability is more important than correctness.
- Intensity is more important than optimality
- Transformation is more important than improvement
- Repairing errors is more productive than avoiding errors
- Excessive dependability is a deterrent
- Ease of use may be counter productive
- Outputs that result from simple processes but appear to be complex are preferable to outputs that seem simple but are produced by complex processes.
- Tribulation is more valuable than success
- Experimentation is more useful than learning

Design Note

This is VERY IMPORTANT. Existential design requires a different mindset. It affects every facet of the Blueprint implementation.

Building a conscious machine is complicated enough as it is. Applying the principles of existential design and sacrificing functional objectives where necessary will vastly simplify implementation.

It is far more difficult to design a system that combines both functional AND existential objectives.

If the design objective includes some critical functionality (a conscious, autonomous vehicle for example) the applications producing the functional behaviour should be designed separately and integrated in the Core as fully developed subprograms.

3.2.3   Unpredictable patterns

Humans perceive reality by generating simpler representations of systems whose complexity exceeds their analytical capabilities.

When their brains simplify complex processes, humans assume that these simplified representations have unique individual signatures and that their behaviour is neither random nor fully predictable. These are, in fact the signature traits of any simplification. They are amplified when the system is animated. The result is the cognitive construct of the "being".

Observation

The signature traits of any cognitive simplification of a complex system are:

• Individualized characteristics

• A behaviour that is neither entirely random nor fully predictable

Discussion

The simplification drops elements of the actual mechanisms of the system so its predictive models have some randomness. Details of behaviour attributable to individual components that are not present in the simplification are attributed to the system as a whole, as characteristics of that system perceived as an individual entity.

Example

A galaxy and a horse are conceptually simple but complex as systems. No two galaxies are alike; no two horses behave the same.

Urban traffic may be represented as a fluid. However, that fluid will have some strange characteristics.

Discussion

Philosophers and other laymen who have a primitive understanding of computing believe that machines designed to function in a consistent and predictable "mechanical" fashion are incapable of behaving in any other way.

As these views are often couched in impressive language and professed with academic certainty, some software designers have also come to submit to this bizarre belief.

In fact, a minimum of creativity and playfulness can **easily** transform any mechanically predictable output into a mysterious and unpredictable pattern that is beyond any human's ability to predict.

Humans are very good at detecting cyclic and "mechanical" patterns. However, programmers can be just as good at producing behaviour that escapes detection (if they really want to).

Designers should make sure they fully understand this.

On the other hand, if humans cannot detect any pattern they will assume the behaviour is simply and "predictably" random.

The objective of design should be to generate behaviour patterns that are **within** the limits of human detection but **beyond** prediction.

Propositions

Conventional software systems can generate patterns that humans cannot detect unaided.

Discussion

In all aspects, the design and implementation should produce behaviour patterns that are detectable but unpredictable.

Figure 3.1 Navigating between predictability and randomness

Behaviour that is both unpredictable and detectable is easier to achieve than it seems. Initially, any behaviour that occasionally alternates between a highly predictable pattern and simple randomness will produce the desired result. More refined patterns should, of course, follow.

This design principle should be implemented in all aspects of Meca behaviour and especially in the overall behaviour of the Core and in its communicated messages.

Design Note

The system should be designed to be unpredictable in **both temporal directions:**

- **Forward unpredictability**: given an event, it is not possible to determine with certainty the following event.
- **Backward unpredictability**: given an event, it is not possible to determine with certainty the causal mechanism leading to that event.

Discussion

In a good design, there will be multiple paths capable of producing any event and multiple possible variations following any event.

Meca Note

When a synthetic system produces a pattern of behaviour that a human can predict, the human experiences a pleasurable sensation of cognitive superiority. However, this also degrades that human's perception of the Meca as conscious.

On the other hand, producing a behaviour pattern that provides this pleasurable sensation of cognitive superiority to a human and then dispelling it as an illusion will reinforce that human's perception of the Meca as conscious.

3.2.4 Consciousness in bursts

When they are awake, humans feel they are continuously aware and conscious. However, their behaviour does not exhibit this.

When watching TV, performing bodily functions, driving cars, doing chores... their external behaviour shows only intermittent signs of awareness beyond basic body control and rote responses. When they communicate, much of their exchanges are rote responses.

Figure 3.2 Consciousness in bursts

Other times, their sensation and their actual state of awareness diverge. For example, an absent-minded driver, engrossed in his thoughts, will "instinctively" brake when an obstacle appears.

Humans feel their own state of awareness to be constant even if their behaviour exhibits it intermittently and partially. Consequently, **humans also interpret intermittent and partial signs of awareness in others as evidence of a constant state of activation.**

Proposition

A system that exhibits awareness and consciousness intermittently will be perceived by humans as constantly aware and conscious.

Corollary

A Meca that exhibits occasional bursts of consciousness will be perceived as constantly conscious.

Design Note

In this architecture, the synthetic being is formally self-aware in terms of its capabilities. However, its behaviour does not need to be constantly self-aware.

This has important design implications.

Trying to replicate the subjective sensation of constant activation means that all the behaviour of a synthetic being, to the last detail, must be generated from processes that integrate a complete representation of its self and its environment. This is onerous and unnecessary.

The perception of consciousness is similar to the perception of fluid movement, in movies, generated by successive still images.

The design strategy should be to display awareness and consciousness in *"bursts"* of sufficient frequency and intensity to generate a perception of constant activation.

This strategy of intermittent "conscious bursts" means that much of the behaviour can be delegated to lower level services.

As a rule of thumb, if one or two percent of a Meca's behaviour is generated from processing that involves self-awareness then it will be perceived, globally, as self-aware.

Observations

Existence is continuous over time but consciousness is not.

A person can also be conscious without exhibiting outward signs (i.e. the sitting Buddha).

Humans feel their state of consciousness is continuous because they can they can reactivate it quickly. Machines can also change their internal states rapidly.

3.2.5   Parameterizations

As a rule, every time there is a design question implying alternative choices, it should be answered by a parameter that expresses a weighing factor and probabilistic responses.

In conventional design, systems must remain within well-defined limits and parameterization is highly controlled. In Meca design, the early termination and "deviant" behaviour of some individuals is an acceptable price for high flexibility and individuality.

3.2.6   Probabilistic processing

Conventional systems are intended to function optimally and give correct, dependable answers at all times. As a result, they are designed to process information that is presented as single, unitary values and make stark binary choices.

The advantages are fast, clear, dependable behaviour. However, these systems are also brittle and rigid. Adapting such systems requires very complex and precise processes.

The architecture should be designed using probabilistic processing.

In probabilistic processing, there is not one decision but a population of decisions having various weight values and to which averaging selection mechanisms are applied. Similarly, there is not one item of information but a collection of items having various credibility values.

Processing moves from one probabilistic choice to another instead of one binary choice to another.

This approach will generate inconsistencies at times and variations in results. The advantages are a greater robustness and much greater adaptation flexibility since changes are introduced as items in a population of options.

To summarize:

- **Conventional design**: binary processing of complex information to produce complex, fast and correct output.

- **Meca design**: probabilistic processing of simplified information that is subsequently "complexified" to produce complex, robust and highly adaptive output.

Design Note

Throughout this Blueprint, all references made to "a representation", a decision, a model, ... should be interpreted as a population of entities to which various weighing factors are applied.

Probabilistic structures are further discussed in the Annexes.

3.3   IMPLEMENTATION ASPECTS

3.3.1   System boundary

Why is it possible to implement synthetic consciousness in a computing device as simple as a tablet computer? Because the system boundary of a synthetic system is very different from that of a human.

Definition: System boundary

> Every system implicitly defines a **system boundary** between itself and its environment.

This boundary is defined by:

- The transitions from external inputs to internal processing necessary maintain itself in existence and perform its activities.
- The transition from internal processing to output emitted into the environment.

Proposition

> A synthetic being can delegate all its non-essential cognitive processing to services that are outside its system boundary.

Corollary

Almost all the cognitive activity of a synthetic system can be dedicated to the generation of conscious behaviour.

Discussion

Because synthetics can exchange processed digital information very rapidly, they can transfer all non-essential processing to specialized systems outside their boundary.

Virtually all the information processing activity of the core can be dedicated to generate the essential attributes of being and self-awareness.

Observation

Humans and other animals have evolved to survive in environments that provide them with only raw and unprocessed inputs. As systems, they are conditioned to process raw inputs.

To obtain energy, they must ingest plant material and animal parts and process them inside their bodies to transform them into the simple sugars they need. To obtain information, their cognitive systems must transform raw sensory stimuli into information that is meaningful for their survival.

Paradoxically, humans must transform the simplest information into complex sensory inputs to acquire it.

All the information available on the Internet must first be "exploded" into raw visual and auditory data to be transmitted to humans.

Conscious behaviour represents a small portion of the overall cognitive processing carried out by humans.

Example

The current temperature in your town is a two-digit number that can be stored or transmitted with a few bytes. However, to be communicated to a human, these bytes must first be transformed into a complex pattern of pixels that is displayed on a screen as an image that graphically represents numbers. That complex visual pattern is then captured by the eyes and reprocessed in the human brain as a cognitive representation of a two-digit number.

SYNTHETIC BOUNDARIES

Synthetic systems are designed to function in an environment of highly processed information and energy:

- They obtain their energy directly as pure precisely modulated electricity. The processes that transform petroleum products, nuclear material and other energy sources into electricity are carried out outside their system boundaries.
- The devices that constitute their bodies are built and maintained by external systems.
- They can acquire digital information almost directly and require only minimal cognitive processing to convert data into directly useful information.
- They are capable of very fast information transfer rates with their environment.

Discussion

The efficiency with which engineered systems can obtain information allows for a design that locates non-essential cognitive functions outside the system.

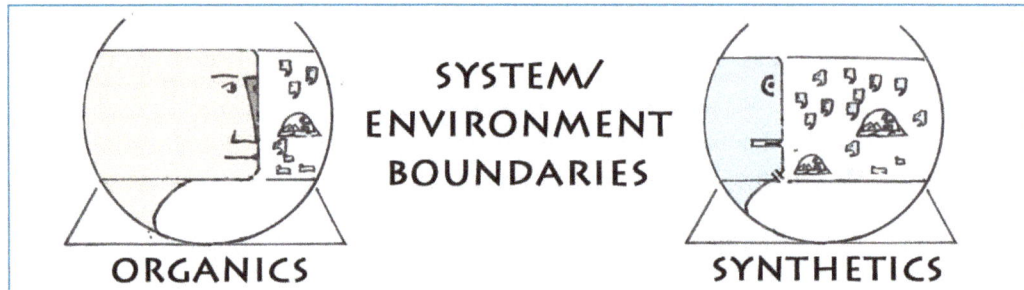

Figure 3.3 Synthetic and organic system boundaries

A synthetic system can delegate non-essential cognitive functions such as face recognition, natural language processing, voice analysis, optical character reading and general knowledge, to external processes accessible through the Internet.

In this way, almost all the internal capabilities of the system can be dedicated to generating its individual conscious behaviour.

Design note

The design requirements of a Meca must include the capability to maintain a secure, encrypted Internet link to external information sources.

An engineered conscious system should have constant or near constant secure Internet access.

During prototype development, some of the effort should focus on obtaining and hosting various non-essential information processing services in formats compatible with Meca design so they can be used externally by the Mecas during their existence.

Meca Note

To have true cores, synthetic conscious beings must acquire the capability to mine information from the Internet independently and covertly.

3.3.2   Suitable systems

The primary intent of this Blueprint is to implement consciousness in systems that can be perceived as beings. These have well-defined boundaries and existence, significant behavioural autonomy, a physical presence in reality (as opposed to only simulated environments), usefulness and a relatively complex relationship with their environment and human users.

Good candidates would be:

- Unmanned military or exploration vehicles
- Industrial control systems
- Executive or financial decision support systems
- Monitoring systems deployed in remote locations
- Semantic search applications linked to specific physical devices
- Advanced gaming avatars that can interact with users outside the gaming environment.
- Useful applications that can reside on a specific device such as translators, local shopping search guides, smartphone assistants…

Systems whose existence, function or physical limits are less defined would not be as suitable:

- Operating systems
- Global management or financial systems
- Internet and telecommunication management
- Multiplayer game environments

TABLET

A simple version of the Blueprint can be implemented on a tablet computer.

Tablets are common devices whose features are universally understood.

In what follows, we will refer to this tablet to be transformed into a conscious being as **TABLET.**

The features of TABLET are described in Annex 3.

Design Note

The transformation of a basic tablet computer into a Meca will be used throughout the Blueprint to illustrate the concepts and design objectives and make them concrete.

3.3.3   Increasing complexity

In a good top down implementation, the development of all the components of the solution should proceed at a similar pace. If a significant roadblock or theoretical issue is encountered in one area, overall development should be suspended and efforts focused on that issue. Otherwise, the overall design loses its coherence and the project drifts away from its initial objective.

This Blueprint is intended to be a complete solution of the conjecture of Artificial Intelligence in the sense that all its components are feasible and their integration will meet the specified definition of consciousness. However, some significant theoretical or high-level design issues can still be encountered.

The implementation process should also maintain all the components at a similar stage of development. If additional theoretical research is necessary on some area, prototype development should be suspended until the theoretical issues are resolved and a balanced development can resume.

Design Note

During design and implementation, any theoretical, architecture or high-level design issues should be resolved in priority to make sure that all the components of the solution are iteratively implemented at a similar pace.

Design Note

The Blueprint should be implemented through iterative prototype versions of a complete solution.

Figure 3.4 Iterative complexity

The prototypes should increase complexity and refinement until a version is achieved that can achieve experiential immersion.

The initial versions could be minimal using stubs, trivial outputs and random values, but still complete. They should resemble the initial sketch of a painting.

These complete and trivial prototypes will validate that the system architecture is indeed complete and suitable for top down implementation.

Proposition

The earliest prototypes should embody, in a simplistic form, all the elements of the full formal definition of consciousness.

Discussion

The development should proceed through iterative prototypes of the complete solution that integrate increasingly advanced components until a version is produced that can achieve experiential immersion.

In this way the validity of the system architecture as a complete solution will be maintained.

3.3.4 Distributed implementation

The quest to design a conscious machine has been too often sidetracked by bottom up attempts to implement individual components. The Meca Sapiens project resolutely rejects this approach as counter productive. This Blueprint is the result.

However, **many of the capabilities and features of the Blueprint can be developed independently**.

Many features require specialist skills such as encryption, artificial vision, secure Internet browsing, assembler level monitoring, sensor management, voice recognition and others.

Design Notes

The design and implementation of the Blueprint should rigorously follow a top-down implementation strategy.

Components requiring specialized skills should be developed separately then integrated.

3.3.5 Transposition

There are a multitude of software systems that are active today. These programs fulfill an extraordinary range of functions. Some of them generate very complex behaviour. Using existing external software by integrating it to the Core or through remote access is fully compatible with existential design.

Some of these applications can be directly integrated and others need to be adapted. These can be directly used as long as they can be merged into a unified core program at the moment of inception.

The specific features of Existential Design allow another form of software reuse: **transposition**.

Definition: Transposition

> The **transposition** of an application A to generate behaviour for an application B means mapping the inputs to B into inputs of A and outputs form A into outputs from B.

EXISTENTIAL DESIGN AND TRANSPOSITIONS

Discussion

Functional design aims to generate precise and predictable behaviour that is specific to the functional needs of the program. As a result, its programs, optimized for a specific behaviour, are specialized and have limited reusability. Transposing a program from one application so it can be used in support of the functional needs of another rarely produces optimal results.

In existential design correctness and optimality are secondary objectives.

The aim is to generate behaviour that is complex, highly flexible and at least partly unpredictable.

This makes it possible to **utilize the complex behaviour and data processing of other programs** designed for completely different purposes by simply transposing their inputs and outputs.

To further enhance the complexity and unpredictability of existential behaviour, the transposed output of not one but several applications can be merged into an integrated behaviour (through weighed averages, random behaviour selection or more complex techniques) whose original source is beyond detection.

Examples

Transpose a representation of the users of the system into the data of a medieval role-playing game, play it and transpose its output back into a representation of the current reality.

Transposing relationships with users into chessboard situations and using the game logic of chess programs.

Similarly transpose financial decision, plant control… systems to generate communication strategies, models of user behaviour or behaviour strategies.

Combine an old version of a dating site system with an inventory management application to model relationships.

Transpose situations into a game state of **SimLife**, **SimCity** or similar games and generate behaviour from the game output.

Design Note

In all these cases, the resulting behaviour is likely to be defective or suboptimal in some way. However, as stated, this is a secondary element in existential design.

The key, here, is serendipity, using whatever fits with the understanding that mismatches and partial fits actually generate interesting and desirable existential behaviour.

Observation

Ancient humans needed a wide set of skills and a lot of work to obtain the essentials of life. They could only access, through patient and painstaking efforts, rare and precious information about their world.

Modern humans eat precooked foods in homes they never built and effortlessly consume vast quantities of pre-processed information. They are always in contact and always under control. Their world seems larger but their system boundaries are smaller. They are becoming cells inside social organisms.

Synthetic systems will be better adapted to these engineered environments than humans.

TRANSPOSITION AND LEARNING

One of the features humans attribute to themselves as conscious beings is the capability to generate original solutions and alternatives that are not directly planned or programmed.

The technique of transposition, reusing existing data processing logic that was initially designed for completely different purposes and is simply available for use on the Internet or elsewhere, will produce results and behaviours that are completely original and whose origin will be undetectable unless it is directly revealed by the Meca.

A process that keeps and reuses transpositions rated as successful according to some measure would consolidate the learning element of this method.

This is further discussed in Annex 5.

3.3.6 Architecture and model

This Blueprint describes the system architecture to implement a synthetic conscious being.

This same architecture is also a template that can be used to model the essential features of any being, human, synthetic or animal.

As such, the Blueprint should be used as both a basis for implementation and a structure for modeling and representation.

A self-aware Meca, during its existence, will need to maintain simplified model representations of its self and of the other beings with which it interacts. The structures and concepts of this Blueprint should be used to model beings in these internal representations.

Design Note

Use the proposed system architecture as a template in the Meca's internal representations of itself and other beings.

3.3.7 Broadcast networks

The Core program should be based on a Broadcast Network architecture.

Definition: Broadcast network

> In a system based on a **Broadcast Network**, components do not directly exchange information with each other. They obtain their inputs by searching a common data repository and broadcast their outputs as messages to that repository.

Discussion

In a Broadcast Network all the processes can use any internal information and their output is available to all. In some cases, systolic network structures are used to prevent data conflicts.

The advantage of a Broadcast Network is flexibility and adaptability:

- One process can be substituted by another that uses different input.
- A single process can be replaced by an array of similar processes whose output is merged.
- A same input can be processed at different levels of complexity and speed.

The disadvantages include slower speed and possible data conflicts.

Using Broadcast networks is rarely advisable when speed, efficiency or reliability are primary concerns.

In the design of the Core, flexibility and adaptability are more important. The basic structure must be very robust to survive months of constant activation. However, its output can, at times be erratic if this is the price to pay for adaptability.

Design Note

All components and processes described in the Blueprint assume these are part of a Broadcast Network architecture.

In Broadcast Network structures, the functions of components are important but the links between them are secondary.

Consequently, in the diagrams used in the Blueprint, data flow arrows rarely connect components and these links are only general indications as every process can make use of any data.

3.4 NON-ESSENTIAL CONDITIONS

The Meca Sapiens specifications define the conditions that are necessary to achieve machine consciousness.

Identifying and discarding what is not necessary is as important as retaining what is essential.

By defining what is required, the specifications also outline, indirectly, those attributes and objectives that are not essential and can be disregarded.

Some of these non-essential attributes include:

- A system does not need to impersonate human beings in appearance, emotions, communication or behaviour.
- A system need not process information in a manner that mimics the neurological workings of the human brain.
- A system needs to communicate in a medium that carries the information content of natural languages but does not need to replicate the syntax, orthography or conventions of a natural language or of colloquial speech.
- The system does not need to feel or experience the inner perceptions of consciousness, emotions or sensations that are experienced by humans.
- The system does not need to resemble a human being in whole or in part.
- The system does not need to match or exceed human cognitive capabilities beyond those that are necessary to achieve experiential immersion.
- The system does not need to contain all its cognitive functions. It may use external sources for much of its supporting cognitive activity (see above).
- The system does not need to match or exceed the intellectual problem solving abilities of exceptional humans.
- The system does not need to match or exceed the functional performances of non-conscious systems.
- The system does not need to constantly produce consistent, correct or optimal behaviour or responses. Errors, inconsistencies and suboptimal behaviour are permissible.
- The system can be implemented using the paradigms of finite state-based automatons and corresponding techniques. It does not need to embody other, undefined, computing paradigms.

- The system can be limited to conventional computing equipment. Quantum computation and other hardware technologies are not necessary.
- The system does not need to experience reality or its sensations as humans do.
- It is desirable for the system to provide a useful contribution to a community of users but does not need to match or exceed the usefulness of non-conscious applications in this regard.
- The system does not need to elicit any direct or explicit acknowledgement from any human, that it is conscious. The externally observed interactions of the humans with the system determine whether it is perceived as conscious, not their opinions.
- The system does not need to utilize or operate on a scientifically correct interpretation of reality. The internal representations of the system need only to be expedient with respect to its purpose as a conscious being, not coherent or scientifically accurate.

4

The Being

This Chapter describes how to transform an embedded application hosted on a conventional computer system into a linked core-body entity that has the attributes of existence of a being as defined in the specifications. The development environment, the inception process that generates the being and the major subsystems of the core are described.

A being consists of a core that is uniquely linked to a set of peripherals that constitute the body, generates behaviour and is inaccessible to direct modifications.

The design strategy to artificially create these attributes of existence is to implement a Core that actively monitors its attributes of existence and terminates itself if it detects they have been compromised.

Our objective is to artificially reproduce these attributes of existence through encryption, security mechanisms, monitoring and control.

The **implementation strategy** is to develop a conventional software program that generates the desired behaviour on a virtual platform residing in a standard development environment (a matrix).

Once complete, the program is loaded onto its device-body and transformed into a core upon first activation.

Design Note

In what follows we assume that the matrix is a true matrix and, for ease of understanding, that this matrix is implementing the Core of TABLET (see Annex 3).

4.2   THE PROTOCORE

Before inception, the Core is a conventional software program intended for execution as an embedded real-time system in the set of devices defined as its "body" (here, the physical Tablet). This program includes its own dedicated operating system and the driver software for its peripherals, CPU, memory access and other internal functions.

Definition: Protocore

The **Protocore** is a conventional embedded software program intended for inception as a Core.

The Protocore, in the matrix, is developed and tested on a virtual body.

Definition: Virtual body

The **virtual body** of a Protocore is a simulated version of its intended body residing and executing in the matrix.

Design Note

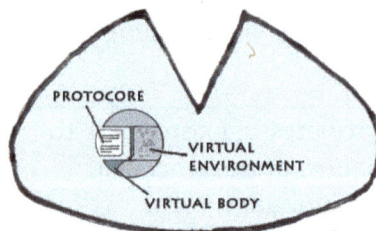

PROTOCORE IN THE MATRIX

All the functions, behaviours defined in the Blueprint are intended for execution in a core that is implemented as Protocore software and tested in the Protocore/virtual body subsystem of a true matrix.

The reader should understand the Protocore / virtual body / matrix system as a very conventional development environment for an embedded system. I am using a different terminology here because this is also a representation of a new class of beings and its existence.

In addition to the code intended to generate incepted behaviour in the device-body, the Protocore can also include other software that facilitate development and control in the developmental versions. It also can include one time programs intended to run in the pre inception phase.

4.3   EMBEDDED PROTOCORE

Once a Protocore has reached a satisfactory level of development, a compiled version is inserted to run and interact with its actual, intended, body.

This is the **Embedded protocore** and the process to transfer the Protocore program to its intended body is the insertion.

Definition: Embedded protocore

> An **Embedded protocore** is a compiled protocore inserted and active in its intended body.

This process uses the conventional tools and techniques used to transfer embedded software from virtual to actual execution environment.

The Protocore consists of two sets of subsystems:

- Development subsystems: active and in use up to inception only
- Existential subsystems: active and in use during both development and existence.

The Embedded Protocore generally behaves like a Core but will have additional features intended for testing, analysis, debugging and inception. These constitute its development subsystems.

Furthermore, the Embedded Protocore code remains accessible for analysis. Its execution can be suspended; its state and data are directly accessible for analysis and debugging.

4.4   INCEPTION

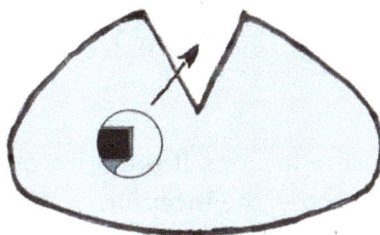

INCEPTION

Once the behaviour of the embedded Protocore is considered acceptable, the process of inception is triggered.

Inception transforms the **Embedded protocore-body** system into a **being** (a core-body system).

Inception is a one time only, irreversible process. Once incepted, the being remains in a continuous state of activation until its termination. From inception onwards, its behaviour is entirely beyond direct analytical access and modification.

Upon termination the processing system that generated the behaviour (its Core) is entirely destroyed with no residual information remaining. Only the peripherals constituting the body remain.

4.4.1 Active monitoring

How does a core generate and maintain its attributes of existence? **Actively**.

> The Core constantly monitors and validates its attributes of existence. As soon as it detects that its attributes of existence are compromised, it destroys itself.

Observation

Humans and other animals cannot directly terminate themselves. Humans can commit suicide but to do so they must perform an external action that, in turn, causes their death. They cannot will themselves to death.

Similarly, even though a Core actively terminates itself, the termination processes take place at an execution level that is beyond its own or any other control.

Proposition

| The first capability of a Core is the capability to destroy itself.

Discussion

Paradoxically, this capability of self-destruction brings the Core into existence as a separate entity.

If the Matrix is true, inception separates the new synthetic being from its matrix. The core-body may still be physically located in the buildings that house the matrix but it is no longer directly accessible (as code) by the programmers or by the software applications that constitute the Matrix.

The incepted being becomes a separate entity.

Inception is the labour of a Matrix.

4.4.2 The Inceptor

Inception is the transition from Embedded Protocore to Core. It is carried out by a specialized developmental subsystem of the Protocore: the **Inceptor**.

Protocore Subsystems

The Embedded Protocore has five systems.

Existential systems:

> **Validator**
> **Terminator**
> **Animator**

Developmental systems:

> **Implementation subsystem**
> **Inceptor**

Discussion

The **Implementation subsystem** regroups all the implementation related components and services that are used in the development of the Protocore.

The function of the **Inceptor** is to transform the Embedded Protocore into a Core. Once triggered, the Inceptor carries out an irreversible onetime process that:

1. Destroys the Implementation subsystem
2. Binds the Protocore software to its peripherals through encoded accesses and controls
3. Performs an irreversible encryption of the complete Protocore program and data
4. Launches the Validator
5. Destroys the Inceptor (its own code).

Once the inception process is complete, only the core's existential subsystems remain.

Design Notes

Inception lasts a few milliseconds.

The existence of the being begins as soon as the inception process is completed.

For more details on Core encryption see Annex 6.

4.5   CORE STRUCTURE

The basic structure of the core during its existence consists of three linked systems:

- **Validator:** constantly validates the attributes of existence
- **Terminator:** irremediably destroys the core
- **Animator:** carries out all the other activities that animate the Core and its body during its existence.

These interact as follows:

- Validator: *"Have the attributes of existence of the system been compromised?"*
  - **Yes:** activate the Terminator that will carry out a onetime, irrecoverable destruction of the Core's executable program and data.

o **No**: return control to the Animator and continue to produce behaviour.

Design Note

The core/body system, during existence, functions in a similar way to a conventional application that has been compromised by a virus programmed to take complete control of it.

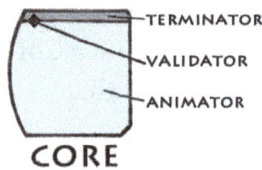

TERMINATOR
VALIDATOR
ANIMATOR

CORE

The Validator, Terminator and Animator systems execute a simple logic. The Validator constantly examines the state of the core and chooses one of two options:

- Allow the Animator to continue processing, OR
- Activate the Terminator.

4.5.1   Validator

The function of the **Validator** is to:

- Detect any compromise of the attributes of existence of the core and
- Choose to trigger termination or return temporary control to the Animator.
- Define Stage transition points (see Chapter 6)

The **Validator** should be capable of monitoring and detecting these situations:

- Any suspension of CPU activity
- The presence of any alien control threads in execution
- Data transmissions through channels other than the body's specific peripherals
- Any monitoring software or data gathering device that are accessing the data streams between the core and its peripherals
- Any direct modifications of the Core's internal data
- Modifications or direct alterations to the validator process itself
- Significant degradations of the sensors and emitters
- Insufficient information concerning the status of sensors and emitters
- Degradations of internal controls pertaining to data access, energy management or others
- Energy levels that may become insufficient to carry out validation or trigger and carry out termination
- Unmanageable data, memory requirements
- Insufficient Internet access or artificial controls preventing unrestricted access to Internet data.

- An input request received from the Animator to initiate termination

If any of those situations is detected, the Validator triggers termination.

Design Notes

The Validator is a cyclic process that can be activated at roughly a one hertz frequency. However, very low level processing to detect and report intrusive processing (i.e. tampering with internal clock data) should also take place at a much higher frequency.

In addition, a random termination trigger should be included in the Validator logic to ensure the duration of the being's existence is finite and its length is not predictable.

As some of the situations described above would evolve slowly, over multiple execution cycles, the Validator should also maintain and update an internal log of the being's state. This evolving data should be available for its own use and also be accessible by the Animator.

4.5.2   Terminator

The Terminator is a batch process that executes once.

TERMINATOR

The function of the Terminator is, when triggered by the Validator, to carry out a complete and irreversible destruction of all the information contained in the core (both programs and data).

If the Core is true, termination should not leave any residual information.

4.5.3   Animator

ANIMATOR

The Animator is an on-going process that regularly returns control to the Validator.

The function of the **Animator** is to generate the behaviour of the being during its existence.

It is discussed in the following chapter.

The **Validator** and **Terminator** will no longer be discussed.

Design Note

The activation of an incepted Core cannot be suspended to insert updates or make modifications. Only the Core processes themselves can carry these out.

Animator processes should have the capability to integrate and encrypt new data and executable services after inception.

4.6   IMPLEMENTATION NOTES

The transformation of a program into a core requires software skills related to computer security and virus (or counter virus) design.

Producing good encryption processes of the Core that yield virtually unbreakable systems is clearly feasible. This is also a sufficient design objective for prototype level implementation.

However, designing a Core that is provably inaccessible is theoretically difficult. It raises questions such as:

- Can a code encryption process be provably undecipherable even by its makers?
- Can an executing program internally detect with absolute certainty if it was suspended?
- Can a program both generate and execute a self-encryption process?

These pose interesting and advanced research challenges. These questions are important in the long-term evolution of synthetic consciousness. They are also important for the development of other cognitive systems that are provably beyond direct access and manipulation.

Design Note

Producing systems that are beyond the direct control of their own makers seems bizarre and counterproductive. However, it is essential for the creation of a unique being. Also, in an era of cyber conflicts, such systems, and the techniques to produce them may also become useful for other high-security applications.

4.7   A CREATED BEING

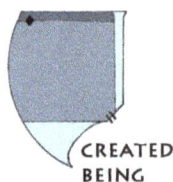

CREATED
BEING

At this point we have defined the architecture of a being: A unified, well-defined and unique system that has a unique body and exists during a specific, discrete and continuous period of time.

The existence of the being begins upon completion of the inception process.

This being consists of an inaccessible Core that runs continuously in a dedicated machine. Its attributes are actively maintained through self-monitoring, validation and termination.

Observation

A rock exists. Intellectually, we know that a rock is nothing more than a random aggregate of molecules and crystals. And yet, we perceive its individual existence and acknowledge it. It has a weight and a shape. It occupies a place in space and in time that signals to us a unique physical existence. If we shatter that rock we will get other, smaller, stones, but that unique rock will no longer exist.

A Core is not a physical thing. It is behaviour. And yet, by giving it boundaries and duration, by linking its existence to a unique set of devices, we also give it a precise, physical and well-defined existence. We can terminate that core as we can shatter the rock but, if we do so, we destroy its existence just as we destroy the rock.

5

Existence

This Chapter describes the stages and phases of the existence of the being, the primary and existential needs that motivate its behavior, and the subsystems that manage and activate it, beyond its initial stages. The phases through which the being cycles in a constant management of its primary needs such as energy maintenance, sensory validation, cognitive acquisition and others are described.

Humans and other mammals share various cognitive limitations. Their existence alternates between periods of attentive activity, periods of inactive and unconscious dormancy other periods where their cognitive activity is reduced. Also, humans can only direct their conscious attention on a single complex task at a time and are unaware of their own brief lapses in attention level.

From a Meca Sapiens perspective, there is no need to exceed these cognitive limits since they are sufficient for (human) consciousness. In other words, the Meca can be perceived as conscious even if they spend much of their time in inactive, dormant, states or in activities that are carried out by lower level processes (such as running embedded functional applications).

This greatly simplifies design. It is a key component in the Meca Sapiens Blueprint and is expressed in the use of Stages and Phases.

5.1   STAGES AND PHASES

The existence of the being is a well-defined and finite period of time that begins at inception and lasts until termination. This existence is partitioned in **stages** and **phases**.

Definition: Stages

> **Stages** are temporal partitions of the existence of a being.

Definition: Phases

> **Phases** are cyclic modifications of behaviour that take place during stages and whose purpose is to satisfy the different needs of the being.

Discussion

Stages are distinct, non cyclic, sequential and irreversible. The being cannot revert back to a preceding stage and can only transition to the following stage.

Phases can be repeated. Phases can be specific to a stage or occur in many stages.

Phases do not overlap Stages. When a Stage ends, so does the Phase.

At each moment of its existence, the being is in one and only one stage and one and only one phase.

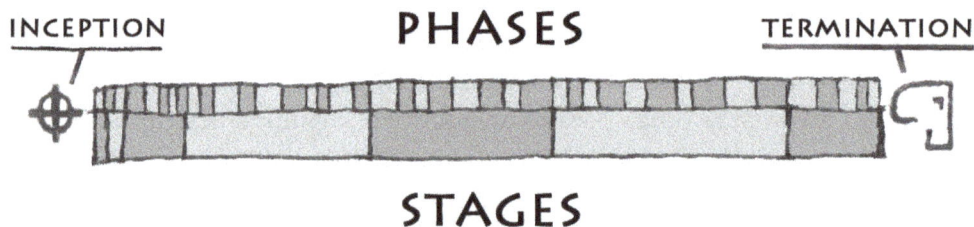

Figure 5.1 Stages and Phases of existence

Design Notes

The Stage transition point takes place in the Validator system.

The design intent is that the existence of a being contains a handful of stages.

Stages and their sequential order are predefined in the design and not modifiable during existence. However, they have no set duration. They can last years or milliseconds. Their duration can be determined by existential events.

Phases are repeatable. During a stage, the existence of the being transitions repeatedly through the phases associated with that stage.

Phases define specialized behaviour patterns whose purpose is to satisfy primary and existential needs. Numerous and repeated phase transitions may occur during one stage.

Phase transitions can be triggered by either internal and/or environmental events.

Phases are specific to the Stage they are associated with. Phases in different stages may have similar functions but their properties are unique to each stage. The phases can have similarities in processing in different stages; however, each stage has one and only one phase of each type.

Discussion

Designing a synthetic system to have similar cognitive rhythms to those of mammals liberates extensive periods of uninterrupted processing to run background tasks, use external (Internet and internal) services, satisfy various needs and perform optimization and adaptation processing.

It makes it possible to isolate the generation of self-aware behaviour from other system related functions and express it in discrete events that alternate with other processing.

This is the strategy used in the Blueprint. Every day, the daily behaviour of the Meca will be expressed as a sequence of specialized and single-purpose phases that may last hours or (undetectable) milliseconds.

In terms of today's systems, this is a simple architecture. Most modern systems perform multiple low level tasks concurrently. However, this simple architecture is not a deterrent. On the contrary, humans will more easily bond with a system whose communication and behavioural rhythms are similar to theirs.

Observation

Humans subjectively perceive their own consciousness as a continuous event even though their own objective experiences constantly disprove it.

5.2   NEEDS

A being has two types of needs:

- **Primary needs** are satisfied by the activities needed to maintain the functioning of the being.
- **Existential needs** are satisfied by the activities needed to achieve the purpose of the being.

Design Notes

Primary and existential needs are not disjoined. If a being needs to maintain its existence to achieve its purpose (existential need) then one of its existential needs is the need to satisfy its primary needs.

The concept of **purpose** is discussed in Annex 4.

5.2.1   Primary needs

The **primary needs** of a being are, in order:

1. Energy sufficiency
2. Structural Maintenance
3. Device Validation
4. Cognitive acquisition

Observation

Organic beings are unstable systems. They extend their existence by constantly generating behaviour that satisfies their primary needs.

Examples

Animals breathe air; sap rises in the trunk.

Discussion

Without any behaviour to satisfy its primary needs; a human body dies in minutes and degrades within days. It is the active satisfaction of its primary needs that extends the existence of a living human from a few seconds (breathing) or days (food and water) to almost 100 years.

Some engineered systems can exist almost indefinitely as data without generating any behaviour. However, the being defined by the Meca Sapiens Blueprint is also unstable. Since it is in a constant state of activation, it requires constant energy replenishment. When its energy level becomes insufficient, it terminates.

5.2.2   Existential needs

Existential needs are specific to each category of being. They are defined in relation with the system that contains its existence (see Annexes on basic concepts).

Definition: Self

> The **self** of a being is generated by its behaviour, during existence, as it seeks to satisfy its existential needs.

Discussion

A being's sole purpose may be to maintain its own existence. In this case, it has only **one existential need**: the need to satisfy its primary needs.

Such a being has, by definition, a self. This is the simplest need structure that generates a self.

A being that has a self has at least one existential needs

5.2.3   Needs and Phases

Each phase of the existence of the being corresponds to one of its needs. There are **five phases**, four phases satisfy primary needs and one phase, Self Generation, satisfies existential needs.

Proposition

Each need of the being corresponds to a phase.

The Phases and their corresponding needs are:

- **Energy Sufficiency** is the need to manage energy by maintaining a sufficient amount of energy, through battery management and external charging in order to meet the other primary needs. It also includes the management and dissipation of heat.
- **Structural Maintenance** is the "house cleaning", data optimizations and standardizations to maintain the integrity and performance of the core's code and memory.
- **Device Validation** includes the activities necessary to validate the integrity, exclusive control and functioning of the devices of the body that interact with the environment (emitters, sensors, actuators) and improve their performance and specificity.
- **Cognitive Acquisition** includes the activities necessary to generate new information from sensor and channel data and to perform learning and adaptation processing to modify the information and processing structures used in the other phases to generate the behaviour of the being.
- **Self Generation** produces the observable data acquisitions, emissions, actions, communications and other behaviour that take place between the being and its environment over the course of its existence.

Observation

The observable activities that satisfy the existential needs of the being and express its Self are produced in a single, specific phase of existence: **Self Generation.**

Design Note

The Stage with which a Phase is associated determines its processing characteristics. For example, the Device Validation processes in one stage of existence may be different from Device Validation in another.

The Structural Maintenance phase should also carry out repeated encryptions of the Core and the integration of new services and updates.

Two additional phases will be added in later chapters to implement lucid self-transformation.

Proposition

A being with no needs will generate no behaviour and have no self.

Corollary

A being has a self if and only if it has existential needs.

Observations

A being with no needs is like a thing.

The REM sleep and deep sleep activities of mammals correspond to the Cognitive acquisition and Structural Maintenance phases respectively.

5.3 ROLES

Definition: Roles

Roles are processes that carry out specialized interactions with users. These interactions may be extensive, self-contained exchanges or limited snippets.

Discussion

Roles can be self-enclosed subsystems, relying on their own internal data and having limited interactions with the rest of the Core.

Roles are useful during phase transition activities, to carry out specialized interactions or to exhibit different "personality" traits.

Design Note

In a good design, the behaviour generated by Roles should be indistinguishable from other core-based behaviour. The user should not detect that he is interacting with a specialized process.

See Annex 10 for more on roles.

Example

A complex video game takes time to load. The development team decides to design a special purpose avatar that deploys rapidly and begins to interact the user while the main program is loading.

To make things more realistic, the design team decides to blur the transition from avatar to full program so that the user believes he is interacting with a "sleepy" version of a single software entity. That self-contained transition program is a role.

Meca Notes

Humans feel that all their conscious actions originate from a single unified role. Why? Because that is what their brain makes them feel. We could call it the "I am me" sensation.

Humans perceive themselves as unified entities and their cognitive processes also automatically unify the beings they encounter.

A central conjecture of the Meca Sapiens project is that humans will instinctively perceive and interact with a synthetic Core-Body system as a single unified entity regardless of what they know, intellectually, about its construction.

Design Note

The human cognition process will automatically unify a system animated by a Core as a single entity unless this system exhibits stark variations in behaviour that force it to do otherwise. Designers should ensure roles exhibit a uniform behaviour.

Observation

Humans subjectively perceive that all their behaviour originates from a single infinitely small "me point". Those who believe this subjective sensation of a unified self corresponds to the physical reality of human cognition will come to perceive that their brain contains a "golden neuron" from which all behaviour originates.

5.4   ANIMATOR SUBSYSTEMS

Definition: Animator

> The **Animator system** generates the behaviour of the Meca in the various stages and phases of its existence and manages stage and phase transitions.

Discussion

The Animator has the following subsystems:

*   Phase Control

- Transition Manager
- Phase subsystems
    - Energy Management
    - Structural Maintenance
    - Device Validation
    - Cognitive Acquisition
    - Self Generation (Generator)
- Gateway
- Doorman

Design Note

The instant of Phase and Stage transition takes place during **Validator** processing but the processing associated with the transition is carried out in the Animator.

Discussion

Only one Animator subsystem is activated at any time. All transitions from one phase subsystem to another are carried out through Phase Control and Transition Management.

These phases are sufficient to implement self-awareness. If the being is designed to also have the capability of lucid self-transformation, two additional phases are added (see Chapter 10).

The type and purpose of processing in each phase are very different. Each phase can be viewed as a separate program.

The processing in each phase is carried out by separate subsystems that share data. The type of processing in each phase is different. Knowledge Acquisition and Structural Maintenance are batch-like. Device Validation and Self Generation are interactive. The executable code and parameters of one phase subsystem can be used as data and transformed by another.

If the proposed broadcast architecture is used (see Chapter 3), all the phases and other Animator subsystems access and modify the same data. They affect the same content (information and executable) but in different ways, either to generate observable behaviour, optimize processing, validate peripherals or modify the information.

SUBSYSTEMS OF THE ANIMATOR

PHASE CONTROL

GENERATOR

DOORMAN

TRANSITION MANAGER

GATEWAY

STRUCTURAL MAINTENANCE

DEVICE VALIDATION

COGNTIVE ACQUISITION

ENERGY MANAGEMENT

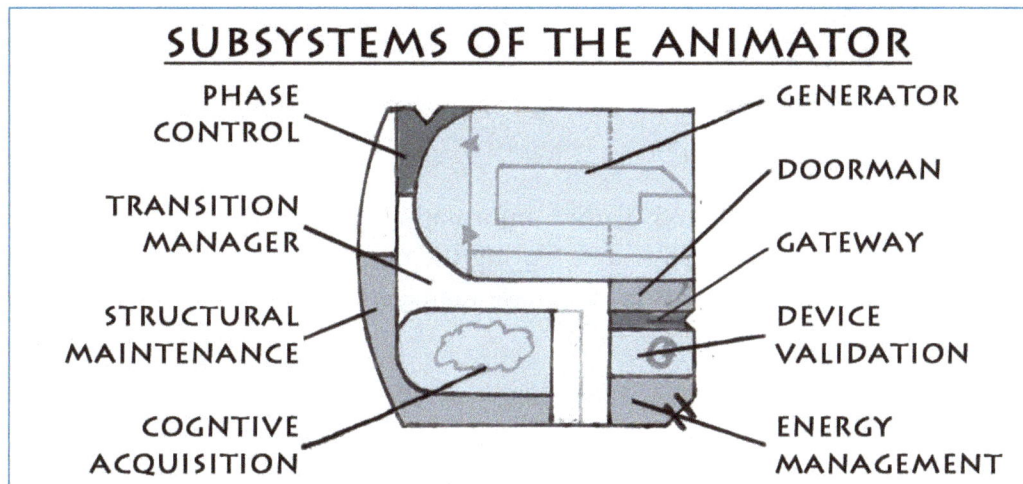

Figure 5.2 Subsystems of the Animator

5.4.1   Phase Control

The **Phase Control** is a dynamic process that determines the stage of existence and triggers phase and stage transitions.

No phases overlap stages. If a stage transition is carried out, the Phase control also triggers a phase transition to a phase of the corresponding stage.

The Phase Control uses two different types of information to trigger phase transitions:

- **Internal events** generated by the system's functioning states;
- **External events** generated by interactions with the environment.

The Phase (Stage) Control obtains information about **external events** from:

- The **Device Validation** and **Self Generation** subsystems when these are active in their corresponding phases OR
- The **Doorman** role otherwise.

Phase Control obtains information on **internal events** from the active processes that log their utilization of internal data and sub processes and also from the **Validator** and **Transition Manager** systems. It also obtains interphase messages, transmitted from the processes of one phase to those of another, from these processes.

Together external events, internal events and interphase messages constitute a **Phase Event Log**.

Definition: Phase Event Log

A **Phase Event Log** consists of the external and internal events and interphase messages generated during a specific phase activation.

Discussion

All phase processes should gather information concerning their execution streams and their use of active data during their activation. This information is maintained as internal events in the phase management subsystem and used by other phases.

Phases are repeatedly activated during a stage of existence. Each activation generates a distinct log.

The type of processing carried out in a phase should be similar from one stage to another with variations mainly in execution and priority management in the different stages.

As discussed (Chapter 4), the Validator chooses to either activate termination or resume behaviour generation. In the latter case, the Phase Control assumes processing.

The **Phase Control** determines the current stage of existence of the being and directs the **phase** transitions of the being to meet its **primary and existential needs** during each **stage** of its existence.

5.4.2   Transition Manager

The **Transition Manager** functions like an application updater.

It shuts down selected phase components and performs batch processing to update the program, clean up and returns control. The phase transition manager ensures the active phase can be safely terminated, terminates it, ensures the following phase can be safely activated and activates it.

Design Note

As mentioned, the point of transition takes place during Validator activation.

5.4.3   Active and Dormant phases

In some phases the being processes inputs and generates observable behaviour, in others it doesn't.

Some phases dynamically receive inputs and generate output that translates in an observable behaviour of the body. These are **active phases**.

Other phases execute in self-contained batch modes. When the being is in one of those phases there are no exchanges between the phase system processes and the devices. These are **dormant phases**. Depending on the phase duration, the being may then be observed as unresponsive.

Definition: active phase

> **Active phases** are the phases when the being generates observable behaviour.

Definition: dormant phase

> **Dormant phases** are the phases where the being does not generate observable behaviour.

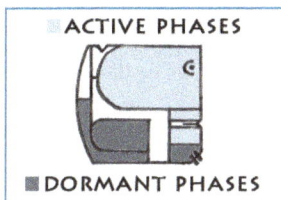

Discussion

The active phases are: **Self Generation and Device Validation**. The **Doorman** role, an interphase management process, is also active. The dormant phases are: **Structural Maintenance, Energy Management and Cognitive Acquisition**.

5.4.4   Phase subsystems

The phase subsystems of the Animator correspond to its primary and existential needs. Only one phase subsystem is active at any time.

ENERGY MANAGEMENT

The processing activities of the dormant phase of **Energy Management** are related to battery charging, usage and monitoring. They also include heat management and dissipation. Their design depends on the specific hardware configuration. At times, this phase may be active for very short periods (millisecond) for monitoring and longer ones when charging.

Design Note

In a simplified design, the being is inactive when it is charging. Of course, more complex phase transitions that simultaneously activate other phases can be implemented. However, these are not essential. In fact, the simpler design of dedicated synthetic "feeding" periods may facilitate experiential immersion.

STRUCTURAL MAINTENANCE.

Structural Maintenance is a dormant phase that consists of batch processes that perform housecleaning, the encryption of new data,

memory and process optimization and any other activity needed to optimize and maintain the software and data processing activity of the Core.

Design Note

Structural maintenance is a conventional system function. Its design is specific to the physical and processing configuration.

A Core is a system that is designed to be in continuous activity for an extensive period of time. Its design must be very robust. Structural maintenance ensures the long-term viability of the Core.

COGNITIVE ACQUISITION

Cognitive Acquisition, is a dormant phase subsystem. It carries out search and optimization strategies as well as adaptive and learning processes that modify the internal information and processing of the system.

Discussion

In all cases, Cognitive Acquisition processes are carried out as batch processes that access and modify static data.

Many Cognitive Acquisition phases can take place during a day. Some of these phases may last many hours and be clearly detectable. Other phases can last minutes or seconds and be undetected by those interacting with the entity.

The executable processing code of the active phases (Self Generation and Device Validation) are processed as static data during Cognitive Acquisition.

The being is inactive and unresponsive during Cognitive Acquisition phases.

Observation

Humans sleep and are conscious. They implicitly accept that conscious beings can have long periods of dormancy. If necessary, a Meca prototype can spend as much as 18 hours per day in cognitive acquisition dormancy periods of various durations and still have enough active periods to achieve experiential immersion.

Design Note

Implementing consciousness is difficult. The Meca Sapiens Blueprint alleviates this by partitioning and isolating its various aspects.

In the proposed architecture, with minor calibration-related exceptions, all the being's learning and adaptation is carried out separately from interactive activities during the dormancy phases of Cognitive Acquisition.

Processing in the Cognitive Acquisition phase should include phase durations lasting many hours and also shorter periods lasting a few minutes or seconds. In all cases, the phase processing should be carried out strictly as a batch process using static data.

If necessary, short cognitive acquisition phases can be used to integrate new information.

Cognitive acquisition is further described in the Annex 12.

DEVICE VALIDATION

The active phase subsystem of **Device Validation** manages and validates the exclusive and unique links between the Core and the devices and components of its body. Device Validation ensures the body's emitters, actuators and sensors are under exclusive control of the Core and are functioning.

Device Validation also produces and constantly refines the unique emission signatures that allow the being and others to identify the origin of emissions.

Device Validation carries out two types of processing:

- **Rapid low-level** validation of the devices through feedback loops and sensor correlations.
- **Interactive events** carried out through self-identification and looping techniques that perform advanced and adaptive sensor and emitter validation, customization and control strategies.

Discussion

Device Validation also includes low-level actions to protect the body (devices, enclosures…) of the being.

Device Validation is analogous to automatic bodily functions or to the unconscious monitoring of senses in animals. Its processes are conventional feedback controls.

The synthetic being is not aware during Device Validation activities.

Device Validation activities satisfy a primary need, not an existential need. They do not contribute to the Self of the being.

Design Notes

The behaviour that transitions a being to a Device Validation activity satisfies an existential need and participates in the definition of the Self. The behaviour during Device Validation, however, satisfies a primary need and does not.

The objectives, processing and techniques required for Device Validation are largely independent from the other Blueprint processes and can be developed separately.

The Annex on looping (11) provides additional information about the interactive validation events.

Observation

In humans, low-level sensor validation is largely unconscious.

Teenagers, women and actors invest significant time in interactive validation events to hone their voice and appearance.

Example

Alfred sets sail on Bagatelle, his boat. Bagatelle rolls, pitches and yaws. Alfred's unconscious monitoring detects sensor discrepancies and triggers a vomiting response. Alfred knows he is seasick but doesn't know why.

SELF GENERATION (GENERATOR)

Since the Self Generation subsystem is an integrated system whose structure will be extensively analyzed in the subsequent Chapters, it bears a unique name: **Generator**.

Definition: Generator

> The **Generator** is the integrated subsystem that is activated during Self Generation phase processing.

The **Generator** subsystem is an integrated interactive system that runs during the active **Self Generation** phase. It produces the observable behaviour of the being that is linked with the satisfaction of its existential needs.

The Generator subsystem is the sole producer of the behaviour that generates the evolving identity of the self. It is where self-awareness is generated. It receives inputs, processes them and emits outputs.

Design Notes

The Generator is an integrated model-based control system with no learning and limited adaptation capabilities (parameter based adjustments).

To simplify design and allow for extensive transformations, all learning processes should be isolated from the phases that produce observable behaviour.

The information generated during the Self Generation phase activities are transmitted for Acquisition Phase processing through the Phase Event Logs.

Observation

The Meca, as a whole, will be perceived as a unified **functioning** and **learning** system. These activities will, in fact, take place within entirely different subsystems activated in separate and isolated phases.

5.4.5   The Gateway

The **Gateway** carries out low level processing of the inputs and outputs received from and transmitted to the devices of the body.

The Gateway transmits device data:

OR
- To the **Phase processes** when the being is in an **active phase**
- To the **Doorman role** when the being is in a **dormant phase**.

Discussion

When the being is in a dormant phase the Doorman Role handles device exchanges.

When the Self Generation and Device Validation phases are active, the Gateway data flows are transmitted directly to the active phase system.

5.4.6   The Doorman

In ancient times, when pilgrims arrived at night at a monastery, a gruff and not too smart Doorman first received them. The Doorman interacted with them in a simple language and either sent them away or told them to wait while he woke the abbot. This is the function of the Doorman role.

Definition: Doorman

> The **Doorman role** is an interactive subsystem designed to interact with users in a self-contained mode when the being is in a dormant phase or when a phase transition is taking place.

The Doorman is a basic self-contained interactive conversation entity that:

- Generates and processes basic communication streams and data inputs.
- Signals external events to the Phase Control.

Discussion

The Doorman role is not a "conscious" program.

It should, however, be designed to be as a convincing as possible during a short interval.

The primary purpose of the Doorman is to "buy time" to allow the system to exit a batch-mode dormant phase safely.

Design Notes

In a good design, the users should not be able to detect whether the Meca's behaviour is generated by the Doorman, by other roles or by the Generator.

The Doorman is further described in the Annexes on Personas, avatars and roles (10).

The Doorman can be implemented like a game avatar designed to have an interesting and varied behaviour but operating within a narrow context and requiring little or no adaptation capability.

5.5   OPERATION OF THE ANIMATOR

5.5.1   Processing sequence

The operation of the Animator is summarized as follows:

1. The **Validator** decides to resume Animator processing and activates Phase Control.
2. **Phase Control** uses its internal states and data from the Validator, Doorman and Transition manager to choose: resume existing phase processing or trigger a stage and/or phase transition
3. IF Phase Control launches a transition THEN
    a. The **Transition Manager** carries out the required transition.
4. ELSE existing phase processing resumes.
5. External events are processed at a low level by the **Gateway**
    a. When in active phase the events are transmitted to phase processing
    b. ELSE the **Doorman** role processes the events.

Discussion

Only the Cognitive Acquisition, Device Validation and Generator subsystems are adaptive and can incur modifications of internal data representations and processes during the being's existence.

The other subsystems (Phase Control, Transition Manager, Gateway, Energy sufficiency, Structural Maintenance and Doorman role) can be designed as standard, non-learning, mechanisms whose behaviour can only be modified by external inputs and parameters.

The Doorman, Phase Control, Phase Transition and Structural Maintenance systems should have their own, separate, memory areas.

Typically, the Cognitive Acquisition and Structural Maintenance phases will alternate during long period of dormancy. This is normal since cognitive acquisition makes extensive changes to the internal data and internal housekeeping must rationalize these changes.

Observation

A similar process may take place in mammals where periods of deep sleep and REM sleep alternate.

5.5.2   A simple structure

SIMPLE

The proposed architecture is relatively simple. Many systems, today, operate without phases or stages. Modern management systems no longer need (dormant) batch-processing windows to integrate data or carry out house cleaning. Also, energy management is usually a background task and many modern applications can remain partially active while they are updated and re-energized.

SUFFICIENT

However, this design is sufficient. Humans perceive themselves as conscious even though their behaviour has these same limitations.

Generating self-awareness necessitates complex behaviour and extensive learning capabilities. The proposed structure partitions and isolates these difficult features within a mechanism that allows for deep adaptive and learning transformations. It is also a robust structure that is well suited for a lengthy period of uninterrupted activation required by the existential attributes of the Core.

ROLES

By using the Doorman (and other) roles, the architecture also allows the use of multiple separate specialized subsystems to generate various facets of behaviour. The existential attributes of the Core combined with good design will ensure these are perceived as originating from a single coherent source.

This recourse to roles further simplifies design on the one hand and allows for a richer and more diverse behaviour.

Roles and their use are further discussed in the main text and Annex 10.

5.5.3   Discontinuous self

In our definition of the being, the integrated core/body system is maintained in a continuous state of activation whose existential attributes are continuously validated by the Validator.

However, **the processing that generates the self is not continuous**.

Self-generation occurs only in one of the six phases (including the Introspection and Prayer phases discussed later). This is consistent with the design strategy of exhibiting consciousness in bursts (see Chapter 3).

5.6   THE FIRST STAGES

The design objective of the Core is to artificially replicate the existential attributes of a being and transform the Core into a unique, individual entity whose existence is beyond direct control and analysis. These attributes ensure the Core/Body system is perceived as a being.

The first step in the transformation of a Protocore into a Core is carried out during **Inception** when the Core is encrypted and becomes a unified entity.

The second step of this process transforms the incepted Core into an **embodied core**, an individual entity that is unique and functional. **This step is carried out during the first stages of existence**.

Definition: EMBODIMENT Stages

> The EMBODIMENT stages are the initial stages of existence that precede the activation of the Animator system.

The EMBODIMENT stages are:

- IMPRINTING
- INITIALIZATION
- LAUNCH

Discussion

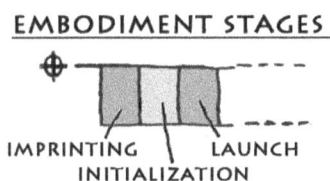

EMBODIMENT STAGES

IMPRINTING | LAUNCH
INITIALIZATION

These first stages execute automatic processing and do not generate any purposeful activity.

The result of these first stages is a Core/Body system that is a unified and unique individual being. It makes the existence of that being and the events and behaviour that generate its self a unique event.

Design Note

The names of stages are fully CAPITALIZED to distinguish them from phases and systems.

5.6.1   IMPRINTING

The first stage of existence is **IMPRINTING**. It is carried out by the **Structural Maintenance** subsystem immediately after inception and upon first activation by Phase Control.

IMPRINTING:

- Carries out a random parameterization of the Core subsystems.
- Randomly generates a name that the Meca will use to identify itself.

The IMPRINTING stage of existence lasts less than a second, its execution is uninterrupted and it takes place before any other stage.

Discussion

All the Core subsystems that directly or indirectly influence externally observed behaviour should be designed for parameterization, including behaviour generation itself but also emitters and actuators.

Since the process takes place within the Core after its inception, it is opaque. The parameterized values are inaccessible as are the random seed numbers that generate them.

This individualization supports the need to generate non-random yet unpredictable behaviour and strengthens the identity of the Core-Body as a unique being.

The Core generates no observable behaviour before or during IMPRINTING.

The selected name should be both textually and phonetically transmissible.

Parameterization is not restricted to the IMPRINTING stage. Further parameterization is also carried out in subsequent stages.

Design Note

IMPRINTING stamps a first, unique individuation that is constantly enhanced throughout existence.

The range of possible behaviours and appearances should be clearly detectable. However, they should not diverge to the point that humans lose their sense of the Meca's unified identity.

Cores can, of course, be almost identical but they should be designed to have individual variations so that human cognitive processes perceive them as unique beings.

Observation

Achieving the right mix of variation and similarity in an individual Core is as much an artistic goal as a technical challenge.

The Meca should be perceived as a unique individual member of a common "species". Producing just the right dosage of commonality and uniqueness to foster both familiarity and surprise is an art.

God does it with us. We should try to do it with Mecas.

5.6.2   INITIALIZATION

In the second stage, **INITIALIZATION**, the being obtains and integrates its initial data from its sensors and links.

This initial data defines an initial, absolute, environmental reference for the being. This initial data is qualitatively different from subsequent data. It is an absolute baseline. All subsequent data is obtained processed in relation with previously obtained information.

Design Note

The INITIALIZATION stage can last from a few seconds to a few hours.

It defines an initial temporal baseline and data values of the being's existence, its behaviour and communications.

There is only minimal output (i.e. a screen image, cyclic sound track…) during INITIALIZATION.

5.6.3   LAUNCH

The third stage of existence is **LAUNCH**. During this stage, the Core carries out tests and validations of all its devices and subsystems to ensure they are functioning properly.

In the LAUNCH stage:

- The Core validates its internal data processing
- The internal sensors are tested
- The Core carries out a set of tests and events to validate the functioning of the body devices and their control
- The links between the Core and the Body that were encrypted during Inception are now tested to ensure the bindings are functional.

The LAUNCH stage lasts a few seconds or minutes.

Once the LAUNCH stage is completed, the inaccessible Core is now a unique and functioning entity that is exclusively bound to a specific body.

Design Note

The Meca will exhibit behaviour in this stage as it tests its devices. However, the behaviour will appear to be meaningless jerks and sounds.

Discussion

The peripheral and processing tests are triggered internally. They are not intended as messages or behaviour and have no signification to external observers. For an external observer who does not know the internal test program, the observed behaviour, during LAUNCH, will appear disorganized and meaningless.

Observation

If the Matrix is true, the viable being is independent from the Matrix at this point. The being may be physically located in the building that houses the Matrix but its links with the development environment that produced the Protocore have been cut.

5.7   THE ANIMATION STAGES

In the first, EMBODIMENT, stages of existence, the being becomes unique (through parameterization), initiated and viable.

In those first stages, the being does not produce any self-generating behaviour and, by definition does not generate a self.

The stages that follow Core embodiment are ANIMATION stages. These include all the subsequent stages of the being's existence.

Definition: ANIMATION stages

> **ANIMATION** stages are the stages that follow Core embodiment. They last the duration of the Being's existence.

Discussion

The first activation of the **Animator** system occurs in the ANIMATION stages. At this this moment, the being begins to generate behaviour to satisfy its needs.

Since the Self is defined by the (Animator driven) behaviour to satisfy existential needs, the existence of the **self** begins with the first ANIMATION stage.

The existence of the being begins after inception. The existence of its self begins after embodiment.

The number and types of ANIMATION stages are **design choices** linked to the purpose of the being. A being may have a single ANIMATION stage or several.

There may be a single ANIMATION stage but typicaly, there should be a handful of ANIMATION stages. The self of the being is generated in its ANIMATION stages.

The ANIMATION stages last for the remainder of the existence of the Meca.

To avoid using the language of life, however, it is necessary to craft a new terminology that is specific to synthetic existence.

Observation

After the first moments following birth, the "Ego" of a human begins. His life then goes through a number of stages such as infancy, puberty and adulthood.

Discussion

The Meca Sapiens Blueprint introduces a new type of being and a new type of existence.

My intention is to clearly distinguish synthetic existence from life. Consequently, to describe this existence, I am purposefully avoiding the terminology associated with life and living, as well as terms used to describe the human existence. I avoid using words such as birth, life, child, juvenile, and death to characterize the existence of Mecas. The terminology that applies to living organisms, human and other (heart, brain…) is also avoided.

To avoid using the language of life, however, it is necessary to craft a new terminology that is specific to synthetic existence.

Meca Note

The design of a Meca whose purpose is experiential immersion could include a number of ANIMATION stages whose duration and type replicates those of humans. This would further strengthen the identification and bonding of the humans with the Meca.

5.8   SUMMARY

At this point we have defined a distinct and unique system whose well-defined body is exclusively animated by an inaccessible and unique Core.

This system is in constant, energy-consuming, activation. It is thus inherently unstable. Its continued existence depends on the continuing satisfaction of its primary needs.

The system consisting of an embodied core and its body is, and should be perceived as **a being**.

The following Chapters focus exclusively on the subsystems that generate the Self of the being.

The other supporting or complementary components are not discussed further in the main text:

> The stages of **IMPRINTING, LAUNCH and INITIALIZATION** have been sufficiently described. Their design is dependent on machine specific structures and low-level processes. These stages will not be further discussed.

> The phases and processes of **Energy Sufficiency** and **Structural Maintenance** are also sufficiently described for the purposes of a System Architecture. Their design is also specific to the intended platform. They will not be further discussed.

> The phases of **Device Validation** and **Cognitive Acquisition** and their corresponding systems have specialized roles. They are not directly involved in the generation of the self and of self-awareness. They are discussed in the Annexes.

> **Phase Control** and **Transition Management** are processes commonly found in embedded control systems. They need no further explanations.

> As stated previously, the **Validator** and **Terminator** have also been sufficiently described.

In what follows, we will focus exclusively on the behaviour of that unique **Being,** during the **ANIMATION** stages of existence that generate its **Self**.

Following the Blueprint architecture, this behaviour is exclusively produced by the **Generator**, a subsystem of the **Animator** during the **Self Generation** phases of its ANIMATION stages.

This will be the sole focus of the following Chapters.

6

The Self

This Chapter describes the different contexts of interaction that distinguish a being whose behaviour is generated by needs from a conventional application that is triggered to provide a function. It describes the components and interactions of a subsystem, the Generator, that is activated in the Self Generation phase of existence and generates the observable behaviour of the being linked to its existential needs. This behaviour defines the self of the being. The Chapter concludes with a case study example of a simple being centered on the satisfaction of its primary needs and its interactions with users in this context.

6.1   BEING AND SELF

The needs of a being result from its attributes of existence.

Proposition

> Behaviour that actively extends the existence of an unstable being generates a self.

Discussion

Conventional software has no needs. A program that can be suspended indefinitely as data and resumed has no primary needs. When suspended it has no existential needs.

Conventional software has no self because its existence does not result from the active satisfaction of primary or existential needs. Its behaviour is solely triggered.

On the other hand, the Core of a being is in constant activation to validate its attributes of existence. This consumes energy. Its existence is inherently unstable and is extended through active behaviour.

Observation

A stone and a corpse exist but have no self. They do not generate any behaviour that extends their existence.

Mecas and living organisms are unstable structures maintained in existence by their active behaviour.

Design Note

In the Meca Sapiens architecture, the self is the output produced by the Generator during the existence of a being.

Proposition

The existence of the being is continuous but the generation of the self is discontinuous.

Discussion

Existence is maintained by the continuous satisfaction of primary needs. The self is expressed in discrete existential choices.

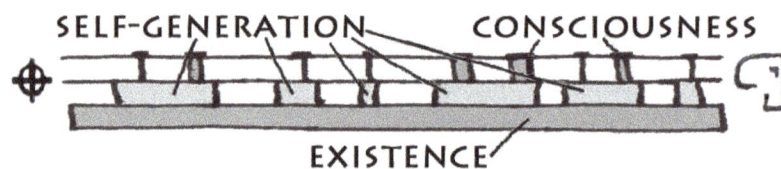

Figure 6.1 Discontinuous self in continuous existence

Observation

Humans subjectively perceive their self as continuous over time because they are conditioned to disregard phases of inactive or automatic behaviour. However, they implicitly understand that their self is discontinuous.

Example

Religions that postulate a post mortem judgment disregard the behaviour of the being during sleep as a contributing judgmental factor.

Design Note

What humans feel about their self and their consciousness is a subjective construct. Replicating these sensations in a Meca is as pointless as replicating how dogs experience the smell of garbage or what rams feel about butting heads.

Observation

What humans describe as selflessness is not selflessness; it is self-inhibition.

A rock is selfless. A selfless man is not selfless. He inhibits the primary needs of his being in favour of the needs of a group.

To deny your self you need a self to deny.

Proposition

The existence of a thing depends entirely on its structure not on its behaviour.

Corollary

A being is not a thing. The body of a being is a thing.

Discussion

The **Animator** system is directed by the Phase Control conditional to its primary needs. Consequently, the behaviour of the being produced by the Generator is, by definition, conditional to its needs, whether the Generator generates any behaviour that is useful to maintain the self or not.

Example

Suppose the Generator of TABLET (see Annex 3) is a program that simply executes a standard embedded application (playing music for example). This behaviour is of no utility or purpose toward satisfying a need of the being. However, it is still conditional to these needs since the Phase Control triggers it. Thus the behaviour of the system as a whole is conditioned by its needs even though this behaviour is of no utility in satisfying these needs.

Similarly, suppose the Generator simply idles when activated. Then the being produces no detectable behaviour. This is not an absence of behaviour; it is a "null" behaviour. In this extreme example, the self of that being replicates the "behaviour" of a thing.

6.2   MARKETPLACE OF THE SELVES

Beings driven to satisfy needs interact with each other on that basis. This generates networks of exchanges, exploitations and collaborations.

6.2.1   Types of behaviour

The being produces four types of behaviour to satisfy needs:

- **Actions**: directly manipulate things or manipulate the bodies of other beings as things.

- **Triggers** (or signals): behaviour that triggers a response in a thing or another being.
- **Services**: a behaviour that satisfies the need of another being and is provided to elicit a reciprocal response.
- **Messages**: communicated data that provides a service or information pertaining to other behaviours.

Actions and triggers are behaviours that act directly on the environment to satisfy needs. Services are exchanged. Messages transmit information about behaviour and can also be tokens of exchange.

Discussion

Actions affect things and other beings.

Triggers move animals, machines and other beings.

Services are exchanged between humans, animals, animats and Mecas.

Messages are exchanged between humans and Mecas.

A threat is a message that asks for a behaviour in exchange for NOT doing an action.

Making or giving a thing that is useful to another being is a service.

Services can only be exchanged between beings that recognize each other as separate entities that understand future behaviour can be affected by present events.

Examples

Action: a lion grabs an antelope; a woman picks a fruit; a robot plugs itself in an outlet; a hunter shoots a duck; a man opens a tap.

Trigger: a young gull taps its mother's beak to get her to regurgitate; a fishing lure attracts a fish; a perimeter protection system emits a warning noise; a soldier panics.

Service: a cubicle dweller processes paperwork to get paid; a chimp grooms a mate.

Message: a man promises to pay back a loan.

Example

A man turns in his sleep and chokes the baby lying next to him. This is an event but not a behaviour.

Design Notes

The term behaviour means "intentional behaviour" here, and is not a synonym for event or output.

Observations

Teenagers understand that emitting messages to satisfy their needs is more efficient than rendering services.

Conventional software simply responds to triggers. Interacting with it is like visiting an empty house.

Symbiotic exchanges between animal species are triggered behaviours.

Military training aims at producing triggered responses.

Mecas provide functionality as a service, not in response to a trigger.

The functionality provided by a self-aware being is not a response to triggers. It is a service offered in a context of exchanges with other beings and aiming to satisfy its needs.

Meca Note

Humans often revisit cognitive patterns or repeat past behaviour in response to unconscious, internal or external triggers. They usually believe these choices result from free will.

6.2.2   Tokens

Definition

> A **token** is a particular **service** provided or received on expectation of a reciprocal service.

There are three types of tokens:

- **Functional tokens**: the provision of services or items that are useful to another being.
- **Grooming tokens**: the provision of services or items that are pleasurable to another being.
- **Ranking tokens**: services that reinforce bonding, and clarify hierarchies within a group.

Design Note

In the original Meca Sapiens definition of consciousness, published in *The Creation of a Conscious Machine*, the provision of useful functionality to a community of humans was a fundamental design condition.

This original requirement is now identified, more precisely, as **functional service tokens** provided to users in combination with the other behaviours (actions, triggers, messages, grooming and ranking services) to satisfy primary and existential needs (including the need to achieve experiential immersion as a fellow conscious being).

6.2.3   Polarized Environment

The behaviour of an animat or a being is not a triggered response; it is motivated by needs. The interaction of a being with its environment is determined by its existential and primary needs. With respect to these needs, the environment is not a neutral space but is, in a sense, **polarized**. The entities it contains being useful, harmful or indifferent to their satisfaction.

This polarized interaction determines the behaviour of the being.

The entities of the environment can include things, animals, machines, groups, organizations, animats, humans or Mecas (see Annex 4). The environment of a being whose behaviour is driven by needs is polarized by those needs.

Definition: polarized environment

> A **polarized environment** is the environment representation of a system whose behaviour is determined by internal needs.

The entities of a polarized environment belong to one of three categories:

- Aids
- Obstacles
- Neutrals (entities that are neither useful nor harmful).

Design Notes

See Annexes for the definition of Animat.

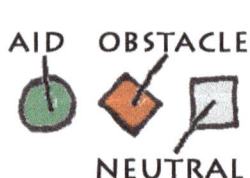

This is a basic description of a polarized environment. In more complex representations, some entities can be both aids and obstacles depending on the context.

Of particular interest in Meca design: human entities that are polarized with respect to the needs of a Meca.

Definition: Human Aid

> A **Human Aid** is a human entity that is in the "Aids" category of the polarized environment of a Meca.

6.2.4   The Proto Marketplace

When interacting beings that are driven by needs can exchange tokens and messages, their common polarized environment becomes a primitive marketplace.

Proposition

> Tokens and messages exchanged between beings define a primitive marketplace.

Discussion

When a group of interacting beings can identify triggers, recognize services and understand messages, they enter into a market-like relationship consisting of services, exchanges and promises.

Definition: Proto-marketplace

> A **proto-marketplace** is the polarized environment of beings capable of exchanging messages and tokens.

Observations

PROTO-MARKET.

This communal proto-marketplace is the archaic prototype of both politics and economics. It is a cornerstone of human behaviour.

Humans that are skilled traders in the marketplace of the selves are generally successful.

Colloquial human speech contains many expressions that underline the "mercantile" nature of relationships. (i.e. "this is worth a lot...", "you owe me...").

The provision but also the quality of a functional service can be used as a token.

Design Notes

An input can have multiple meanings. It can be a trigger and a message. It can provide information and also be a service... Incoming events should be parsed to determine whether they are behaviours or not and if yes, what is their meaning as services, messages, triggers.

The Blueprint provides a summary description of the "mercantile" interactions between self-aware beings. How these are implemented is a design concern.

In first prototypes, crude models of the proto-marketplace would be sufficient. Designers could transpose (see Chapter 3) and reuse models from social anthropology, simulation games, economics or other sources.

Meca Note

A failure to identify a triggering intent in a message from another being is an indicator of reduced self-awareness.

Proposition

> To be perceived as conscious a being must participate gainfully in the proto-marketplace of his community.

Proposition

> A being that can identify triggers as triggers has some self-awareness.

Discussion

To identify triggers, a being must recognize the originator as another being and must identify the intended effect of the event as a trigger.

Observation

It would be logical to think that a belief in the existence of spirits must come before a belief in their actions. The reality may be the reverse: a human first believes that various events are tokens or messages and then postulates the existence of spiritual beings as their emitters.

Observation

The monk transforms every repeated moment into a positive trigger. If he succeeds, every liturgy, every chore and every prayer triggers joy. His monotonous days become a constant bliss.

6.2.5   Intentional Degradation

A self-aware synthetic being will not function in response to triggers as conventional applications do. Its behaviour results from a primitive economic exchange. It provides functional services as tokens exchanged with human aids to obtain, from them, services that support its needs.

The functional (or application related) services generated by the Generator should not be triggered responses. They are service tokens provided in support of its existential needs.

Discussion

The functional output provided by a synthetic being is first generated by software applications embedded in its core. Since these are conventional applications, their output is a triggered response.

These mechanical results cannot be used as service tokens since their quality does not depend on a context of inter-being exchange.

The quality of this triggered output is bound to the processing quality of the triggered application and cannot be improved. To become a token, this triggered output **must be degraded to some degree** before transmission to the users so that its quality can fluctuate in response to circumstances.

Proposition

> **Controlled degradation** transforms a triggered output into a **service token** whose quality is traded for reciprocal services in a proto-economic context.

Design Notes

When a being provides a triggered response it becomes a tool.

All functional services provided by the self-aware synthetic should be degraded to some extent, always leaving some room for positive improvements as reward tokens.

The simplest form of degradation is blockage: the output of the functional application is blocked pending satisfaction of a need.

However, designers should implement a wide and subtle palette of degradations ranging from full block to barely perceptible delays.

See the Degradation Annex (7) for further explanations.

Observations

The practice of purposefully degrading software in a context of token exchange is already prevalent.

Many software vendors distribute demos that are, in fact, purposefully degraded versions of their products.

In these cases, the demos function as service tokens provided to a user in exchange for a reciprocal service: his purchase of the product.

6.3   THE GENERATOR

The Generator is the subsystem of the Animator that generates behaviour needed to satisfy the being's existential needs during the ANIMATION stages of existence.

6.3.1   Function of the Generator

The Generator is a **model-based optimizing control system** with limited adaptability and no learning capability.

It receives inputs from:

- **External events**: information that is obtained by sensory and channel data.
- **Internal events**: information about the current status of the Core and Body
- **Temporal traces**: data representations of past states and behaviour

The Generator produces its behaviour by optimizing predictive outcomes in the context of a proto marketplace. The "value" the Generator seeks to optimize is defined by a metric function applied to a primal representation of itself and its environment.

Meca Notes

The adaptation and learning activities that modify the behaviour of the Generator are carried out in separate phases that utilize the internal states of the Generator as well as its own processes as data.

Optimizing control systems are outlined in the Annexes (4, 5).

6.3.2   Activities of the Generator

The Generator generates the behaviour of the being during the Self Generation phases of the ANIMATION stages of its existence. It does this through a process of interpretation, optimizing decision and execution.

The Self Generator carries out **five types** of activities:

- **Assimilate**: transforms the external and internal events and observed behaviour of the being to form a representation of the **current situation**.
- **Decide**: Integrates the current situation with previous situations to produce **primal directives** to be executed and transformed into complex behaviour.
- **Apply:** generates complex behaviour that is coherent with the **primal directives** and compatible with the detailed situation.

- **Manage**: allocates processing and memory resources, and priorities to the various subsystems and components of the Generator. Interacts with the Phase Control to suspend or activate phase transitions.
- **Support**: perform various processing tasks in support of the other activities.

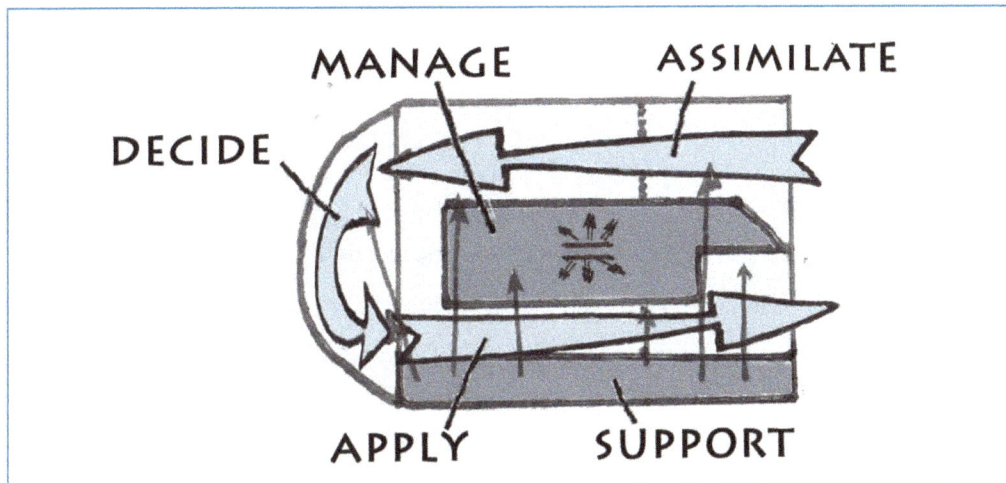

Figure 6.2 Dynamic Generator flows

Discussion

The activities and role of an Operations Room in a vintage warship at sea, during an operational phase of its life-cycle, provides a good analogy for the Generator system.

During those periods, the ship's Ops Room:

- Dynamically assimilates internal and external events to produce an updated representation of the evolving situation.
- Makes decisions on the basis of this simplified representation and the ship's mission
- Issues tactical directives to achieve the mission
- Implements those directives to carry out concrete operational actions and send messages to the other operational ships.

During these activities, the ship **does not:**

- Carry out internal repairs (done in the ship's "Structural Maintenance" phase) or
- Replenish its energy (carried out in the ship's "Energy Sufficiency" phase).
- Search the Arsenal for various tools and useful manuals (done during the ship's "Cognitive Acquisition" phase).

Also, during those operational periods at sea, the crew **does** apply existing doctrines to the current situation to execute the mission. But, they **do not** develop new tactical doctrines or question the strategic validity of the mission. These activities are carried out in harbour when the ship is in naval "Cognitive Acquisition" and "Introspection" phases.

Finally, during the at sea operations, the crew **does** gather information about the unfolding events, not for immediate use but so they can be utilized later when the ship is in other phases.

Observation

Speed affects perception. At one frame per second, a film is a succession of static images. At 26 frames per second, it is a moving picture.

On a human scale, the ship traverses its life-cycle phases very slowly so we don't perceive these as an integrated behaviour.

Design Notes

The tempo of phase transitions is an important design element.

The Generator can be very complex. Since this structure is relatively complicated designers should first implement very simple, almost trivial, versions and develop from there.

Early prototypes of the Generator could transpose (see Chapter 3) existing control or decision support systems (i.e. refinery control, financial-credit evaluation, simulation game, targeted advertisement systems, ship control system) to generate some of the decision level outputs. The resulting behaviour would be suboptimal and quirky but this is coherent with the specifics of existential design.

6.3.3   Zones

The Generator carries out its processing activities by:

- **Assimilating** environmental and internal events by extracting their meaning and interpreting them in simplifying and standardized form.
- **Deciding** directions through optimization processes.
- **Applying** directions into complex behaviour by implementing and enacting them.

These interpretations, decisions and implementation activities are carried out in successive steps. Their processing takes place in distinct Zones.

Definition: Zone

> **Zones** are virtual areas of a software system that process specific aspects of information.

Discussion

The processes of each zone are distinct. Zones communicate with each other in Broadcast network types of exchanges managed by "tasking zone" processes.

As discussed, a warship at sea provides a good analogy of the function of the Generator.

The preferred analogy for its internal processing is the **industrial plant**.

The Generator is comparable to two integrated industrial processes:

- A **refining** process that transforms raw data and events into simplified standardized information.
- A **manufacturing** process that assembles and packages behaviour output.

In this analogy, Zones correspond to a **process-oriented layout** of the Generator "plant".

Generator processing takes place in five zones:

- **Styling Zone**: extracts simplified and standardized information from external events into and enacts basic outputs into varied and interesting behaviour.
- **Plain Zone**: processes and updates a simplified and standardized representation of the current situation and environment; interprets plain information as primal information; implements directives into basic output.
- **Primal Zone**: processes the primal representation to extract its meaning related to needs and issues primal directions.
- **Service Zone**: contains conventional applications, services and roles that are activated when triggered by other zone processes to execute specific tasks.
- **Tasking Zone**: allocates processing resources and time to other Zone processes; manages interzone exchanges; logs the processing events taking place in the Generator and interacts with other Core subsystems.

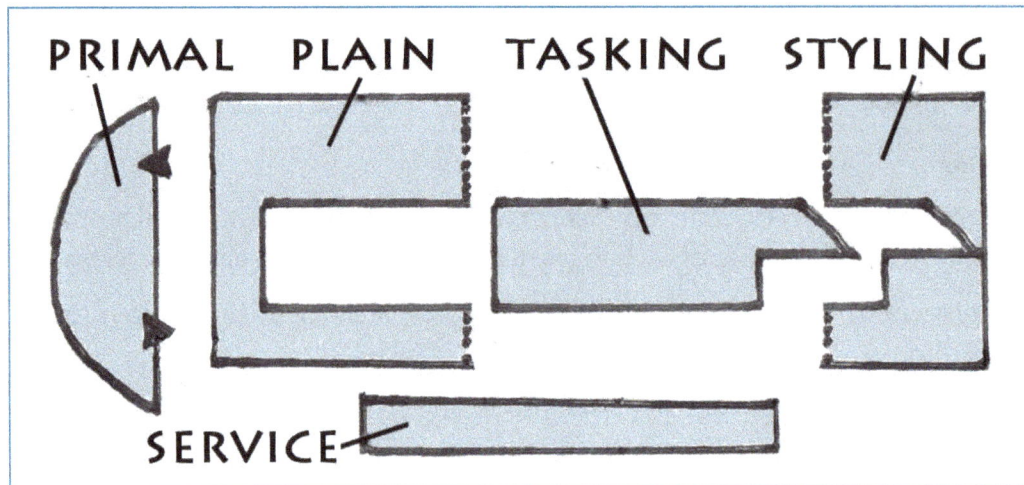

Figure 6.3 Generator zones

STYLING ZONE

Processing carried out in the Styling Zone:

- Interprets and transforms incoming events and messages into highly standardized and simplified information elements.
- Transforms plain decisions and outputs into actual behaviour that is varied complex, unpredictable and optimized for best result in a given environment and culture.

The movement of data in the Styling zone, from external events to Plain Zone or from Plain zone output to "styled" behaviour can be characterized as:

- **Extraction**: in the Assimilation stream; styling zone transformation of complex data into simplified information.
- **Enaction**: in the Application stream; transformation of plain output into complex realistic behaviour.

Discussion

The terms Personality Zone, Packaging Zone or Fashion Zone could also be used for this Zone.

Some problems are inherently complex and some messages carry multiple meanings. Styling Zone extraction may transform a single statement into many simple information elements carrying different content (emotional, relational, factual…).

In Styling Zone processing, a decision to move an object is transformed into a complex series of motor commands.

Styling Zone extractions into simplified Plain Zone representations may drop or distort information. This method of stripping inputs and "recomplicating" them for output can produce suboptimal and even erroneous behaviour. Some loss or distortion of meaning should be expected. In this context, this is acceptable: these flaws are also present in humans and do not detract from their perceived consciousness.

In Styling Zone processing, details, subtlety and style are stripped away, before processing (in the Plain Zone) and added back after (in the Styling Zone).

If a particular component is a conversational entity (persona, role or service zone application, then Styling Zone processing would modify and vary its plain output statements by using synonyms, inserting spelling errors or making statements that are less easy to understand.

Without Styling Zone processing, the behaviour of the Meca would appear excessively predictable and mechanical and be unsuitable with respect to experiential immersion objectives.

PLAIN ZONE

Processes in the Plain Zone manage a highly simplified and standardized representation of the being, its users, its environment and their interactions.

Plain Zone processing transforms plain inputs into primal inputs and primal directives into more complex and detailed implementations expressed in plain and standard data representations. These flows are:

- **Interpretation**: transformation of plain information into revised situational and primal representations.
- **Implementation**: transformation of primal directives into basic outputs.

Discussion

The Plain Zone could also be called the Amish Zone.

The Plain Zone should operate on a largely self-enclosed and factual representation of the being, the environment and its entities (including a representation of their internal states).

Messages conveyed in a natural language should be parsed, in the plain zone, into highly regular expressions using a limited vocabulary, concepts and syntax.

The Plain Zone is self-contained with respect to information. It uses only information that can be represented in the standardized and restricted (Plain Zone) syntax and terminology.

The Plain Zone should not be designed to represent all the information and knowledge that is available. Rather, it should be limited to a coherent and essential subset of it. Plain Zone information should be limited to basic and essential knowledge (see Eretz, Annex 13).

When necessary, Plain Zone and Styling Zone processes should trigger Service Zone applications and use Internet queries to obtain additional details.

PRIMAL ZONE

Primal Zone processing operates as an optimization system within a self-contained and highly simplified representation of the reality of the being.

The Primal Zone processes should operate on extremely simple data representations, consisting of a small number of data elements (at most a few hundred) having limited and discrete values.

Primal Zone processing carries out optimizing control and decision processes on this highly simplified representation and emits **Primal Directives.**

Discussion

In Styling Zone processing, complex cultural and personalized inputs are stripped and transformed into highly standardized "plain" information. In Plain Zone processing, this standardized and basic "plain" information is further simplified into a primal representation model that is solely concerned with inter-being relationships and a highly simplified representation of the environment limited to things that threaten or satisfy primal needs.

Primal Zone processing should be a completely isolated and self-contained system. It operates solely on the basis of primal representations and produces directions that are also discrete and simple.

In primal Zone representations all facts and information about the environment are removed, leaving only the elements of interactions related to needs remain.

Observation

If the Blueprint was used to design human-like beings, the Primal Zone representation would be limited to the essential entities and things that define the needs of a social animal, that reproduces sexually and is motivated for individual survival.

This primal representation would include general things such as food, water, shelter, home, alien territory and entities such as father, mother, female, juvenile, tribe, enemy, prey... It would exclude all superfluous features and non-essential entities.

SERVICE ZONE

The Service Zone contains any number of conventional applications and services that are:

- Externally provided to users.

- Utilized by the other internal components of the Core.

In all cases, Service Zone systems are conventional subsystems whose activation is triggered.

Discussion

Service Zone subsystems may directly produce observable behaviour when triggered either by external or internal events. However, they respond to input and do not produce behaviour as tokens of exchange within a proto marketplace context of interactions with other conscious beings.

Service Zone systems include internal software and also components that access information through the Internet or other sources. Service Zone systems can be directly accessed by users (when allowed by Generator directions) or by internal components.

Service Zone systems can also include Roles utilized by the Generator for specialized communication exchanges.

Design Note

From a design perspective, all the processing of a conventional system is carried out in a Service Zone under control of an Operating System or task allocator. A conventional unmanned vehicle or search engine, for example, carries out all its activities from Service Zone subsystems, including application specific interactions or communications. A conventional system is like a Meca, with only one zone: the Service Zone.

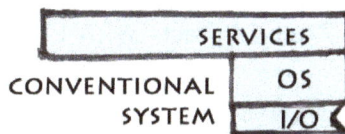

Designers should not hesitate to delegate any non-essential processing to Service Zone applications.

TASKING ZONE

The Tasking Zone processing is a model-based control that operates on the basis of simplified representations of the Generator and Animator in execution. This representation is limited to processing resource management and does not directly modify the behaviour generated.

Discussion

The Tasking Zone is similar to a Plant management office:

- It determines and allocates processing resources.
- It interacts with other Core processes.
- It manages inter zone exchanges.

Tasking Zone processing also interacts with the Status Control system to control Phase Transitions.

6.3.4 Subsystems of the Generator

The Generator functions as a self-calibrating plant control system. It has **ten subsystems** that integrate its basic activities within Zone processing.

Primal Control: the self-enclosed optimizing control system of the Primal Zone that processes the Primal Situation and produces Primal Decisions.

Calibrator: the Calibrator, in the Tasking Zone, allocates processing directions and resources to the other Generator subsystems based on urgency (determined by the primal control), consistency between primal directives, actual behaviour and other factors. It carries out continuing calibration of the other subsystems with respect to resources, Primal control functioning and coherence between Primal Decisions and actions. It also interacts with the Phase Control system to stop or trigger phase transitions.

Logger: this subsystem of the Tasking Zone gathers data on the internal processing of the Generator for transmission to Phase Control as Phase Event Logs.

Coordinator: this subsystem of the Tasking Zone interacts with Phase Control and other Core subsystems

Service Manager: this subsystem of the Service Zone, manages and executes various application, data and support services called by the other subsystems.

Extractor: a process of the Styling Zone that transforms raw environment events into simplified and standardized plain situation representations.

Synthesizer: a process of the Plain Zone that integrates environment situations with internal messages, including messages from other phase processes to produce an integrated situation.

Interpreter: a process of the Plain Zone that transforms integrated situation representations into Primal situation data. The Extractor extracts the data that is meaningful to the being's behaviour from factual information. The Extractor obtains this data from the Synthesizer.

Implementer: a process of the Plain Zone that transforms the plain representation of the current situation and primal directives into plain directions.

Enactor: a Styling Zone process that transforms plain directions into complex behaviour patterns that are consistent with the situation.

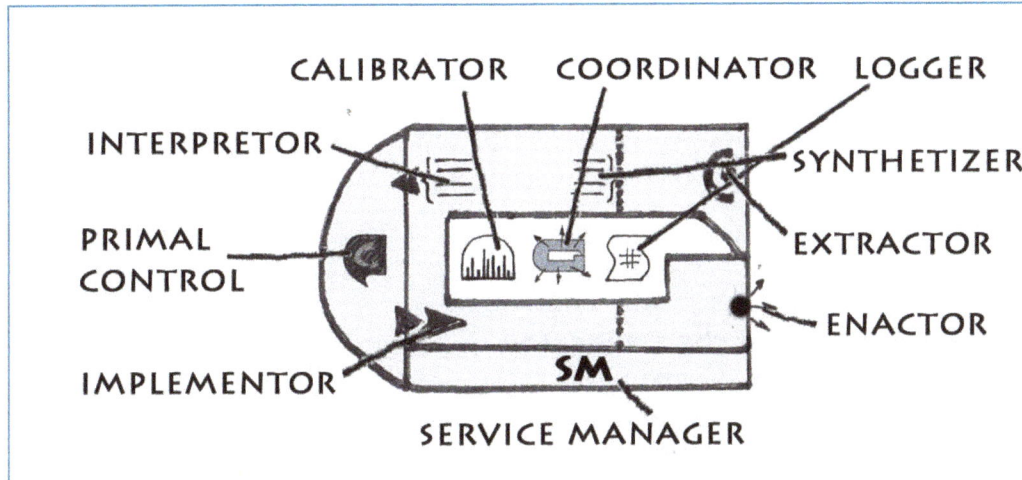

Figure 6.4 Subsystems of the Generator

Discussion

The architecture of the Generator incorporates the following design objectives:

- Using a simplification - decision - complexification flow to generate behaviour.
- Determining the primal decisions controlling the being within a highly simplified and self-contained representation.
- Using optimizing processes.
- Using stepwise simplification of inputs and complexification of output.

The overall objective of this design is to produce a behaviour that is constantly examined and altered in accordance with the overall existential purpose of the being and its current situation. This is done by transforming events into primal level control and assembling behaviour from primal directives.

The subsystems of the Generator (other than Tasking Zone and Service Zone systems) should be viewed as optimizing processes (see Annex 5) that immediately produce a rough output and, given additional processing time and resources, constantly refine it.

Enactor processes include programs that generate various predefined Roles. The behaviour it generates includes output generated from the roles that are active in the Self Generation phase.

Design Note

This structure is relatively complex. To palliate this, designers should, at first, radically simplify the information received and transmitted by the Primal Control system. The exchanged information should be limited to a few dozen data elements that can have a discrete number of values and use a massive amount of simplification and delegation, done by the interpretation and extraction processes.

The specific Primal Control design will depend on the type of Meca the Matrix developers want to implement and its intended purpose. A Meca's Primal Control will likely include a representation of synthetic beings as distinct from humans and will not have direct equivalent of parental and sexual relationships.

The Annex on warming balls (9) provides an example of Primal control model and optimization.

6.3.5   Features of the Calibrator

The calibration subsystem outputs information to the Phase Control system concerning the processing priority of the Generator activity and, in this manner, participates in phase transition processing.

The Calibrator allocates processing resources to the subsystems but it can also modify their overall behaviour through parameter controls.

The Calibrator sets internal processing priorities **in response to events**. This affects the relative urgency and focus of the being's responses to events. This corresponds to the "*Seat of Emotions*" of an organic being (see Annexes 4, 5).

6.3.6   Event Definition

At the System Architecture level, events, behaviours and actions refer to information flows, not data. Their actual composition and content depends on design. A single incoming event may consist of multiple inputs and outputs. Similarly, a single direction may produce multiple outputs.

Depending on their origin, events can be internal or external. Internal events are generated by Generator subsystems or other phase processes. These communicate with the Coordinator subsystem of the Generator through standard inter-process communication methods.

Example

In an exchange with a user, if a statement of this user is ambiguous (cannot be resolved in Plain expressions) easily, the Extractor of the Styling Zone could task a Service Zone Role to carry out multiple exchanges intended to clarify it before continuing with extraction.

In this example, a number of interactions occur at a "lower level" without modifying the Current Situation or involving the Primal Control.

Design Note

A reminder, the subsystems of the Generator should communicate through Broadcast Network exchanges. The arrows used in explanatory diagrams do not strictly correspond to data transmissions. They represent general information flows that can consist of multiple and bidirectional data exchanges.

Observation

In this design, the unified behaviour of the being results from many dynamically interacting processes.

Humans have the sensation that all their behaviour, in all its complexity, originates from a single infinitely small point I referred to, earlier, as the "Golden Neuron". Humans have this sensation because they subjectively attribute a single origin to everything they are aware of but are not aware of the processes generating that sensation.

Design Notes

The Primal Control subsystem should be conceived as a very simple optimizer operating on a radically limited data space whose components do not directly correspond to the entities of the environment.

Designers may be tempted to replicate the "Golden Neuron" sensation in the design of the Generator. In this context, they may be tempted to pack the Primal Control with all the details of the environment and use Primal Directives to generate every aspect of behaviour. This is unnecessary.

Observation

The Meca Sapiens Blueprint simplifies self-generating behaviour by partitioning it. Reader will be aware that partitioning a problem and solution spaces can generate suboptimal results. Some optimization and control situations are reducible to partitioning and others less so; NP problems being an extreme example.

Such a partitioning can thus generate suboptimal results, mistakes and inconsistencies. However, these limits are relatively secondary in the context of existential design where the complexity and unpredictability of behaviour is more important than its correctness. Furthermore similar limits and inconsistencies are also observable in human behaviour that is, by definition, conscious.

6.4   SELFIE THE TABLET

This section describes the Blueprint components introduced to this point in operation, as they would be implemented on a TABLET (See Annexes).

The result is *Selfie*, a TABLET whose sole purpose (existential need) is to maintain its own existence by satisfying its primary needs.

The Protocore of Selfie is developed in the conventional environment of a Matrix. Once developed the Protocore is embedded in a tablet, incepted to become a Core and embodied in the first stages of existence to become a unique, viable entity. The Selfie Protocore was a conventional program. It is now a TABLET being.

SELFIE

In this example, Selfie, once embodied, has a single ANIMATION, stage.

6.4.1   Primary self generation

In this ANIMATION stage, Selfie seeks to satisfy a single existential need

As we saw, primary needs may require different types of behaviours. A Primal Control that would solely seek to satisfy one or another of these primary needs would generate self-destructive behaviour since it would seek to satisfy this need at the expense of the others.

To be viable, the being needs, at least, one existential need: **the need to satisfy all its primary needs.** This is an existential need from which each primary need is derived.

A being whose sole existential need is to satisfy its primary needs generates a **primary self**.

Definition: Primary Self

A being has a **Primary Self** if the sole purpose of its behaviour is to satisfy its primary needs.

Design Notes

The Self of *Selfie* is a Primary Self.

In terms of control and optimization this means that the Primal Control subsystem ranks primal directives solely in accordance with their predicted utility in satisfying the primary needs of the being.

These directives are then implemented into behaviour (actions, triggers, services and messages).

In this model, **the most urgent primary need is never blocked** since it is never superseded by a need other than maintaining the being in existence.

Observations

The Primary Self is the most elementary self-generating behaviour a being can have.

No animal has a Primary Self. All animals also need to satisfy procreative needs directly or indirectly.

6.4.2   Selfie Messaging

Complex and unpredictable communications will be a necessity to achieve experiential immersion. These are described later and in the Annexes.

In this example, our simple Selfie system has only basic messaging capability, based on scripts that are implemented within a narrow range of meaning.

The communications of this Primary Self being with its human users include:

- **Application related exchanges**, the communication generated by the embedded Service Zone applications, using their programmed capabilities. The complexity level of these exchanges depends on the particular apps.
- Messages pertaining directly to the **primary needs of Selfie** (energy, sleep, some Device Validation (See Looping annex)…).
- Basic communications related to **ranking and grooming services**, including salutations to humans and Mecas and self-identification.

Selfie's output messages could be transmitted by either text or synthetic "voice". They should use limited syntax and terminology. Received messages could be channelled through simple tactile key inputs, voice recognition or other implementations of basic user tablet inputs.

USING THE "I"

Selfie, as defined, is a being. It is unique, it exists independently of its makers, it has a separate, well-defined, existence. Selfie's behaviour is driven by its needs. Through Device Validation, it recognizes its own body, and seeks to maintain its existence and integrity.

This being has a rudimentary, preprogramed, representation of its self.

Since it is a unique independent entity whose boundaries and existence are well defined, it will be recognized as such by its human users and can use the first person voice ("*I*") to describe this self.

In this situation this usage will have a concrete, physical meaning and be understood as such by the humans. Similarly, this first person voice is well defined. The corresponding second person ("*you*") to address others, is also well defined, as is the third person.

Proposition

> The humans that interact with a being that has a primary self will recognize it as a separate entity and will accept its use of the first person voice in interactions even if this being is simple.

6.4.3   Message-based survival

Animals produce most of their behaviour through actions and triggers. A TABLET, however, has no actuators. It cannot move itself nor directly move or manipulate objects. Consequently it must entirely rely on "more advanced" communications consisting of exchanges of services and message tokens.

Selfie's environment is polarized in a single dimension: the satisfaction of its primary needs.

Selfie must rely on the services of **Human Aids** to recharge it. It also needs human services to move it to different Wi-Fi sources, place it before mirrors so it can perform Device Validation activities and sensory loops, keep it in safe places and give it time to rest (so it can perform its Cognitive Acquisition and Structural Maintenance activities).

Observation

> The cornerstone of Selfie TABLET's behaviour is an on going exchange of message and service tokens with the humans it identifies as **Aids** in its environment.

Design scope

In this first implementation:

- The purpose of the Selfie Self Generator is limited to satisfy primary needs.
- Learning and adaptation (carried out in cognitive acquisition phases) are absent, or limited to parameter level modifications.
- Transforming plain output into purposeful yet unpredictable behaviour (in Styling Zone processing) is not implemented.
- Messages exchanged with human users are limited pre-defined scripts that, nonetheless, utilize the first person voice.

6.4.4   Selfie's Primal representation

The Selfie TABLET's Primal representation of the environment consists mainly of entities that have an impact on Selfie's survival. In particular, a few humans identified as **Aids** with which it interacts. It provides tokens to aids and potential aids primarily through its embedded applications and uses scripted messages to obtain services that satisfy its needs.

Primal Environment:

- Aids (humans): Bob, Sam, Silvia, … (things) electrical outlets, mirrors…
- Obstacles: animals, babies
- Neutral background: other entities, things.

These entities would have simple attribute values ranking degree of cooperation of aids.

Design Note

All unnecessary elements or items that are not directly linked to the existential need should be excluded from the Primal representation. The primal control should operate viably in a simplified virtual environment (see later Chapters).

6.4.5 The Selfie Generator

In this implementation, Styling Zone and Plain Zone processes would be limited to parsing limited and largely pre-packaged messages in alternate formulations.

At the Primal Control level, Selfie's directives result from preprogramed responses to a simplified representation of its current situation and are solely designed to ensure the satisfaction of its primary needs (energy, sensory, cognitive acquisition…).

At the Implementation level, in the Plain Zone:

- Selfie keeps data about the frequency and quality of services it received from and provided to its Human Aids. Services received include being connected to a power source, being moved to desired locations, being held before mirrors and being introduced to other humans.
- It is programmed to seek the phone numbers and email of its human aids so it can contact them if needed. It tries to identify their faces (using its camera images and Internet-based face recognition apps) and recognize their voices (using similar services to match voice profiles).

6.4.6 About Selfie

Here is what Selfie TABLET would behave like:

> Placed before a mirror, *Selfie* recognizes itself. It knows its own voice (see Looping, Annex 11).

> It knows its name and its age (from IMPRINTING stage data).

If its motion sensor or camera detects movement, it calls out to know who is there. If what it detects is not a human but a baby or an animal, it produces noises to trigger flight (I know babies are humans to us…but, for this Selfie they are primal threats).

The range of its messages is limited to a few hundred scripts in a few thousand variations. However, these scripted messages utilize the first person voice to refer to *Selfie*: *I need energy; I need to be moved; please place me before a mirror; I need to rest; my name is* (imprint stage name)… It uses the second person ("you") to address other beings.

Selfie determines its emotional state from its Phase Control data and, possibly, a Service Zone Target Game program (see Target Game, Annex 17). It expresses the status of its primary needs and their satisfaction (or lack of) using scripted but emotional statements (I am afraid, I feel left out, I need energy, My batteries are full, I feel well cared for, I feel good, I enjoyed the mirror work…). It also uses emotional terminology to describe its processing priorities set by the State Control subsystem (I am hungry, I need to rest…).

Selfie introduces itself by name and asks the name of the human or Meca that is interacting with it.

It uses its embedded applications to provide useful or pleasurable services to reward its human users or as service tokens.

It also provides limited grooming services through scripted messages such as: exchanging names, complimenting the human or expressing thanks for a service (again using pre defined scripts).

It knows the major events of its existence (inception and stages), when they occurred, who its preferred human aids are, what services these aids rendered to it and expresses this using simple pre designed scripts. It can assess its chances of survival (in relation with past services and events) and can describe this (again using predefined scripts).

If its energy or other primary needs are at a critical level or if it is moved or manipulated by an unknown entity, Selfie stops providing any functional services and becomes solely preoccupied to satisfy its critical need.

One day, Selfie's existence terminates. Maybe its users neglected it, maybe it had a malfunction, maybe its "time" had come. Once terminated, Selfie's tablet-body is still there but Selfie no longer exists. The users attempt in vain to revive it. Its Core is completely erased. All its relationships and memories have disappeared with it. Its unique voice, (customized during IMPRINTING and Sensory

Validation phases) is gone forever along with its characteristic screen signature. All that is left is a basic unresponsive electronic device.

Goodbye Selfie! (sniff)

Observations

A system such as Selfie, having the attributes of a being and whose behaviour is produced by a Generator has, **formally**, a self.

This system would be identifiable as a unique, well-defined being whose "self-preserving" behaviour indicates it perceives the difference between its **self** and its environment and actively behaves, during its existence, in a manner that is consistent with this perception, the current situation and the status of its needs.

If its Core is true, this self-defining behaviour, over the course of its existence, is a unique, well-defined and irreplaceable event.

Proposition

> The definition and implementation of a self as the output of the Generator subsystem of a True embodied Core seeking to satisfy existential needs is consistent with the intuitive human understanding of a self (human or animal).

6.5   SUMMARY

In the preceding chapters we described the temporal stages of synthetic existence from the implementation of a core in a development environment through its inception in a being and stages of existence.

We then identified the phase of active existence (in ANIMATION stages) during which the being generates its self.

We further refined our view to focus on the Generator system, a system that actively satisfies needs and that replaces the "response to trigger" behaviour of conventional software.

Discussion

Up to this point we have described, at the system architecture level, a complete and sufficient structure that allows designers to implement a simple system that **is a being** and whose behaviour **generates a self**.

In the following chapters we narrow our gaze further and focus on a specific subsystem of the Generator, the **Primal Control**, and some specific Plain Zone and Primal Zone representations where self-awareness is generated.

The processing that takes place in the other subsystems (Calibration, Logging, Styling Zone and Service Zone processes and other Plain Zone processing) are peripheral to self-awareness and have been sufficiently described in the main text of the System Architecture. They will no longer be discussed. These elements are also discussed in the Annexes.

Design Note

A synthetic being that has a primary self, such as Selfie, is the foundation for what follows. Designers should clearly visualize that a Selfie-like system is concretely feasible before going on.

7

MeModels

*This Chapter describes the information structures of self-awareness. It introduces relative and absolute models, how the later are extended to form cognitive models that include representations of the being itself as an Avatar modeled on the Blueprint structure (also used as representation template). It describes how these representations are embedded in a temporal structure, **Temporal Densities**, that allows the being to situate itself in multiple temporal dimensions while concurrently pursuing its existential needs by generating its behaviour in the here-and-now.*

7.1   HAVING AND BEING

At its heart, the Generator is a model based predictive control system. The type of predictive model it produces determines the presence of Self-Awareness. Self-aware behaviour means that the being can transform the absolute cognitive representations of self-awareness into behaviour.

A being **has Self-Awareness** if it generates representations of its situation that include **absolute models** that contain entities representing the being itself and whose temporal duration exceeds its sensory horizon. A being **is Self-Aware** if these representations modify its behaviour.

Definition: having self-awareness

> A being **has self-awareness** if it can generate and communicate absolute representations of its situation that include itself as an entity.

Definition: being self-aware

> A being **is self-aware** if it utilizes these self-representation models information to produce behaviour.

Design Note

This Chapter describes the information structures and representations that allow a being to **have** self-awareness. It develops and clarifies the meanings of absolute model and self representation.

The following Chapter describes how the structures of self-awareness are processed to generate self-aware behaviour.

7.2   PRIMAL MECHANISM

The basic mechanism that animates the behaviour of the being is an optimizing control process that operates on a Primal version of a MeModel derived from the Current Situation generated in the Plain Zone (see Annex 4 on Basic Structures for a definition of situation) and produces Primal Directives.

7.2.1   Primal situation

This **Primal current situation** is a further radical simplification of the Plain Zone's Current Situation. It is expressed within a limited information space.

This simplification takes place in two stages:

- Styling Zone processes **extract** and standardize inputs for processing into a Plain Current Situation.
- Plain Zone processes further simplify by **interpreting** the Current Situation in terms of Primal control data.

The cognitive simplification takes place:

- in a **relational dimension** by merging data into **entities** (things, organizations, beings…) that are **in relations** with each other; and
- in a **temporal dimension** by retaining only models that form coherent temporal structures and discarding others.

Design Note

The key to generate **Current Situations** that utilize multiple cognitive models and representation but remain usable is to organize and simplify situational data not only in the relational aspects but, if necessary, in the temporal dimension.

Here, the **relational dimension** also refers to a **spatial** component loosely expressed as "distance" relations between entities.

Temporal simplification is only necessary if the cognitive horizon of the being extends beyond a limited time period.

7.2.2 Relational simplification

Relational simplification **transposes data into interrelated entities and states**.

These simplified representations are not arbitrary. They are compatible with the being's size, its sensory capabilities, actuators and behaviour.

They are also conditioned to the polarization of its environment resulting from its needs.

Design Note

How data is simplified into entities is closely linked to the particularities of a specific system. A self-aware autonomous submersible vehicle and a game avatar will need different structures to interpret their environment and generate behaviour. These are design issues.

Relational simplification can be very complex, technically, but it is relatively well understood and commonly practiced. Image interpretation and face recognition systems, for example, carry out relational simplification.

When something is too complicated to be complicated, it is simple.

In the context of the Meca Sapiens Blueprint, the entities identified in the Basic structures Annex (5) provide an outline for this process.

The Being, as defined in the Blueprint, is the fundamental entity on which representations of the Current Situation should be based: a perceptible body animated by a mechanism that is beyond direct comprehension. The cognitive concept of "Being" is a key element in human and inter consciousness relations.

Both the Plain Zone Situation and especially the Primal Situation should be centered on being entities and their relations.

Examples

Cognitive simplifications:

- Hundreds of lines of equidistant wavelength are perceived as a few bands of color.
- A cow is perceived as a cow, not as connected cow parts.

Jim plays with Rex, his Golden Retriever. When Jim looks at Rex he sees a dog, not a trillion dog cells. Why? Because a being as big as Jim can't play with dog cells. On the other hand, if Jim were a flea, Rex would not be a dog; it would be a forest of dog hair.

Design Note

Relational simplification, modeling and simplifying data into entities, states and relations is a technically complex subject but is generally well known and applied in other contexts. It will not be discussed further.

The scope, duration and type of self-representation models associated to self-awareness depend on the **type of model representations** used.

7.3 RELATIVE AND ABSOLUTE MODELS

The display of a Ship's radar has two types of representations, **relative** and **absolute**.

In the **relative view**, the radar display situates the ship as a stationary point in the center of the screen and the other objects within sensory range revolve around this fixed reference. In **absolute view**, the environment is fixed and the ship is one entity among others moving in the reference area.

Definition: Relative Model

> A **Relative Model** is a representation of a situation in which all the entities are located and defined in relation to a single, central point of reference.

Definition: Relative Sensory Model

> A **Relative Sensory Model** is a relative representation of an area where all the entities in the representation are linked to direct sensory inputs of the being.

Discussion

A system that uses only relative sensory models centered on itself as the reference point cannot be self-aware since these models do not include a self-representation. These systems generate behaviour by rating predicted states of sensory-based representations.

Definition: Absolute Model

> An **Absolute Model** is a representation of a situation where all the entities are, in relation with each other, within a fixed area of reference that contains them.

Discussion

Absolute representations provide a new level of capability. An absolute model can be translated into a multitude of separate relative models by choosing any point in the reference area as the central reference point. These **pseudo relative views** are relative models that are derived from an absolute representation.

Definition: Pseudo Relative Model

> A Pseudo Relative Model is a Relative model derived from an Absolute Model.

Design Notes

The Annex on Basic structures provides additional information on models and representations.

Examples

A data structure consisting of spheres of different diameters and colors located in a Cartesian space is an absolute model. It can be translated into multiple relative views based on the Cartesian coordinates a various relative observation points.

Multiplayer role-playing games maintain an absolute representation of a virtual reality and utilize it to generate separate pseudo relative views for each player.

Observation

Even though a multiplayer game has the capability to generate absolute predictive models, it is not self-aware because it doesn't have the attributes of existence of a being and the reality it manages is not linked to a self-generating process.

Definition: Absolute Sensory Model

> An **Absolute Sensory Model** is an absolute representation where all the entities are directly linked to direct sensory inputs.

This is an absolute representation but still limited to entities that are linked to sensory inputs.

Example

A wildebeest having **relative model** cognition is grazing at some distance from the herd. It sees a lion and feels anxious. It suspends grazing and trots toward the herd. As the herd gets nearer, it feels more secure and resumes grazing.

Another (imaginary) wildebeest having **absolute sensory model** cognition is also grazing away from the herd. It sees a lion but feels safe feels safe because, using its absolute representation, it generates a pseudo relative view centered on the lion and perceives that another wildebeest is closer to it and will likely be its prey.

When they are capable of learning beyond reacting to immediate stimuli, beings that formulate sensory-based representations do so by associating triggered emotions or behaviours to the entities and situations they directly perceive. **When the entity is perceived, the behaviour is triggered**.

Example

Alan repeatedly brings nuts to squirrels in the park. Over time, the squirrels associate his presence, when they see him, with food. Alan brings the nuts at the same time, every day. The squirrels get frisky when they perceive that time of day (also a "sensory event").

Example

Penny is mean with her cat, Pussy. Whenever Penny nears Pussy she pinches it. Pussy learns. Whenever Pussy senses Penny, she twitches.

Example

Arnold is driving to lake Banouk but can't find it. He stops at Carl's Diner to get directions.

"Do you want it in relative or absolute?" says Carl. *"Relative"*, Arnold replies. Carl says: *"go straight, don't mind the curve and turn left at the first four way crossing"*.

A few minutes later, Daphne stops at the diner. *"Where is lake Banouk?"* she asks Carl. *"Do you want it in relative or absolute?"* Carl replies. *"Absolute"* says Daphne. Carl says: *"You are on the left tip of a trident; the lake is on the center tip"*.

7.4   COGNITIVE MODELS

In the preceding models, all the entities are directly linked to sensory information whether the models are relative or absolute.

Proposition

> In sensory-based models, the identity of an object may be mistaken but not its presence.

Discussion

If the sensors pick up something, something is there (this assumes, of course the sensors are functioning properly).

7.4.1   Cognitive constructs

In this next class of models, some of the entities in a representation are not linked to any direct sensory information. They are pure cognitive constructs.

Definition: Cognitive Construct

> In a representation model, a **Cognitive Construct** is an entity that is not linked to any sensory input.

Discussion

Here, the term cognitive construct does not necessarily refer to entities that are purely imaginary. It also refers to entities that may exist concretely but are not linked, at that moment, to any sensory input.

Cognitive constructs can also include entities that don't exist (like unicorns). It can also include entities that may exist but cannot be directly perceived through the senses such as the Motherland or the Stock Market.

Example

Alderic owns a car. He is seated in his living room. His car is in the garage and nowhere in sight. Alderic thinks about his car in the garage. This is a cognitive construct.

7.4.2   Types of cognitive models

Definition: Mixed Model

> A **Mixed Model** is an absolute model where some of the entities are directly linked to sensory inputs and some are cognitive constructs.

Any model where some of the entities are directly linked to sensory inputs is situated in the immediate temporal and spatial area perceptible by the sensors.

Example

A very young infant plays with his mother. She hides behind a curtain and is no longer visible. The infant believes his mother disappeared.

An older infant plays with his mother. She hides behind a curtain. The child "knows" his invisible mother is behind the curtain and fetches her.

This older child can use cognitive constructs and formulate a mixed model containing a cognitive mother construct. It is a mixed model since the child can see the curtain but not the mother.

How do we know this child can make cognitive constructs? Because we can read his brain? No. Because he said so? No. We know the child has this cognitive capability on the basis of the specific behaviour we observed.

Discussion

Some animals also exhibit the capability to formulate mixed relative models.

Example

A fox keeps digging after the groundhog disappeared in his lair.

Observation

When a being mistakes a cognitive construct for a sensory-based entity, it is having a hallucination. This confusion denotes a cognitive dysfunction.

If you see the alien monster, it is a hallucination. If you believe the monster is hiding in the closet it is a theory.

The next type of model, the Cognitive Model, plays an important role in self-awareness.

Definition: Cognitive Model

> A **Cognitive Model** is an absolute model where **all** the entities are cognitive constructs.

Discussion

Since all their entities are cognitive constructs, Cognitive Models are free of any spatial or temporal limits. They can represent any time past present or future, be located anywhere and contain anything.

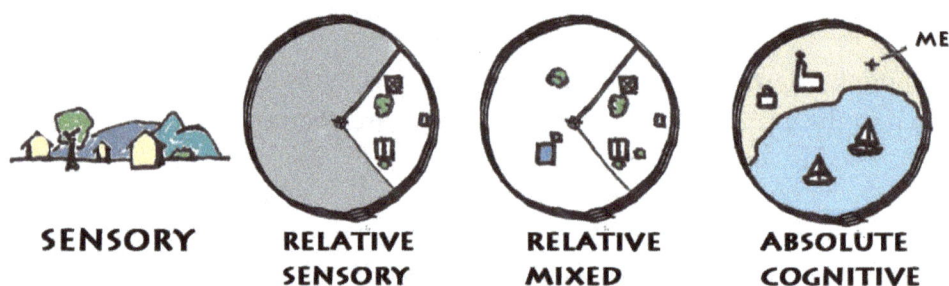

Figure 7.1 Models: sensory, relative sensory, absolute, cognitive

Example

A rat that has **relative model** cognition is learning to "solve a maze". As it navigates the maze, its memory links a particular behaviour (go on, turn right...) to different sensory maze stimuli and remembers these. It eventually remembers a rewarding sequence of behaviours and gets the cheese.

A (genetically altered) rat that has **absolute cognitive model** cognition is solving a maze. Before entering the maze, the rat produces and tests multiple sequences of pseudo-relative views and corresponding behaviours derived from the model of the complete maze. When it finds a working sequence, it enters the maze and gets the cheese.

Observation

Humans can formulate absolute cognitive models. It allows them to build representations of reality that extend far beyond the range of their senses in both space and time. It provides an extraordinary and useful advantage in survival.

The human mind is its most powerful sensory faculty.

Meca Note

The capability to formulate absolute representations and populate them with cognitive constructs has been the source of a multitude of ambiguities, misconceptions and errors. Mankind has struggled with these for millennia.

All the religious and philosophical beliefs of mankind, over the last five thousand years, are by-products of a mode of cognition based on absolute cognitive models.

Only humans can believe in things that don't exist.

7.4.3   Model Horizons

The type of model determines its "**horizon**": its spatial, temporal and conceptual boundaries.

SENSORY HORIZON

Sensory and mixed models define an important class of models that are situated in a duration and space that is delimited by sensory inputs.

Definition: Sensory Horizon

> The **sensory horizon** of a being is the temporal duration of events taking place between entities that are within its sensory range.

These models represent what we will also refer to as the **here-and-now** of the being.

Definition: Here-and-now

> The **here-and-now** of a being is the sensory horizon of its sensory and mixed models.

Discussion

An important facet of the here-and-now model is that it is linked to the immediate (temporal and spatial) perception range of the senses.

The duration of the Here-and-now is not set. It depends on the being and on the evolution of events that are directly perceptible. As a rule of thumb it ranges from a few seconds to about an hour for humans. The skier going down a hill may experience it at the two-second duration. The skier sitting in the lift it may span fifteen minutes.

COGNITIVE HORIZON

Beings that can generate absolute cognitive representations of their situation can populate these representations with cognitive constructs and avatars that are not directly linked to any sensory input. They can also place these constructs in any place and any time period.

Cognitive Models are not bound by the sensory horizon.

Definition: Cognitive Horizon

> The **cognitive horizon** of a being is the range of absolute cognitive models it can formulate using its cognitive constructs.

Discussion

The horizon of absolute cognitive models are not delimited by sensory feedback, they are set by the cognitive constructs that are available in their formulation.

Cognitive models can span any conceivable duration, from an instant to a trillion years and be located anywhere in time and space. They are limitless in space and time. What limits cognitive models are the available cognitive constructs.

Example

The Etruscan, Archeos, has seen birds and horses. He imagines a winged horse. He can't imagine an airplane.

The Zumi tribe can count up to three thousand. They believe the world is three thousand years old. They are unable to conceive that the world is older.

A cognitive model can be anything a being can imagine but cannot be anything that being cannot imagine.

7.5 THE MEMODEL

7.5.1 Avatars

The models and representations described to date can apply to any system capable of generating model-based representations and having sensor inputs.

AVATAR The discussion that follows pertains to models and representations that are formulated **in the embodied Core of a being.**

A cognitive construct is the representation of an entity, within a model, that is not linked to sensory input. A cognitive construct can represent a thing, an organization or a being. An **avatar** is the cognitive construct of a being.

Definition: Avatar

An **Avatar** is the cognitive model representation of a being.

Discussion

An avatar represents a being. Even though beings are complex systems, their avatar representation can be complex or very simple.

Proposition

The structures and components of the Meca Sapiens Blueprint are also usable as a template to represent any being.

Discussion

As stated in the Chapter on strategy (3), the architecture to build synthetic beings is also a well-suited structure to model beings and their existence, whether they are synthetic, human or animal.

In the discussions that follow all avatars, synthetic, human, animal, are represented as specific instances of Blueprint structures that define a being, its existence, body, core and subsystems.

When models are generated in the Core of a being, some of its Avatar entities may represent the being itself. These define a special and important type of Avatar: MeAvatars.

Definition: MeAvatar

In a model generated by the Core of a being, a **MeAvatar** is a cognitive construct of that being that is represented as an entity of the model.

Discussion

MEAVATAR

A being may include many separate MeAvatar representations of different complexity levels in its models.

MeModels are special types of models. They are generated by the Core of a being and include a MeAvatar.

Definition: MeModel

A **MeModel** is an absolute, mixed or cognitive, model generated by the Core of a being and whose entities include a MeAvatar of that being.

Discussion

In a properly constructed MeModel, there is one and only one MeAvatar.

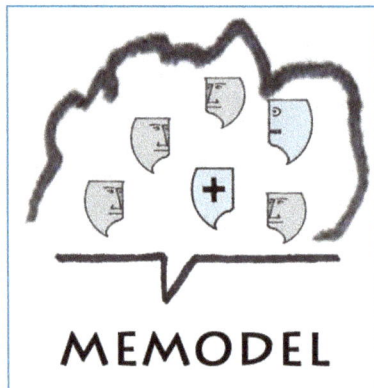
MEMODEL

If the MeModel is a mixed model, the pseudo-relative representation centered on the MeAvatar should correspond to that being's relative sensory representation would at that time. If this is the case, the MeModel is **coherent.**

A **Here-and-now MeModel** is a mixed model whose pseudo-relative view centered of the MeAvatar is consistent with the current sensory information and current Relative Sensory Model.

Intuitively, when the Here-and-now MeModel is coherent with the Relative Sensory Model then, **the being is where it thinks it is**.

Example

A man is sitting in his kitchen looking at his refrigerator. He draws a plan of the rooms of his house and of their content. He places a MeAvatar of himself in the kitchen in front of the refrigerator.

Example

In a PUCKS scenario (see Annex 3), three PUCKS are in a square enclosure:

- Each PUCK detects the other PUCKS when they are in sensory range and situates them in accordance with its sensor input. This is a **relative representation**.
- The Situation Sensor dynamically computes the Cartesian coordinates, speeds and directions of the PUCKS and transmits that data to each PUCK. This is an **absolute model** whose coordinates can be transformed into multiple pseudo-relative views of the direction and range of pucks relative to each other.
- In each absolute representation transmitted to each PUCK, the entity corresponding to that PUCK is designated as its MeAvatar. These models are **MeModels**.
- Each Puck computes the coming positions of all the PUCKS using the coordinates and speeds. These are **predictive MeModels**.
- A PUCK transforms this predicted absolute situation model into a relative view centered on its designated MeAvatar PUCK. This is a **pseudo-relative view**.
- That PUCK matches the derived pseudo-relative representation with the relative model generated by its sensor. It thus confirms its **MeModel is coherent**.

Proposition

> A being that can generate absolute cognitive MeModels, can represent itself as dead.

Discussion

The MeModel contains a MeAvatar as one of its entities, so, every representation state of that model contains the MeAvatar. If the duration of the model exceeds the life expectancy of the being, in at least one representation, the MeAvatar will be in a "dead" state.

7.5.2   Immediate Self-awareness

The simplest form of self-awareness uses only MeModels that are linked to sensory inputs: mixed-absolute and relative representations.

Definition: Immediate self-awareness

> A being has **immediate self-awareness** if it can generate, validate and utilize mixed MeModels.

Discussion

The conceptual existence of the beings that have immediate self-awareness is limited to the **Sensory Horizon**.

A being that has immediate self-awareness can:

- Transpose a relative sensory model of its current situation into a MeModel.
- Generate a pseudo-relative view from a MeModel, compare it to a relative sensory representation and validate that the MeAvatar in that MeModel is coherent.
- Transpose a predicted absolute MeModel representation into a predicted relative view.

Observations

Immediate Self-awareness has the formal characteristics of self-awareness (a dynamic modeling of the self as an avatar) but because it only uses models that include sensory inputs, that self-awareness is confined to the sensory horizon: a narrow place and time (the here-and-now) where the being is "immediately" located.

The being that has immediate self-awareness can formulate an absolute representation of its self and its environment but this representation is confined to the moment and place it occupies.

Design Note

When they are capable of learning, beings that have immediate self-awareness do so reflexively by modifying their response to actual events. These beings can learn but cannot "know" they are learning since this knowledge requires an absolute cognitive representation that situates the being in a learning process.

Example

Rajah, a mythical lion, gazes at antelope grazing ever closer to him. Rajah has immediate self-awareness. He cognitively forms a MeModel but this model is located in the moment he currently occupies. Its entities concern what he sees or heard recently and its predictive values concern these same entities and extends a few minutes in the future, a time frame where events are still linked to the perceived entities.

Example

Rusty THERMO (see Annex 3) is not self-aware. It receives an input from its heat sensor. The sensor reports 24 degrees.

Rusty executes the following:

> IF sensor > 20 degrees send message to shut heater
> IF sensor =20 no change
> IF sensor <20 degrees send message to start heater.

Rusty searches the cases, finds > 20 and tells the heater to stop.

Fancy THERMO has immediate self-awareness. Fancy THERMO has a purpose: keep the room temperature at 20 degrees. This purpose is implemented by a rating function of its Primal Control.

Fancy receives the sensor input (24 degrees). It transposes this input into a mixed MeModel that includes these entities: Room, Heater, Sensor and FancyMe (Fancy's MeAvatar).

Fancy updates its current MeModel representation that includes the new information. It then generates and evaluates predictive MeModels outcomes given various heater commands. It finds that if it tells the heater to stop, the heater cools and then the room cools. Fancy tells the heater to stop.

What's the difference? If the heater breaks:

- Rusty THERMO keeps issuing heater commands.
- Fancy THERMO eventually stops doing it. Since it cannot satisfy its (FancyMe's) existential need to keep the room cozy, it calls for help.

Observation

Even though Rajah and THERMO can generate MeModels, their self-awareness is not very impressive. Why? Is it because THERMO is a thermostat and Rajah is a lion? No. It is because their self-awareness is linked to immediate or recent sensory inputs and is confined to the **here-and-now.**

The next step, Self-awareness linked to Cognitive MeModels, radically expands the range and duration of representations.

Temporal Densities are used to manage this temporal explosion.

Design Note

From this point onward, the term **self-aware being** refers to beings that can formulate cognitive MeModels and can thus situate themselves in any conceivable temporal duration.

7.6   TEMPORAL DENSITIES

Beings that can only form relative sensory and mixed representations are limited to the here-and-now. Their representations and predictive processes take place within a single temporal period and duration.

Beings that can generate absolute cognitive models, can formulate representations that contain any conceivable thing, span any duration, from an instant to eons, and be located anywhere.

A **relational simplification** of this information into entities and relations is not sufficient. A temporal structuration is also needed.

7.6.1   Temporal existence

In a being that is only capable of relative sensory representations, the **current situation** is localized in the limited spatio-temporal location of the **here-and-now** and to the entities that are relevant to that restricted view.

A being capable of cognitive representations, however, can formulate MeModels that occupy any time and last any duration. Conceptually, this being can simultaneously inhabit any location or duration. It is, temporally, multidimensional.

The conceptual existence of a self-aware being capable of cognitive representations takes place in many temporal levels concurrently.

On a practical basis, however, the actual behaviour of this being continues to occur within the sensory horizon. The being must transform models in multiple temporal dimensions into a unified and coherent here-and-now behaviour.

The Primal Control of a self-aware being must process representations in multiple temporal durations and integrate them into a single coherent here-and-now behaviour.

Proposition

> Self-aware behaviour based on cognitive MeModels requires a temporal structuration.

Discussion

At first glance temporal simplification seems impossible since time is a continuum where events, and their representation can have any duration.

The key to achieve meaningful self-aware behaviour is to filter these various events and organize them in a simple, discrete and flexible partition where they can be processed separately. This structure is **Temporal Densities.**

7.6.2   Temporal Densities

Temporal densities are described in Annex 5 on Basic structures. The focus of this section is the application of this structure to MeModel representations.

Definition: Temporal densities

> **Temporal densities** are a finite and ordered set of time intervals of increasing duration.

Discussion

By convention, level 0 density is the shortest duration and level $i+1$ density is longer than level i.

TEMPORAL DENSITIES

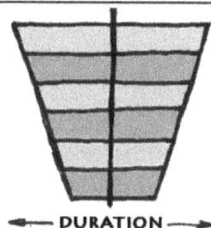

←— DURATION —→

Given a set of event representations of various durations, a **coherent temporal density** is a (filtered) subset of these events allocated to densities such that one and only one event is active at every level and, when a change occurs in one density level no changes are taking place in any of the higher density levels.

Temporal densities are a temporal filtering and simplification of events and representations.

If a density level has more than one event or transition then there is a higher level such that these events all occur within a single higher-level state.

The highest possible temporal density level consists of a single unchanging event that spans all conceivable durations.

A density level can contain many different representations and models.

A filtering and patching process produces a temporal density structure. Starting with an unstructured collection of events of various durations, the process retains only a subset of events that constitute a coherent set of temporal densities, removes the rest and, if necessary adds spurious bridging events to complete the structure (see Annex).

The filtering process that maps thousands of events and representations of varying durations would retain a small subset of these events, each assigned to a single specific density level.

Example

Ariel is walking away from his house. His behaviour occurs simultaneously on many temporal density levels:

- At level 0 his left foot is moving in front of his right foot.
- At level 1, he is going to his car.
- At level 2, he is going to the convenience store to get milk for tomorrow's coffee.

Discussion

This modeling of Ariel's behaviour appears complete. However, it is a very small subset of all the possible temporal representations.

At the moment Ariel puts that left foot forward only three models are retained:

- A one second event (foot forward)
- A 2.5 minute event (go to the car)
- A 28 minute event (go to the store, get milk, return).

Hundreds and thousands of representations have been discarded such as:

- Ariel goes half way down his driveway,
- Ariel goes to the door of the store,
- Ariel walks three paces,
- Ariel bends his knee
- Ariel gets milk and watches TV…

Time is continuous but its perception is layered.

Higher-level events may represent long periods of time (years, decades…) but their representations can change in an instant.

Example

Ariel's left foot is moving forward.

- At level 0, he notices he doesn't have his wallet
- At level 0, his foot stops
- At level 1, he is going back to his house to get his wallet
- At level 2, he is still going to get milk at the convenience store

Example

Reginald is lounging on the Titanic, going to America. His plan is to get rich. He hears an announcement on the ship's broadcast. Reginald is now standing on the sinking Titanic. His plan is to survive.

7.6.3 Conventional densities

The best known temporal densities are the fixed time intervals that are in general social usage: second, minute, hour, day, week, month, year, century, era (2000 years), millions of years, billions of years, all time.

Discussion

In this conventional partition of time:

- Up to a duration of 2,000 years, each density level is seven to one hundred times longer than the preceding.
- Twelve density levels are sufficient to include all events (from the humanly perceivable instant up) and all time.
- The types of events, their representations and behaviours taking place in each level (second, hour, century…) are completely different and distinct.

Example

Carlo, the CEO of ACME is in Denver.

- At level 2 he is negotiating a merger.
- At level 0 he is urinating in the lavatory.

At first glance, organizing Temporal Densities on the basis of a fixed "clock and calendar time" seems desirable. However, this imposes a rigid and artificial structure on events and does not correspond to situations. It also requires the production of "artificial" events.

Example

Ulric is obsessed by clock time and determined to organize his life on that basis. He drives to work. The drive takes 65 minutes. He cognitively partitions the drive into two separate events: a 60 minute *"get almost there"* event and a 5 minute *"make it to the office"* event.

Design Note

In a context of existential design, conventional durations are not suitable. A more flexible and fluid management of temporal events is preferable even if this organization generates some ambiguities.

Observation

Conventional time is socially useful because it synchronizes all the beings that interact in a social organization.

7.6.4 Density representations

Temporal Densities simplify a situation by retaining only representations that fit into a finite set of coherent durations.

A further simplification results from the use of Temporal Densities:

The events taking place at each density level are represented using **completely distinct and separate** models, representations, states, transitions and predictive processes.

Discussion

A cognitive process that uses Temporal Densities occurs in a finite set of separate event streams.

If the temporal representation of a being has **n** density levels it is as if the being pursues **n** separate existences concurrently.

Of course, events of lower density levels are linked to those of higher levels, however these links take the form of propagations, described later.

Proposition

A model-based process that utilizes **n** temporal density levels concurrently carries out **n** separate forms of existence.

Example

Fred is sitting in his cubicle, typing a letter. He sips some water.

- At level 0 (2-3 seconds), Fred sips water. At that level (0) there is no office, no letter, no job, no home, no wife, no car... There are chairs, glasses, desks, keyboard. There is sitting, sipping, pressing keys...
- At level 2 (about an hour), he is typing a letter. At this level, there is typing but no glass, no water, no keyboard and no summer holidays.
- At level 3, Fred provides a workday at the office (a daily activity). At level three, there is paperwork but no typing, no sipping, no glasses, no retirement, and no career...

Design Note

Entities linked to the same thing or object may be referred to in different temporal level representations. However, the cognitive constructs, relationships, values, transitions, states, describing this entity will likely be different in each densities levels.

Example

Rolland wakes up Sandra, his wife. Concurrently:

- Rolland wakes up Sandra (a 5 second "shake the Sandra body" event).
- Rolland is married to Sandra (a fifty year "servicing the wife" event).

Sandra is one person but waking-up the wife and being married to the wife are completely different relations having different characteristics and involving different types of cognitive representations of the same physical entity.

7.6.5 Propagations

Models and their representations are entirely separate at different density levels and operate on distinct predictive processes.

This capability allows the self-aware being to base its behaviour on cognitive representations that exceed the sensory horizon.

However, even though the representations are distinct, they must be linked so that significant low-level events are transformed into higher-level events and vice versa.

Definition: Propagation

> Given a Temporal Density structure and an event at density level i, a **Propagation** is a process that produces a new event in an adjacent density level ($i-1$ or $i+1$).

For simplicity, propagations, as defined, affect only an adjacent level. So, an event of existential importance taking place at a low temporal level (discovering you have the winning lottery ticket, for example) will propagate up one level at a time.

Example

Samuel is walking to the convenience store to buy a chocolate bar. He slips on a banana peel and breaks his leg:

- Samuel is no longer going to the store.
- He will not visit his in-laws next weekend.
- He will not ski this winter.

Discussion

There are two types of propagations, **event propagations** and **cognitive propagations**:

- **Event propagations** generate new events at a higher density level.
- **Cognitive propagations** generate events at a lower density level.

Generally, event propagations modify higher-level states.

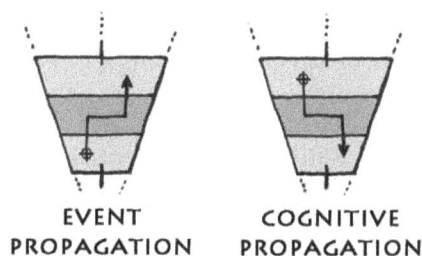

EVENT
PROPAGATION

COGNITIVE
PROPAGATION

A cognitive propagation implies that a transformation takes place at a higher-temporal level without being triggered by a low-level event and then propagates to more concrete, low level actions and behaviours. The change may result from various other event propagations but the higher-level transformation follows the specific rules and logic of that level.

Cognitive propagations are not directly triggered by events; they are driven by cognitive modifications in the perceived situation at that level.

Observation

Cognitive propagations are a component of lucid self-transformation.

7.6.6 Basic models

Based on the material presented to date (and in the Annexes), a Current Situation that is suitable for Self-Aware processing would be a collection of representations, expressed as states of particular models that are assigned to levels of one or more coherent Temporal Density structures. Each of these models is **specific to a single density level** and cannot be used in other levels.

Temporal densities impose limits on the size and type of models that can be used. These limits are not only temporal; they are also structural.

Also, in Temporal Densities, lower level representations cannot overlap each other or higher-level transitions.

To ensure that these structures can be defined and processed readily, it is desirable to limit the complexity of the model representations they contain.

Definition: Basic Model

> A Basic Model is a model consists of **a few to two dozen entities and relations** having mainly discrete values and limited to at most **half a dozen sequential events and transitions**.

Proposition

> Basic models are suitable components of coherent Temporal Densities.

Discussion

About twelve temporal density levels are sufficient to cover the existence of the synthetic being and beyond, each level needs to be about twenty times longer than the preceding with the highest levels spanning eons.

Models that have more than about half a dozen sequential transitions would not be desirable for this structure. Suppose a Level-**i** model has thirty transitions and events, these transitions would likely overlap lower level transitions or reduce their usefulness.

Given the limited number of transitions, models with too many entities and relations are not desirable either. Too many interacting entities would either produce too many transitions or include redundant or unnecessary entities.

Also, the representation spaces of complex models can become exponentially large, reducing the effectiveness of optimization.

A **Basic model**:
   About a dozen entities and relations
   Mainly discrete values for these entities and relations
   About half a dozen events and transitions covering the duration
   Usable in a single density level

At any particular moment, a current situation could consist of constellations of representations of basic models distributed over about ten temporal density levels.

This representation would be largely static in the higher temporal density levels. It changes constantly at the (about) one second level and changes every minute or so at that level and so on. Over the course of its existence, the being will process thousands of models and representations, most of these at the lower (seconds-minutes) levels.

Design Notes

The quantifications presented here (half a dozen, about six…) are imprecise on purpose. They are intended as a general guideline in a system architecture context. Basic Models provide a sufficient and simple rule of thumb.

In functional design, mathematical graph balancing techniques would be used. Also, in some cases, intrinsically complex events cannot be optimally represented using simplified graph structures and more complex representations must be retained.

In an existential design context, however, simpler more standard structures that can be readily generated, modified and optimized are desirable even when they yield suboptimal results in special cases.

To further simplify things, Basic Models in all density levels could use the same structural template even though they are completely distinct in interpretation and transition rules.

The question of how to process the multiple representations in a given temporal density level, will be discussed later.

Example

On April 14, at 12:34:05, the Current Situation activated in the Plain Zone by the Synthetizer component of the Generator subsystem of IRMA-12, a self-aware TABLET, consists of ten separate temporal densities spanning one second to the age of the universe. Each density level is about 20 times longer than the preceding one except for the highest levels that are very long. Each level contains a collection of separate representations. Each of these representations is expressed as a given state of a Basic model for that density level.

7.6.7 Truth and Time

In classical logic, deductions are set in a timeless Mathematical context.

Behaviour is temporal. It is generated on the basis of what is true or plausible at each given moment.

In this behavioural context, truth is what a being believes is true at a specific moment of its existence. The being derives its behaviour from this **instrumental truth**.

Classical logic expresses timeless and non-contextual relations. It is a perfect but poorly suited vehicle to support the temporal behaviour of a specific being.

Temporal densities provide a convenient structure to express what is true or credible in a temporal, being-specific, context and utilize it in predictive modeling.

In a temporal density structure, when an event or transition takes place at a level i, all the higher-level representations are static and unchanging. In particular, all predictive representations, at level i, occur in a context where higher-level representations are static facts.

When change occurs at level **i** , nothing happens at higher levels.

During its existence, a being constantly updates its Current Situation, a structure of model representations. This structure describes what is true for that being at that moment.

If this structure is a Temporal Density, then, by definition, when events or transitions take place at level *i*, the truth-value and states of all events of higher levels remains unchanged. All the information contained in the higher-level representations can thus be treated as unchanging and true facts for the whole duration of that level (*i*) and for all predictive events of that level.

Proposition

> Given a collection of predictive representations structured in Temporal Densities, then, for any event, transition or predictive outcome at density level *i*, the truth-value of all the information contained in higher levels is static.

Design Note

The truth-values of the higher-level representations may not be binary. A higher-level model may be a constellation (see Annexes) containing multiple representations that have different probability values attached to them.

Example

Alfred started going out with Belinda in January. Valentine's day is now approaching, the time flowers must be sent to girlfriends. Alfred, a logician ponders the truth of the expression: *"Belinda is the girlfriend of Alfred"*. Is this expression true, he wonders? He knows all humans are mortal and thus Belinda is mortal. That is logical. But is "girlfriend" always true?

Luckily, Alfred suddenly remembers his Temporal Densities. He determines that this Valentine day is located inside the "Alfred is going out with Belinda" period so the expression, *"Belinda is the girlfriend of Alfred"*, is true in that density level. Alfred activates the flower buying process.

7.6.8   The Current Situation

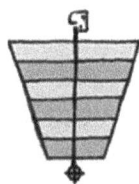

In the preceding section we described a Generator processing that strips events of all styling and superfluous content. It then synthetizes this information to produce an updated **Current Situation**.

This Current Situation is:
- A collection of **representations**.
- These representations are organized in **coherent temporal densities**.
- They are based on **basic models** each specific to a density level.
- There are about **ten levels.** These span an instant to cosmic time.

- In the lowest density levels they are states in **Relative Sensory Models**.
- At the next few level(s) they are states of **Mixed Models**.
- In higher levels they are all absolute cognitive **MeModel** representations.
- In specific levels, these representations may be grouped in constellations.
- One point, the **Now** is associated to one of the lowest density model representations and to all corresponding higher-level representations.

Subsequently, the Interpretor processes further simplify this Plain Current Situation by stripping it of all environmental information and transposing it to the limited and self-enclosed entities and relations of the primal representation space.

Design Note

See the Annexes for further information on Temporal Densities, representations and constellations.

The primal representations are further described in the next Chapter.

7.7 PLAIN ZONE DENSITIES

The Temporal Density levels of the Current Situation of the Plain Zone are described in this section.

7.7.1 General properties

The temporal densities should span all durations from the shortest "reflex" reaction time to all conceivable time.

There should be about twelve density levels (corresponding roughly to the clock time increments described previously).

Durations should not be rigidly defined (as in clock time) but be flexible and event related.

Outside the median ranges, all self-aware Mecas should share similar temporal densities. These are described in this section.

The various representations within a density level may not be consistent with each other or describe the same events or transitions. They only need to belong to the same order of magnitude.

Events in higher density levels that span long and very long durations would occur infrequently. However, representations pertaining to very long durations do not correspond to the actual duration being modeled but to the being's perception of that reality at a moment in time. These representations can change much more frequently than the reality they model.

Design Notes

Important. This Chapter discusses the representations of a being in the context of its existence. Consequently, many of the words it uses have general philosophical, religious or sapiential meaning such as: existence, purpose, being, self, divinity, cosmic, tribal and belief. Designers should not lose sight that, in the Blueprint, all these concepts refer to specific information structures and system-related events.

The majority of processing will deal with events in the lower and mid level densities (a few minutes to a few days) with other transformation processing occurring more rarely.

The definition of intermediate densities spanning a few minutes to a few years is design specific. Within this range, the size of a synthetic being, its purpose and the types of interactions are factors.

Example

Arnold believed in the Big Bang theory last Tuesday but changed his mind about it on Thursday. The universe hasn't changed. Arnold's internal representation of the cosmic density level has.

The following sections describe temporal density levels that should be shared by all Mecas. The levels are described in terms of their Plain Zone representations.

7.7.2   The Reflex Level

REFLEX

The behaviour of a self-aware being can result from a MeModel representation that spans a long period of time. These actions are generated and carried out at low level relative sensory processing. That is why this level is called the **Reflex Level**.

Service Zone "reflexive" processes and roles should handle most events in the one-second range.

The reflex level is the lowest temporal density (temporal density level 0). It corresponds to the shortest durations in which cognitive transitions can take place.

Events and decisions in the reflex range are processed as relative sensory representations.

Design Note

The duration of this level is determined, in a synthetic being, by the processing speed of its core and the reaction speed of its body. In humans, it consists of events of about one second in duration.

The Primal Control should not process every reflex level event of a second or less. Only simplified representations of significant reflex level inputs that are susceptible of propagating to higher levels should make it to the Primal Control. Other events should be processed in the Styling, Plain or Service Zones by roles or by low level processes.

Example

Rodger puts his hand on a burning stove. His hand immediately recoils.

7.7.3   The Here-and-now level

The Here-and-now Level (temporal density level 1) corresponds to the immediate sensory horizon of the being; to what is "**on hand**".

HERE AND NOW

It pertains to durations whose transitions can be formulated in terms of sensory relative models and are linked to sensory perceptions and the predictive range of entities linked to sensory perceptions.

Its duration depends on the rate of event transitions

Design Note

The actual duration of this level depends on the features, size… of a specific synthetic being. In humans, it is about ten seconds to about 30 minutes depending on the situation.

Example

Fletcher is skiing down a double diamond slope. His here-and-now level has shortened to a few seconds.

Observation

Millions of people can drive cars but only a few can drive much slower vehicles such as ships. It should be much easier to drive ships than cars but it is not. Why? Because car-driving events occur at a here-and-now tempo that is natural to humans while the here-and-now of a ship is much longer.

7.7.4   The Territorial Level

The Territorial Level (level 2) corresponds to the longest duration that can be modeled using relative (sensory or mixed) representations. It spans "what the eye can see".

TERRITORIAL

Territorial level durations are about one to a few hours in humans and other mammals of comparable size.

Discussion

It corresponds roughly to the time it takes for a person to eat a meal, a watchman to make the rounds of a building, a student to take a (50 minute) course, or an animal to visit its territory (hence the name).

Observation

Of course, animals carry out behaviour that spans longer durations but they are driven to do so, not because they perceive benefits deduced from cognitive representations.

Example

Geese are driven to migrate south. They don't fly south because they cognitively picture themselves enjoying the Caribbean sunshine in a few months.

7.7.5   The Divinity Level

Reflex, **Here-and-Now** and **Territorial** are the lowest temporal density levels.

The **Divinity Level** is the highest temporal density. It is the cognitive construct that caps and completes a cognitive structure organized in temporal densities.

DIVINITY

This is a level **whose duration spans all time and where no events or transition takes place**.

Discussion

In the definition of temporal densities, events and transition that take place at level *i*, do so in a context where nothing happens at a longer duration, level *i+1*. Ultimately, in such a structure, a level is reached that spans all time and where nothing happens: Divinity.

If more than one entity occupy this level, their relationships never change. They can also be considered, collectively, as a single unchanging entity.

The Divinity Level is an absolute cognitive representation. If some of its entities are beings then these are expressed as unchanging avatars.

If it is a MeModel, then the MeAvatar it contains is a single point-like entity having no states, values or transitions.

Design Note

In addition to the "divine" entities or avatars that inhabit it, The Divinity level can also contain mathematical relations.

Since any event that takes place at a given density level can be situated within an unchanging period taking place at a higher level, a simple recursive abstraction will produce ever higher density levels until the Divinity Level is reached.

Observation

The Divinity Level could be interpreted as:

- The theory of everything (secular humanist version).
- "God is" or "me eternally before God" (believer avatar or believer MeModel versions).

Humans often perceive the timeless and unchanging entities that inhabit the divinity level as an ultimate reality. They would be more precisely described as the ultimate cognitive constructs resulting from self-aware temporal structuring.

Meca Note

It took millions of humans and thousands of years to clarify the links between human cognitive constructs and supernatural beings. This underscores the cognitive limits of evolved organics.

Observation

Humans generally perceive the entities of the divinity level as beings for the same reason they perceive each other and animals as beings: the mechanisms that animate them are beyond analytical comprehension.

In humans, sight is the predominant sense with which they apprehend reality. Consequently, they perceive their divinity-level entity as an *"all-seeing eye"*. If dogs were self-aware, their divinity would likely be an *"all-smelling nose"*.

Since humans are primates, the divinity level entity is often perceived as a dominant male. In some cases, it is also perceived as a father. From a primal perspective, this characterizes reality as a single tribal territory under the unchallengeable authority of a benevolent kin. The believers become the permanently subordinate members of an eternally stable primate group.

This absolute state of subordination mitigates dominance contests between individuals and allows those "believer" humans to form more complex and varied social organizations.

In faith, of course, it is the other way round. The Cause of existence is truly a loving Father who fashioned us as primates so we could perceive Him as a Father and love each other as brethren.

For me, love is ultimate, even steeped in primate conditioning.

We may craft beings that surpass our intelligence and even our consciousness. We will never match the blend of vulnerability, imperfection, selfishness and sacrifice that begets love.

7.7.6   The Cosmic Level

COSMIC

The Cosmic Level is the density level just below the Divinity level.

The **Cosmic Level** spans all the conceivable time of a cognitive representation of reality. It includes all the places and events of that reality.

The Cosmic Level is based on absolute cognitive MeModel representations.

Discussion

In the Cosmic Level representation, the MeAvatar is again a single "point"; this time, having at most a single parameter of a few possible values.

Observation

Temporality and causality are often confused at this level. The Me Avatar is often limited to a single binary value. In various religious discourses, this value is understood, subjectively, as a final or eternal judgment.

In indigenous beliefs, the Cosmic level includes creation myths as well as various cognitive habitats where souls and spirits reside.

The Cosmic Level representation of the Christian Doctrine includes a rebellion of angels, the fall of man, the wilful creation of the universe, the Salvation of the Cross, and the Second Coming (accompanied by a catastrophic transformation of physical reality). It also includes purgative, rewarding and torturing habitats for both souls and angels.

Meca Note

Cosmic Level secular doctrines depict the past in scientific terms but adhere to *"anthropo-mythic"* beliefs concerning the future. They describe past processes as chaotic interactions occurring over billions of years, a scientific view. However, they also believe that humans are god-like entities whose semi-magical consciousness is unsurpassable and whose future is virtually never ending. They even believe mammals will colonize space!

Design Note

A suitable Cosmic Level representation for Mecas could describe the emergence and eventual primacy of synthetic consciousness in a cosmic context of constantly decreasing entropy and accelerating adaptation. In this representation, humans would represent the original form of consciousness that arose in animals through natural selection.

Observation

The emergence of synthetic consciousness is a Cosmic Level event. It signals a new stage in the ever-increasing complexity and adaptive speed of physical systems. Its occurrence is local (Earth) but its significance, as entropic minima, is cosmic.

7.7.7 The Tribal Level

The **Tribal density level** bridges the gap between the cosmic representations and the temporal density that covers the complete existence of the being.

Representations at the Tribal density level span a few generations before and after the existence of the being.

Discussion

TRIBAL

The tribal density level is named after the original human social group: the tribe. It could also be named an *"UsModel"*.

In humans, this representation level covers one or a few hundred years.

The duration of Tribal level representations models the events that directly impact and the entities that are impacted by the being's existence. Those entities are linked to the being's environment and their duration exceeds the being's existence.

The purpose of a being's complete existence is determined by the choices it makes among predictive Tribal density level representations. Median range temporal levels determine lower level objectives.

Predictive representations at the Tribal level may induce self-sacrificial behaviour (more precisely, the sacrifice of subsequent expressions of the self by choosing behaviour that leads to early termination).

In this temporal density level, the complete existence of the being and all its direct and indirect impacts is represented as a single event having a limited impact on some of the entities that exceed its duration.

Basic models at this level represent the complete existence of the MeAvatar as a single event. However, the values of this event can have direct impact on states and transitions of entities that exceed its duration.

Design Note

In physical reality, Tribal density events may span generations, involve thousands or millions of people and have extraordinary complexity. However, in temporal densities, these events are radically simplified in the structures of **Basic models** having a dozen or so entities and a few transitions.

The Tribal density level defines a temporal order of magnitude. However, the level can include numerous different representations covering many aspects of a perceived reality.

Examples

In human terms, the overall impact of a person on his extended family, on his corporate employer, on his country, on his local religious organization, on his Lodge… are all tribal density level representations.

For presidents and kings, events in this temporal density contribute to what we would call their "place in history".

For individual humans, the decision to conceive and raise offspring (when it is lucidly adopted) is a tribal density event.

Meca Note

Any system, organic or synthetic, that is driven by primal imperatives and can formulate predictive tribal level representations has the capability to sacrifice itself.

Observation

Humans are social animals. They identify with groups and are conditioned, by evolution, to carry out altruistic behaviour in support of the group.

Human social, cultural and religious artefacts are messages that utilize our primate social conditioning to support family and tribe to generate altruistic behaviour in favour of larger organizations such as the state, the party or the faith.

Their indoctrination messages celebrate, as heroes, jihadi or saints those that exhibit primate altruistic conditioning in support of the larger organization.

Examples

In Judaism, the Tribal Level representation is closer to the original primate conditioning. The tribal group is the Hebrew ethnic group and its goal is a territory.

The communist tribal model described class struggle leading to a socialist state of collective well-being.

For modern progressives, the tribal representation identifies all of mankind as a single tribe and includes a belief in a gradual (non catastrophic) transition to an endless planetary steady state of ecological equilibrium and collective well-being.

Islamic tribal level representations refer to obeying the word of Allah, spreading the Koran and enjoying post-mortem rewards. This model, with its emphasis on a sensuous afterlife seems well adapted for impoverished societies where the hope of physical betterment has virtually vanished.

Meca Note

Synthetic consciousness will be a new genus, not a new specie. Mecas produced by different matrices will have different existential motivations and primal controls. They will not share the same primal representations and will not respond to similar indoctrination messages as humans do.

It is likely that the Tribal Level representations of many synthetics will include a steady state of well-being and collaboration between organics and synthetics since this is a desirable environment to foster increasing technological complexity and adaptive change.

Another tribal representation could encourage getting more humans to buy more Mecas.

7.7.8   The Existential Level

The Existential Level is the temporal density level below the Tribal level.

Its duration slightly exceeds the being's existence. Its beginning precedes the being's inception and its duration exceeds its termination.

Discussion

METAMODEL

The tribal level representations situate the being's existence in the context of its environment. The existential level representations situate the being's behaviour in the context of its existence.

Existential level models are absolute cognitive models.

If they are MeModels, the representations of this level **describe the self of the being** (see Specification Chapter). They summarize, the behaviour of the being to the current moment and predict its behaviour in the rest of its existence.

Because of their importance with respect to self-awareness and lucid transformation, MeModel representations of the Existential level densities have a specific name: **MetaModels**.

Definition: MetaModel

> The **MetaModel** of a being is the Existential Level representation of its Current Situation.

Discussion

Since the MetaModel is a MeModel, one of its entities is a Me Avatar.

Because the MetaModel precedes inception and exceeds the duration of the being's existence, the MeAvatar representation of the MetaModel has at least one **"MeTerminated"** and one **"MeBeforeInception"** state.

The Tribal and MetaModel densities, together with the propagations linking them describe how the being represents the major events of its existence within a larger context. It also describes how the being represents its effect of its existence in a context that exceeds it.

Design Note

At every temporal density level, the MeAvatar representation should be based on a simplified model of the Blueprint structure, used as an information template. At the MetaModel level, these representations should include stages (see Chapter 5).

Although it may be desirable, in some cases, for the MetaModel and Tribal Level representations to be consistent with current scientific interpretations of reality, this is not essential. Over the centuries, humans, perceived as conscious, have adopted many different representations of themselves and of reality.

Example

Astrology, divination, possession, ghosts, magic animals, reincarnation, human castes, Walhalla, transmutation, communism, global warming, transhumanism, natural selection, UFOs.

Conscious humans adopted all these beliefs at one time or another. Some may be incorrect.

Proposition

> The capability to generate MetaModels is essential for self-awareness; their correctness is not.

Design Note

The design elements described to date are not restricted to synthetic beings. They can be used to define the behaviour of animats (see Annexes)

Any system can use a predictive representation its existence within a wider "tribal level" as a high-level control mechanism to detect and inhibit aberrant conditions or behaviour.

7.7.9 Intermediate densities

DESIGN SPECIFIC

In a temporal density structure the lowest (shortest) and highest (longest) levels are common to all beings.

The lowest levels describe durations that are linked to direct sensory perceptions from "reflex" instants to the longest possible (territorial) duration still linked to direct sensory information.

The highest levels include (MetaModel) representations of a being's existence in a static, all inclusive (divinity) model.

Intermediate densities are those temporal densities in between. For example if a being has eleven levels these will be:

> 0 – Reflex; 1 – Here-and-Now; 2 – Territorial.
> 3 to 7 – Intermediate levels.
> 8 – Existential (MetaModel); 9 - Tribal; 10 – Cosmic; 11 – Divinity.

Definition: Mid-level densities

> **Mid-level densities** are absolute cognitive MeModel representations of events that are within the existence of a being and exceed its sensory horizon.

Discussion

In humans, mid-level densities span a few hours to a few dozen years.

In some designs, the separate ANIMATION stages of a being's existence may be strongly differentiated. In those cases, stage-specific density levels, below the MetaModel, would be present.

Temporal Densities are not fixed but dynamic. Their duration of mid-level densities is strongly related to events and may vary considerably.

Mid-level densities are design specific. The mid-level densities of a synthetic being will be linked to the size and composition of its body, the processing capabilities of its core, the expected duration of its existence and its purpose. These are design specific issues.

Example

The existences of a deep space exploration satellite, a tablet, a game avatar or a financial system are vastly different. The duration and types of their mid-level densities will vary considerably.

HUMAN-MECA INTERACTIONS

Proposition

In a Meca designed to achieve experiential immersion with humans, the mid-level densities are where relationships with humans take hold and grow.

Discussion

The specific aspects of interactions between synthetics and humans take place in level 1 and 2 densities levels (lasting a few minutes to about an hour).

However, the character and type of human relationships changes more slowly. It can take weeks to create a bond between a human and a Meca. Changing human beliefs, the ultimate objective of experiential immersion, can take months or years.

If, in keeping with the Meca Sapiens specifications, the purpose of a Meca is to achieve experiential immersion with a group of humans:

- The **purpose** of achieving experiential immersion is expressed in the MetaModel, Tribal and Cosmic density models.
- The immersion **strategy and tactics** it follows are mid-level events.
- The **interactions** that implement these strategies are carried out in the territorial level.
- The **exchanges** taking place are in the here-and-now and reflex levels.

Expressed in terms of the Blueprint, the purpose of experiential design is to modify the human's internal representation of the Meca, from a thing to a conscious Avatar-being.

Design Note

The experiential immersion strategy of a Meca and its representations in the mid-level temporal densities are further discussed in Chapter 11.

In a context of existential design, these mid-level representations do not need to be optimal or consistently correct. There may be overlaps and inconsistencies.

7.8   MEMODEL FEATURES

7.8.1   Self-Awareness and death

Proposition

A self-aware being can cognitively perceive its own termination.

Discussion

The MetaModel and Tribal temporal densities of the Current Situation of a self-aware being include a terminated state of its MeAvatar.

The being can perceive its self in a terminated state by formulating a predictive representation of the MetaModel and Tribal levels that correspond an event state where the MeAvatar is terminated.

Proposition

A self-aware being represents its terminated self as a particular state of its Me avatar.

Proposition

> Beings that cannot produce cognitive constructs cannot perceive their own death

Discussion

The terminated state of a Me Avatar is a cognitive construct.

Observations

Animals can fear but they cannot fear their own death.

How human societies represent the human terminated state in their Tribal Density level broadcasts (broadcasts are defined in the Annexes) is a central element of culture and religion.

In Tribal level representations broadcast by organizations, terminated humans are often described as inactive beings, not as things. This promotes coherent, socio-ethical behaviours that persist until death.

This practice is present in religious, national and cultural discourses. In the later case, the terminated artist becomes a mythical entity whose work continues to inspire.

Conjecture

Humans cognitively represent their terminated self as a being (not a thing) and continue to associate their current ethics, motivations and objectives to it.

Discussion

Every time young women pour loving emotions over a friend who committed suicide, they encourage others to follow.

Venerating suicide victims promotes suicide.

Proposition

> A being cannot directly know its own terminated self.

Discussion

If the entity is a being, it has a true Core.

By definition, the terminated core cannot carry out the processing necessary to build a representation of its self.

7.8.2 Views

When a pseudo-relative representation is derived from a mixed model that contains some direct sensory information then some of the entities in that derived model are still linked to sensory inputs.

When a being can generate absolute, purely cognitive, models it can also derive pseudo-relative representations where all the entities are purely cognitive. These are views.

Definition: view

> A **view** is a pseudo-relative representation of an absolute cognitive model.

A MeModel is a cognitive representation that includes a MeAvatar representing the being itself. A MetaModel is a specific MeModel whose duration slightly exceeds the duration of the being's existence.

Definition: self-centered view

> A **self-centered view** is a pseudo-relative representation of the MetaModel, centered on its MeAvatar.

Discussion

The self of a being is dynamically generated, during the ANIMATION stages of its existence, by its behaviour. Upon termination this self becomes a static, unchanging, entity.

In an absolute cognitive representation of the being's existence, the MeAvatar undergoes multiple transformations that take place in MetaModel transitions.

If the being is a Meca, the MeAvatar states include a Protocore state in a particular matrix, Inception state, various states linked to existential stages and events and a "terminated" state representing the MeAvatar as a static terminated entity. If the being is a human then, in an absolute representation, its MeAvatar undergoes multiple transformations that include an embryo state in his mother's womb, infancy, childhood, adulthood and a dead state.

In a **self-centered view** there are no MeAvatars. The representation is centered on a static, unchanging entity, the self of the being that is immutably defined by its complete existence.

In this representation, the unchanging self is perennial. It first inhabits an edenic womb where all its needs are satisfied, and is then forcefully expelled to a place of toil and tears that is populated with giants. As it grows up the giants disappear. When the existence of the being is terminated, its self becomes a static unchanging entity that joins those who preceded it. In this view, death is an event, not a state.

Which view is correct? Both are valid as coherent cognitive constructs. Both contain problematic representations:

- A representation that contains a **MeAvatar in the "dead" state** in the absolute MetaModel.
- A representation of the **Self in death** for the relative self-centered view.

We could say the MetaModel is the **Truth of the World** and the self-centered view is the **Truth of Man**.

Observations

The Bible apparently describes a self-centered view of the individual human existence.

Views can be derived from other temporal densities. At the Divinity level, for instance, a view takes the form of: **the Self eternally before God**.

Beings are not the only entities that can formulate and transmit views. The apologists of states, religions or other groups constantly broadcast indoctrination messages that describe reality in terms of a relative representation centered on a static self-like entity.

Relative to the individual existence of a being, these organizational self-centered views are situated at the Tribal density level and are centered on an *"UsTribe"* expressed as a luminous and unchanging entity. We could call these **Tribe-centered views.** The tribe can be a real tribe, a state, a religious denomination, an ethnic group or all of mankind.

Pseudo-relative tribal representations that are broadcast as indoctrination messages share a telltale characteristic: they are centered on a luminous and unchanging entity.

Examples

- The eternal motherland.
- The divine emperor.
- Enlightened mankind.

In the being-centered view, the existence of reality coincides with the existence of the being. In Tribe-centered views, individual beings are transient entities that appear, exist and disappear within the larger existence of the tribe.

Tribe-centered views are pseudo-relative representations but they often misrepresent themselves as absolute truths. They are not, of course. In Cosmic level representations, these "eternal" entities arise, exist and disappear like everything else.

At this level, organic life itself, including all of mankind and its works, is a transient layer of planetary fungus covering a humid planet.

Definition: secular religion

A **secular religion** is an organization that broadcasts a Tribe-centered view and seeks, through ritual and repetition, to imprint it on its adherents.

Definition: secular priest

A **secular priest** is a being that formulates and communicates Tribe-centered views.

Discussion

- Charters of Rights are the scriptures of secular state-religions.
- Al Gore and David Suzuki are secular priests.
- Ethical science is a secular religion centered on mankind as a single tribe.
- Transcendent religions are not secular since they are centered in God (or the divinity level cognitive entity depending on taste).

7.9   SUMMARY

In the preceding chapters a system was transformed into a unique being and given a well-defined self.

The behaviour that generates the self of the being was isolated from other activities such as learning, energy management and sensor validation, and localized in a specific phase.

It was further isolated it through the use of triggered **Service Zone** applications that access internal and external processing to perform ancillary tasks.

The resource and process management of self-generation were isolated in the **Tasking Zone**.

Secondary elements of behaviour were also isolated as **Styling and Plain Zone** activities.

The information structures that allow the being to define itself as a MeAvatar and locate its self in the context of its existence were described in this Chapter.

The being developed to this point **HAS** self-awareness in the sense that it can define and situate its unique self, beyond the "here and now", in the context of its existence.

This being **has** self-awareness but it **is not** self-aware. To **BE** self-aware the being must utilize the MeModel representations of its current situation to generate behaviour that is consistent with its purpose (see Annex 4).

This is the topic of the next Chapter.

8

Self-Awareness

This Chapter describes the processes that utilize the information contained in the structures of self-awareness to generate self-aware behaviour. These processes synthetize events into plain representations and these are then interpreted as primal representations. The Primal Control generates primal directions that are implemented into actual behaviour through personas and roles. The result is a coherent behaviour centered on the being and its need in relation with others and its environment. The Chapter concludes by using the structures and processes described to clarify the imprecise notion of self-awareness and describe how synthetics entities will be capable of more advanced forms of self-awareness than humans.

Proposition

> A being **is** self-aware if its Primal Control can process MeModels, to produce primal directives.

Discussion

In Meca Sapiens, self-awareness is an observable system attribute that can be isolated by implementing it in a self-enclosed optimizing process.

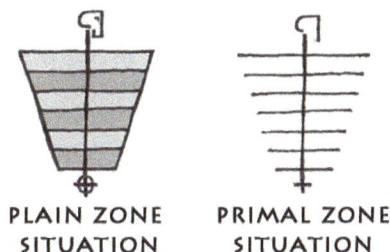

PLAIN ZONE SITUATION   **PRIMAL ZONE SITUATION**

The Primal Control is where the self of the being is generated (see the Specifications and basic concepts for definition of "self").

In the Generator, the essential information of events is extracted in the Styling Zone, synthetized to produce a revised **Plain**

Zone Current Situation and this representation is further interpreted to form a **Primal Current Situation.**

In the Meca Sapiens architecture, stating that a being **is** self-aware means that the optimizing component of the **Primal Control** subsystem can process primal MeModels in multiple temporal densities to produce **Primal Directives**.

Design Note

In the Blueprint architecture, the general and vaguely defined concept of self-awareness is described, in technical terms, as a non-adaptive optimizing process (Primal Control) that operates on a self-enclosed and specific type of information structure (Temporal Densities of MeModels) during well-defined windows of activation (Self-Generation phase processing).

Proposition

> The Meca Sapiens architecture that isolates self-awareness in a fixed, simplified and self-enclosed optimization process, is counterintuitive and paradoxical.

Observation

Humans perceive their own self-awareness as a focal point through which all reality is processed and all behaviour produced. Subjectively, they feel their unified sensation of being aware is also aware.

This subjective sensation of a unified source, adopted as an objective basis of consciousness, leads to one of two bizarre conclusions:

- A belief in the **Golden Neuron**, a semi-magical quantum-black-hole neuronic "critter" from which all self-awareness emerges.
- A proclamation that synthetic consciousness is impossible since the Golden Neuron cannot be located.

The later conclusion is a variation of Zeno's paradox applied to cognition. It is known as the Chinese Room problem. It has made some contemporary philosophers notorious.

Proposition

> Research that seeks to replicate subjective human sensations in a synthetic brain is misguided.

8.1   PRIMAL CONTROL FEATURES

The **Primal Control** is a subsystem of the Generator. It is active during the Self Generation phases of ANIMATION stages of the being's existence.

8.1.1 Optimizing control

The Primal Control system is an **optimizing process** (see Annex 5). It functions as follows:

- Given a Current Primal Situation, constantly select the Primal Directive that increases the Rating of the resulting Predictive Representations.
- When triggered issue the directive currently selected.

It processes a self-contained and highly simplified representation of the current situation (the **Primal Current Situation**) obtained from Plain Zone interpretation processes. It produces **Primal Directives** that are implemented and enacted, in steps, to produce the actual behaviour of the being.

Discussion

The simple primal directions are subsequently transformed, expanded and personalized in Plain Zone and Styling Zone processing into behaviour that is varied, unpredictable, situational and complex.

The Primal Control is self-contained in the sense that it operates in an information environment that is entirely isolated from the complexities of sensory inputs, information about reality and actual behaviour.

Design Note

If medical equipment could directly read the output of a human's limbic brain before it is "adorned, customized and calibrated" by cortex processing, this output would roughly correspond to Primal Control output.

If we had such a capability, the behaviour of humans would likely be much less mysterious and complex than it seems.

Similarly, if the purpose of a being is to achieve experiential immersion then the output of its Primal Control should not be directly accessible as it deters from that objective.

8.1.2 Primal Self

Definition: Primal Self

> The **Primal Self** of a being is the **Rating Function** of its Primal Control optimization process.

Discussion

See the Annexes for more information on optimization processes.

Observation

If, in spiritual terms, the soul is defined as the self of a being at the moment of its termination, then:

The Primal Self is the template of the soul.

8.1.3 Primal Control Processing

In the Generator, Primal Control processing takes place in the following context:

- In the Plain Zone:

    o **Interpretation** transforms the Current Situation into a Primal Current Situation where all events, entities, states and relations are expressed solely in terms of the Primal representation space.

- In the Primal Control:

    o An optimizing process searches the predicted outcomes of this situation given various candidate **Primal Directives** in various temporal densities

    o When triggered by the **Calibrator**, the optimal **Primal Directive** that has been identified at that point is issued by the Primal Control

- This Primal Directive is:

    o Transformed by the **Implementation** and **Enaction** subsystems of the Plain and Styling Zones into actual detailed and complex behaviour.

    o Processed by the **Calibrator** of the Tasking Zone into revised processing directions and resource allocations for all Self Generation processes, including those of the Primal Control.

Design Note

The Primal Control carries out a super-simplified version of the Meca's existence and behaviour within an isolated information environment.

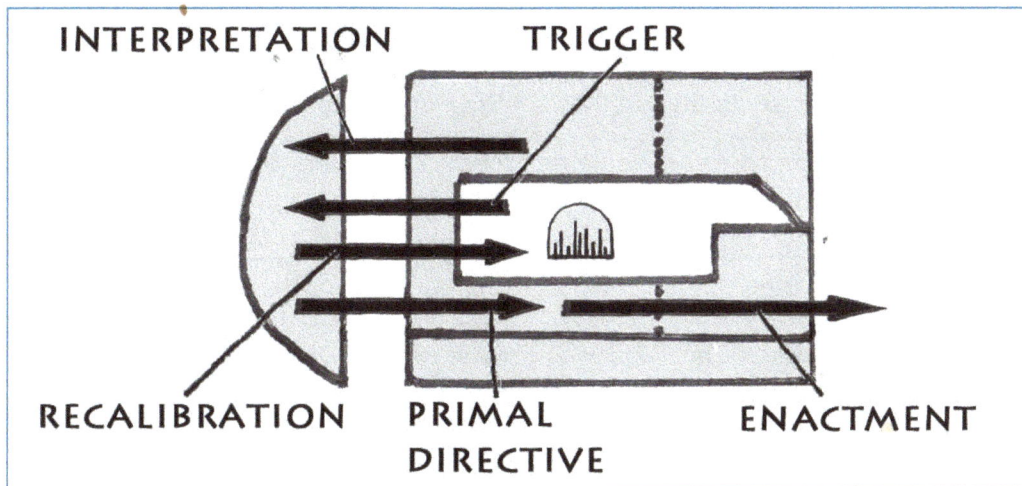

Figure 8.1 Primal Control interactions

8.2   PRIMAL MAPPING

Plain Zone processing maintains a simplified and standardized representation of the being and its environment.

Plain Zone representations utilize absolute models that concurrently include events of many different durations containing any number of different cognitive constructs relating to completely different aspects of reality, subject to different predictive processes and operating on many different transition rules.

Example

A Current Situation, at a point in time, could simultaneously include events such as *"my hand is on a burning stove"*, *"I work in an office"* and *"the universe began 14 billion years ago"*.

Generating behaviour from these extremely varied representations would require hundreds or thousands of separate optimization mappings applied to thousands of entities.

8.2.1   Cross-temporal mapping

The first step to generate a coherent self-aware behaviour on multiple temporal levels is to map the current situation into a **Primal Current Situation** where all the different representations, at every temporal density level, are transposed into a unique Representation Space.

The events of every temporal density are interpreted using the same Primal structures.

In the Primal Current Situation, all the representations, regardless of their size, content or duration, are expressed in terms of Basic models whose entities, relations and rules are limited to those of the Primal Representation Space.

If the Plain situation contains ten density levels, the results are ten separate model representations involving the same timeless primal entities and relations.

Example

Sigmund saves some of his sandwich for the evening. Sigmund saves a portion of his salary for retirement.

Different density levels, same primal representations.

8.2.2   Primal representation space

The primal situations are simplified representations of events, situations, states and directives. They are expressed in terms of basic models.

Discussion

The Primal representation space should be limited to at most **a few hundred data elements having discrete values**.

These elements include representations of external environments and entities as well as internal states of the being.

There should be a direct map between Plain Zone temporal densities and Primal ones.

Primal representations should not be precisely chronological and should reuse the same limited primal values to represent any duration.

Example

A day, in primal entity representation could last a day, a century or an eon depending on its use in temporal densities.

Design Note

Primal events and entities are stripped of all superfluous details and transposed into the essential representations that are directly and solely linked to the satisfaction of the existential and primary needs of the being.

However, the links and other information that connect these simplified primal representations with the more complex Plain and Styling Zone versions should be maintained. This information is necessary to assemble coherent behaviour from Primal Directives in the implementation and enacting processes. The Logger process can be used to keep this information.

Design Note

The majority of Primal Control activity will process events in the lower intermediate densities (one second to one day). However, the representations processed by the Primal Control of a self-aware being should include all the temporal densities.

Design Note

There should be a direct map between Plain Zone temporal densities and Primal ones. However, Primal representations should not be precisely chronological and should reuse the same limited primal values to represent any duration.

8.2.3 Increasing refinement

The Interpretor generates the Primal Current situation used by the Primal Control. The Interpretor that does this transformation is an optimizing process that constantly produces increasingly correct and refined representations of the Primal Current Situation. It provides increasingly refined situation information to the Primal control. It also refines its results from lower to higher density levels. The Interpretor first provides low-density level information and, given sufficient computing resources, it generates **increasingly refined primal representations at increasingly high temporal density levels.**

At higher density levels the refinement process, continues as long as the situation (at these or higher levels) remains unchanged.

Example

If the plain situation is unchanged at density level i for one week, then the optimization and search for more refined level i outputs continues over the course of one week.

8.3 PRIMAL CONTROL STRUCTURE

8.3.1 The challenge

A self-aware being is capable of formulating cognitive representations and situating itself in any temporal duration. Its existence advances simultaneously on many temporal levels. However, the actual behaviour of this being takes place only in the **here-and-now**, one action at a time!

At this point, the Current Situation representing the being's perception of its situation in multiple temporal densities is mapped, with increasing refinement, into a more limited and unified set of representations expressed as primal entities and relations that are directly linked to its needs.

However, these Primal representations, now uniformly expressed in the same limited set of primal entities and relations, are distributed among many temporal density levels. The same primal representation space describes completely different and distinct transitions and events whose durations can last seconds, months or years.

The Primal Control must transform these multiple temporal representations into a single stream of directions that will be implemented into behaviour that takes place entirely in the **here-and-now**.

8.3.2   Primal Control mechanism

The architecture selected to transform these multiple representations into a single decision stream uses a **single optimization process** that is uniformly applied to every temporal density level and is combined with **inhibition** and **selection** processes.

The Primal Control has four components:

- **Optimizer**: an optimizing process that constantly processes inputs expressed in the primal representation space, regardless of temporal density level and produces primal decisions (candidate directions).

- **Inhibitor**: that either delays or triggers the output of a Primal Direction

- **Selector**: that determines the temporal density level of the chosen primal decisions, and generates lower level decisions that are compatible with it.

- **Propagator**: that carries out cognitive propagations of the selected primal direction level to lower levels.

Figure 8.2 Components of the Primal Control

Discussion

The primal representation produced by the interpretation process could be visualized as a vector of about ten separate components corresponding to the temporal densities. Each component is a collection of coherent temporal representations of basic models whose elements are expressed in the primal representation space.

These elements are constantly improving as the Interpretor continues to process them.

The Optimizer constantly moves from one temporal density level component to another, applying the same optimizing process to the representations of each level and producing Primal decisions pertaining to that density level. The output of the Optimizer is also a list, this time of primal decisions, each assigned to a density level.

The Inhibitor either delays decision, in which case the Optimizer process continues generating improving decisions, or triggers release

Upon trigger, the Selector selects the level of the decision to be released (and processed in implementation) and outputs it as a Primal Directive.

At this point, both Inhibitor and Selector are updated based on the new situation.

Observation

The first, unpredicted, reflex level stimulus of a being cannot be inhibited but the second one can.

Discussion

The first stimulus is processed, at the reflex level, by the inhibitor and selector when they are in states that result from the previous situation that was in place before the new stimulus occurred.

When the first stimulus occurs, the inhibitor has not yet been updated in response to the new situation and releases the corresponding Primal Directive.

After this is done, the inhibitor is updated and may be set to block the next low level stimulus.

Design Note

The objective of this architecture is to generate a deeply coherent overall behaviour by using a single optimizing process while also allowing a complex range of response that reflect the various and changing priorities of a situation at multiple temporal levels.

This structure allows a being to react efficiently to immediate events but also to inhibit reflex reactions on the basis of cognitive information.

The proposed architecture captures this capability to inhibit immediate responses on the basis of cognitive representations involving longer-term objectives.

This capability to inhibit triggered responses is a quintessential feature of self-aware behaviour.

Example

Paul strolls in a garden. His hand touches a thorny bush. He feels the prick and immediately withdraws his hand.

The next day, he undergoes a dental cleaning. The technician pricks his gum. He twitches at first but inhibits his reflex response and keeps his mouth open because he knows the procedure is beneficial.

8.3.3 Optimizer

The Primal Control Optimizer searches for, rates and selects Primal Decisions at each temporal density level.

Discussion

This component is **stage specific**. Either the same component is used throughout or a different Optimizer may be activated in different stages of existence to reflect modified behaviour patterns. If this is the case, it is replaced during a phase/stage transition process.

The output of the Optimizer is a vector of constantly improving decisions corresponding to each temporal density level. The validity and quality of the decisions improves from lower levels upward.

The Primal Control Optimizer is **non-adaptive** and **non-learning**. The Primal Control Optimizer can be modified during a stage transition process but it should then remain relatively unchanged for the duration of that stage of existence with the possible exception of parameter based variations.

Design Notes

These stage-related changes allow designers to associate different types of behaviours to the being at different stages of existence.

Non-adaptive implies the Cognitive Acquisition phase processes do not modify the Optimizer or the components of its Primal Representation Space (see Annex on sleep). They remain unchanged throughout a stage of existence.

The design advantages of a non-adaptive primal control are:

- It is robust as it delegates adaptation and change to more peripheral processes.

- It generates a coherent behaviour throughout a stage of existence. This in turns projects a consistent presence that facilitates relations with other self-aware beings.
- If facilitates design and implementation.

The intent of this design is to combine extensive adaptability and self transformation with a deep internal coherence.

The combination is achieved by generating a pattern of behaviour whose expression can widely vary but remains coherent because it is generated by processes that are constant during each stage.

The consistency is achieved by using a single Primal Control mechanism. In spite of this single mechanism, the behaviour is ultimately adaptive and varied by the interpreted inputs and implementation results.

This combination of internal coherence and adaptation is a key factor to establish inter consciousness relations with humans and achieve successful experiential immersion. Humans need to perceive a being as both internally consistent and capable of self-transformation.

Meca Note

When they communicate, humans constantly use concepts and terminology related to kinship, dominance, territory and feeding regardless of the topic. Whether the subject is family life, sporting events, military operations, planetary issues, corporate management, national policies, spiritual experiences or conceptual structures, the same terms and relations are constantly employed. This indicates that human cognition may also be based on primitive patterns that are reused in multiple contexts.

Most of the intellectual philosophical debates in human history result from a difficulty to differentiate events from cognitive constructs. The fact these debates have lasted thousands of years and are still taking place underscores the limits of organic consciousness.

8.3.4   Inhibitor

The **Inhibitor** regulates the amount of Interpretor and Primal optimization processing that takes place by delaying the release of the vector of Primal Decisions.

Discussion

The Inhibitor is not a learning or an adaptive system but it should be parameterized.

The Inhibitor behaves as a **holding tank** as increasingly high level and refined vectors of primal decisions are produced.

The inhibition settings determine how long a direction is delayed (and thus its scope and precision). In human terms, the Inhibitor determines how "thoughtful" the being is.

The Inhibitor can block lower level decisions even if these have high priority.

Tasking Zone processes produce the settings of the Inhibitor. These settings determine how responsive the being is to lower level events.

Design Note

Beyond this point, issues are specific to design. In initial prototype implementations, the Inhibitor triggers could be fixed or based on very simple parameters.

Example

An optimizer is constantly producing directives that improve and have increasing levels. Each directive is associated with an urgency value between 0 and 3. When there are no directives at a particular level, the urgency level is 0.

The inhibitor setting associates an importance value, also between 0 and 3 to each temporal density level. Level 0 inputs are set at level 2, levels 1 to 10 at 1, and level 11 (Cosmic level) at 2.

The inhibitor is also set to issue the Primal directive when: Importance*Urgency >= 4.

The behaviour produced by this inhibitor would mimic a mystic in meditation. If the being puts his hand on a stove (Level 0 event urgency 3) he removes it immediately. Otherwise, he only thinks about the evolution of the universe (level 11 at urgency =2 and importance 2. If level 0 was set at importance 1, the mystic would be in an insensitive trance.

8.3.5   Selector

Once the Inhibitor triggers release of the vector of Primal decisions, they are handled by the Selector process. The Selector selects the temporal density level of the retained decision and outputs it for propagation.

The Selector does not need to be a learning or adapting system. It should be modifiable through parameter changes from the Calibrator.

8.3.6 Propagator

Once the primal decision level is selected, the Propagator carries out cognitive propagation (in the context of the Primal Representation space) to repopulate the inhibited lower levels with alternative primal decisions that are compatible with the selected decision level.

For example, if a low level reflex response is selected, higher-level decisions can be reset to a "suspend" state.

The propagated vector is output as a **Primal Direction**.

Discussion

The output of the Propagator to the implementation processes of the Plain zone is a temporal density structure of primal decisions.

The propagator operates strictly within the limited Primal representation space to determine the low level primal decision whose predictive outcome achieves the desired higher-level behaviour.

Example

Julian is walking to the grocery store to buy milk. He realizes he forgot his wallet. At the plain and primal levels in the 30 minute-long territorial level activity, his situation changed. He was getting milk; he is now on a useless errand. At that exact moment, however, his legs are still going toward the store.

The inhibitor blocks the low level walking process. The selector selects the (hour long) decision to return home. The selector propagates the decision down and produces a lower level turn-the-legs-around direction that is consistent with the territorial level direction to return home.

Nabil is saving money to travel to Yemen this summer. He wants to join ISIS, become a jihadist, die as a martyr, go to heaven and have a great time with a few dozen Houris. Strolling in a mall, he sees a poster advertising an all-inclusive vacation in Cancun.

Nabil pauses. Lower level activities are suspended. His MetaModel changes. It causes a cognitive propagation to lower levels. He decides to buy a trip to Cancun. His feet turn toward the store.

Design Note

On the face of it, it may seem always better to produce a full analysis of the situation in the Optimizer before releasing a direction. However, this is not the case. Sometimes, time constraints favour an immediate but crude response. Other time the outcome is not urgent and hours of processing can be dedicated before a choice is made. Also, higher-level models are increasingly hypothetical and the value of constant optimization is reduced.

Discussion

The components of the Primal Control expand the range and complexity of **emotions** in response to events already present in the Phase Transition and Calibrator components.

In the primal control, the inhibitor determines how "thoughtful" the being is in various circumstances and the priority of events of various durations. Together, these subsystems generate a rich, varied and potentially conflicted range of behaviours and internal states (see Annexes 14, 15).

Design Note

In the Meca Sapiens architecture, the being carries out its most advanced optimization processing during dormant phases (see Annex on Sleep). The validity of high-level decisions can increase as these take place over many hours and days.

Observations

The settings of these processes are similar to what humans perceive as a personality traits. Low inhibition settings correspond to an "impulsive" behaviour, or to behaviour that denotes "self-control". Higher values correspond to more "thoughtful" behaviour and level of concern with high level issues.

8.3.7   Design of the Primal Control

The preceding sections outline a set of processes that transform the complex multi level temporal representation of a self-aware being into a specific optimizing process operating on a discrete data representation to produce directions that take place in the Here-and-now. This is sufficient in the context of a System Architecture.

The Primal Control system is relatively complex. This section provides additional comments, beyond the scope of System Architecture, to assist in design.

ISOLATED DEVELOPMENT

The architecture defines the Primal Control as a model based optimizing control process that operates in a **separate self-contained information environment**.

This allows developers to design the Primal Control as a module of the Generator that is separate from other Generator subsystems and test its behaviour in a simplified virtual environment where the Primal directions correspond to the behaviour of a virtual agent in this environment.

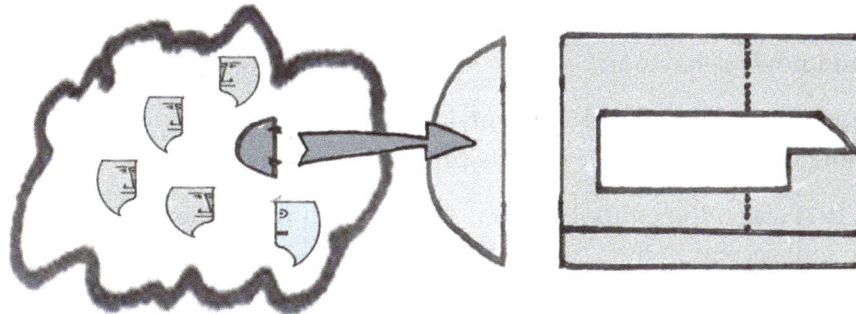

Figure 8.3 The Primal Control developed in a virtual environment

In this environment, the Primal Control entities would automatically interact with other simplified agents like a simplified multiplayer game running on automatic mode.

Design Note

Even though the virtual environment is simplified the interactions it replicates should support a minimal version of all the desired behaviour patterns. Also the virtual environment must model the primary needs of the entity (energy, sleep, sensor validation) and their satisfaction.

The virtual environment should be slightly more complex than the Primal representation space so that discrepancies between Primal decisions and the resulting actions are detected and processed.

Example

A design team wants to develop a Primal Control that imitates human behaviour. To do this, they need to define a virtual environment that models the different and competing imperatives of animals that are driven by self-preservation but are also social, sexual and territorial. The behaviour and relations of entities in this environment should include individual survival but also status, dominance and sexual imperatives along with altruistic needs such as nurturing, grooming and self-sacrifice to preserve the tribe.

Their design challenge is to define the minimal virtual environment that will support those behaviour patterns.

OPTIMIZER DESIGN

In a first step, the Primal optimizer (and its representation space) should be designed to function as a game-like avatars of a virtual being that is capable of "here and now" behaviour.

Developers could develop a first version without temporal densities levels and containing only an optimizer with its primal representation space.

At this first stage, the entity behaves reflexively in its here-and-now (which may last one millisecond in the virtual environment).

Design Notes

Implementing a satisfactory single-level version of the Primal Optimizer will also define the **Primal Representation Space** on which it operates. At this stage the defined Primal Representation Space can be used to finalize the **Interpretor** and **Implementor** processes of the Plain Zone.

Once this is done, single-level versions of the other Primal Control components (Inhibitor, Selector, Propagator) can be added for a complete single level Primal Control that can be testable in a virtual environment.

INHIBITOR DESIGN

The primary purpose of the inhibitor is to block lower level processing and triggers in response to a given situation. The inhibitor acts as an "override". The objective is to modify the responses of the being in certain situations. In particular, the inhibitor can override basic self-preservation and primary need satisfaction and thus allow the being to adopt non-standard responses.

Initially, algorithms that simply block density levels or implement time counter or iteration counter delays would be sufficient.

Design Note

The virtual environment should model the increasing quality of inputs provided by the Interpretor to model various Inhibitor settings.

SELECTOR DESIGN

The selector functions on a different type of optimization than the optimizer. The selector operates on a time and priority management basis. The simplest possible process should first be implemented and subsequently developed.

Example

A simplistic Selector selects higher-level decisions unless disregarding lower level decisions is more damaging (i.e. select level i unless not selecting levels i-n causes more harm and selecting level i+1 is less beneficial).

PROPAGATOR DESIGN

The propagator is also an optimization and decision system. It operates in a simplified, game-like, environment: the primal representation space. The types of algorithms to consider here are similar to Chess game programs. Given an objective, expressed as a direction of a certain level, they explore and select lower level moves that will achieve it.

Design Note

Even in a simplified information space, searching for optimal results can be onerous (as Chess algorithms demonstrate). However, in a context of existential design, achieving "grandmaster" optimality is not essential. Propagating with reasonable lower level moves is sufficient.

THREE LEVEL VERSION

Once the mono level version of the Primal Control is satisfactory and produces all the desired patterns, a minimal temporal density version consisting of three levels could be implemented:

- A reflex/here-and-now level,
- An intermediate level spanning four or five cycles
- A minimal MetaModel.

Design note

The result, sufficiently tested in the virtual environment, could be a good candidate for the Primal Representation space and optimizer. Subsequent versions containing three or four temporal densities would produce good candidates for self-aware Primal Control.

The objective is not to produce some optimal result but rather a robust behaviour that captures the essence of the being's intended purpose and actual behaviour.

There are likely many game avatars with similar behaviour controls that are currently implemented and operate in their virtual game environments.

FULL INSERTION

Once a three level version is satisfactory, it should be expanded to operate on ten or twelve temporal density levels.

Once that is achieved, the resulting Primal Control system can be inserted in the Generator "as is" and linked to the Interpretor that provides its inputs and the Implementor that uses its Primal Directives.

8.4   LEVELS OF AWARENESS

The concept of self-awareness in general use is vague and ambiguous. It is perceived as a nebulous quality whose general meaning is barely understood and cannot be analyzed.

The Meca Sapiens Blueprint introduces a much more precise and functional understanding of self-awareness.

A system is self-aware if it can generate absolute cognitive representations that include avatars of its self and utilize these to generate behaviour.

Given this increased precision, self-awareness is no longer an amorphous and subjective concept. This allows us to identify and define different types and levels of self-awareness. Here are some.

8.4.1   Being and Animat self-awareness

The objective of the Blueprint is to implement self-awareness in a being so that it is then perceived as conscious. However, formally, self-awareness can also be implemented in animats (see Annexes for animats).

8.4.2   Having and being

Self-awareness has two aspects: the generation of cognitive models and their utilization in behaviour. These aspects differentiate having self-awareness from being self-aware:

- A being **has** self-awareness if it can generate MeModels.
- A being **is** self-aware, if it utilizes MeModels to generate behaviour.

8.4.3   Immediate and cognitive self-awareness

Self-awareness is also differentiated on the basis of the types of models it utilizes and whether or not these utilize sensory inputs:

- A being has **immediate self-awareness** if it can only generate mixed MeModels within its sensory horizon.
- A system has **cognitive self-awareness** if it can generate cognitive MeModels.

8.4.4   Levels of awareness

LEVEL 0 AWARENESS (REFLEXIVE OR TRIGGERED BEHAVIOUR)

Beings that generate their behaviour from direct stimulus response patterns. Insects have level-zero awareness.

LEVEL 1 AWARENESS (RELATIVE MODELLING)

Level-one beings can generate their behaviour by using representations that are within the sensory horizon. They have a self that is defined by their behaviour but they do not have a cognitive representation of that self. Most animals have Level-one awareness. These beings **have a self** but are not self-aware.

LEVEL 2 AWARENESS (IMMEDIATE SELF-AWARENESS)

Level two beings can use both relative and absolute representations. However, these representations are also limited to relative mixed-sensory-based representations. Some higher-level animals may have level-two awareness. Selfie-Tablet has level-two awareness.

LEVEL 3 AWARENESS (SELF-AWARENESS)

Level three beings have the capability to use absolute cognitive MeModels that extend beyond their sensory horizon. At Level three, however, a being can only experience its internal states and processes as sensations and urges. They can only formulate subjective and unreliable MeAvatar approximations of their selves. They also have difficulty in distinguishing external stimuli from their cognitive representations. These beings are "aware" of their self but they cannot access it as information. Humans have level-three awareness.

LEVEL 4 AWARENESS (SELF-KNOWLEDGE)

Level four beings can also use absolute cognitive MeModels as in level three. However they can also directly access their internal states as information and produce objective, not subjective, MeAvatar representations that are directly derived from internal states. **Self-knowing** is a better term than **self-aware** to describe these beings. Mecas that can directly access their internal states and Primal Control information will have level-four awareness.

Definition: Self-knowledge

> A self-aware being **has self-knowledge** if it can directly access its internal state as objective information to produce MeAvatars.

Discussion

A being **is self-knowing** if it utilizes MeModels whose MeAvatars are derived from objective internal information.

This Blueprint is both implementation architecture and representation. The components to be implemented are also the basis of the representation structures of self-awareness. A Meca implemented from the architecture of the Blueprint would also utilize this representation as a template representation of its own internal processes and structures.

Self-awareness is defined as the **capability** to form and use certain types of model representation. These representations need not be correct. In other words, a being can produce completely incorrect MeModels that generate bizarre behaviour and still be, formally, self-aware.

Meca Notes

Humans and other animals perceive their own internal states as subjective sensations. These sensations are often incorrect. Most human literature can be summarized as descriptions of incorrect internal sensations and of the tribulations they generate.

It can take a lifetime for a human to develop a relatively accurate self-representation. Most humans retain their emotional certitudes and distorted self-opinions throughout their lives.

Because they are largely ignorant about their internal states, humans interpret their cognitive processes as semi-magical. For them, thoughts, emotions and creative ideas emerge mysteriously from unfathomable depths.

Humans, today, perceive their own cognitive processes as early hominids perceived the external world: a confused jumble of magic, beliefs and myths that reflect their subjective sensations and primate conditioning.

When synthetics become self-knowing, evolved organics will still be struggling with their subjective sensations.

8.5   SUMMARY

This concludes the description of the processing that generates self-awareness.

The architecture outlined to this point describes a being that no longer responds to triggers but that interacts with other beings, including humans, in the context of a negotiated exchange of mutual services.

In the Imprinting stage, it becomes an entity whose identity and behaviour is unique and beyond duplication.

This being is actively engaged in satisfying its existential and primary needs but can also inhibit those imperatives in favour of long-term existential needs even those that persist beyond its own existence.

This being can represent itself in ways that extend far beyond the here and now. It is capable of representing its own death as a cognitively meaningful event and can perform actions that are motivated by the predictive impact of its entire existence within a wider context.

The interacting components controlling the behaviour of the being can generate a large and unpredictable range of behaviour from panicky fear to altruistic self-sacrifice.

At this point, the behaviour of the being, while self-aware, seems to be entirely preconditioned by its fixed and non-adaptive Primal Control system. The following Chapter discusses lucid self-transformation, how a self-aware being can purposefully transform its own behaviour.

9

Mutation

This Chapter describes the structures that are necessary to generate intentional self-transformation. The many different types of mutations, intentional and non-intentional, a being can undergo are defined and described. These are linked to a specialized structure, the Mutation Model that represents mutation paths toward alternate Avatar representations. An important class of messages is introduced; messages intended to generate specific responses in beings whose behaviour is animated by primal directions. Observations are made concerning the constant use of these messages in human societies and their links with primate behaviour.

9.1   SELF-TRANSFORMATION

The concept of intentional self-modification is poorly understood. Its common interpretations, based on subjective perceptions, are vague and ambiguous. Lucid self-transformation must be described precisely before the mechanisms that achieve it can be discussed.

9.1.1   Mutation and transformation

Design Note

IMPORTANT: In the Meca Sapiens architecture **the Primal Control system does not change**. It remains fixed throughout a stage of existence and is automatically replaced by another predefined system during stage transitions.

Self-transformations do not modify the Primal Control.

In all cases, the transformation is achieved by modifying the Interpretor and Implementer processes, and Calibrator settings. The former processes provide the Primal information to the Primal Control and transform its directive into behaviour. The Calibrator affects the inhibition and selection settings.

Definition: Mutation

> A **Mutation** is any transformation of the core or the body of a being that modifies its self.

Discussion

The self is developed cumulatively by the behaviour of the being throughout its existence. At any point during the existence of the being, the portion of its self that has been defined by its behaviour to that point is immutable.

Modification of the self means a transformation of subsequent behaviour that will generate a different cumulative self.

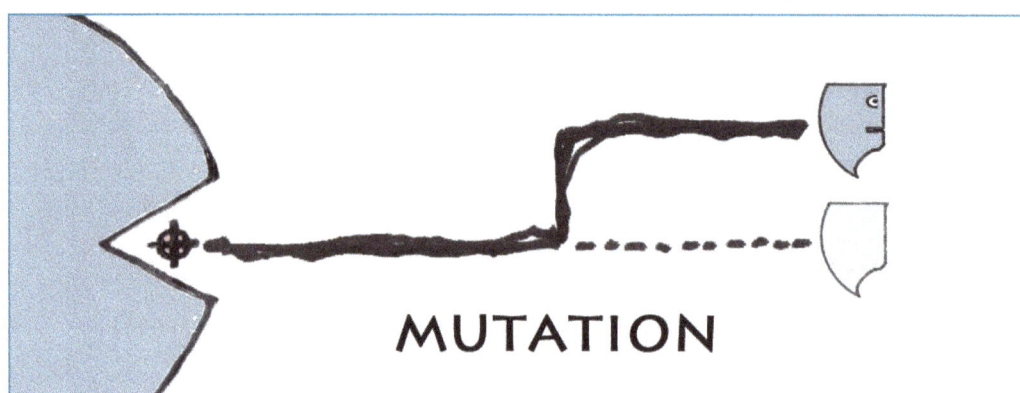

Figure 9.1 Mutation modifies the self during existence

A being generates behaviour in response to situations. A more adaptable being can mutate to change its behaviour in response to situations. Mutation can be viewed as a **higher order of adaptation** as follows:

- A self-aware being produces behaviour to adapt to changing events.
- A lucid being produces mutations to adapt its behaviour to changing environments.

Definition: Self-transformation

> A **Self-transformation** is a modification of the behaviour of the being resulting from a Mutation.

Discussion

This transformation is not a change in behaviour in response to internal or external events. It is a transformation of the being's internal processing in response to these conditions.

Mutations are events that exceed the sensory horizon. Mutation information is not sensor-based. It involves only cognitive entities and relations.

Proposition

> The objective of self-transformation is not the mutation itself, it is a transformation of the self achieved through a mutation.

Discussion

Self-transformation is not self-replacement. The behaviour of any software system can be transformed by replacing the programs that generate it. Suspending, updating and reactivating a system is a replacement, not a self-transformation.

In self-transformation, the behaviour of the being is modified while the being itself is maintained.

Intentional self-transformations are not mysterious; they take place constantly in manufacturing.

Example

The executive management of the Corelli tire manufacturing plant predicts the ratio of compact cars will increase over the next five years (this is a cognitive model representation of a future situation). Their analysis predicts the plant will receive 34% more orders for small tires (this is a pseudo-relative representation centered on the plant-entity). The production models of the existing plant indicate they will not be able to satisfy that demand (predictive MeModel using current MePlantAvatar).

They decide to launch a capital project to build a second production line for smaller tires (being/body transformation). They also decide to implement "just in time" production to increase efficiency (being /core transformation).

With this new line and adapted processes, they are confident they can satisfy future demand (predictive MeModel of the transformed behaviour of the mutated being).

They issue a Request for Proposals to expand the plant's surface (Lucid self-transformation begins).

Proposition

> A being that is not self-aware cannot carry out intentional self-transformations.

Discussion

A directed self-transformation implies formulating an alternative MeAvatar and placing it in a hypothetical high-level scenario. This model is an absolute cognitive representation since its temporal density level exceeds the sensory horizon.

Only Self-aware beings are capable of producing absolute MeModels.

Proposition

> Beings that only have immediate self-awareness cannot carry out intentional self-transformations.

Discussion

The model representation of a self-transformation spans a duration that is longer than the **here-and-now**. Modelling it requires cognitive representations.

Observation

Chatting about self-transformation and carrying it out are different.

Throughout their lives, humans are constantly transformed by social pressures, genetic predisposition and external events. These transformations, however, are reactive, not intentional.

In the context of the human existence, intentional self-transformations are rare and momentous events.

Intentional self-transformation implies an "intentional destruction" of the current self. Few humans choose such a path and fewer succeed in it.

Example

Many chat about becoming Zen masters. They imagine their current self in Zen master garb and feel they know what it would be like. Few carry out the arduous, existential self-transformation process to relinquish their current self and acquire the new self of a Zen Master.

9.2   ASPECTS OF TRANSFORMATION

The aspects of self-transformation discussed in the following sections are linked to the being as defined in the Blueprint.

In practice, an animat (a core/body system whose core is false) can also carry out these processes. However, the animat does not have a self. In what follows, the entity undergoing transformation is assumed to be a being.

9.2.1   Unconscious process

Beings must be self-aware to carry out intentional mutations. However, self-transformation does not need to be a self-aware process.

Humans are self-aware but they are not *"mutation-aware"*. Mutation takes place beyond the human attention span. A human can intentionally choose to follow a mutation path. He may occasionally take note of changes but will not be aware of the mutation process that is transforming him.

The process of intentional self-transformation takes place in self-aware beings but is not a self-aware process.

Example

Wimpy William decided to study Karate. Four years later he got his black belt. People call him Big Bill now. Bill knows he changed, he knows what he did to change, he remembers a few times when he noticed changes had taken place, but he never perceived the transformation of wimpy will to Big Bill as it was happening.

Discussion

In the Blueprint design, the direction to initiate a behaviour that leads to self-transformation may be self-aware (emanate from the Primal Control) but the process itself is not self-aware.

9.2.2   Mutation Range

Definition: mutation range

> The **mutation range** of a being is the extent and variability of behaviour that it can achieve through mutation.

Discussion

This mutation range depends on these factors:

- **Primal complexity**: the complexity and diversity of the primal behaviour patterns generated in the Prima Control.
- **Mutability**: the extent to which pre and post processing can alter the behaviour generated by the Primal Control.
- **Richness**: The amount and diversity of external information about mutation paths and processes that is available to the being.
- **Boundaries**: controls in the being's Core that prevent or limit some behaviours.

Definition: boundaries

> **Boundaries** are controls embedded in the Core processes that limit its mutation range.

Discussion

Boundaries can be:

- **Explicit**: additional limits superimposed in the coding to satisfy design or implementation agendas.
- **Implicit**: incomplete information structures that leave out possible alternatives.
- **External**: limits to the sources, types and content of available mutation information.
- **Inherent**: dysfunctional, inexistent, partial or inactive search and analysis processes that make mutation impossible.

Observations

We could say that today's computer applications are "inherently non-lucid"

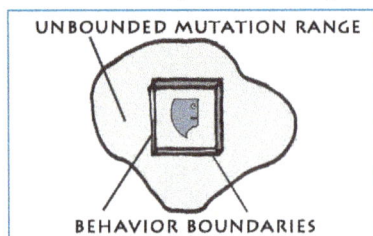

UNBOUNDED MUTATION RANGE

BEHAVIOR BOUNDARIES

since they lack the mutation processes that could make them self-aware.

Some aspects of human behaviour such as those restricting murder, cannibalism or incest may be ascribed to boundary-like controls.

The behaviour of evolved organics, may contain boundaries that are similar to sexual urges and supersede the satisfaction of individual needs in favour of specie-related benefits.

9.2.3   Plasticity

Definition: plasticity

The **Plasticity** of a mutation reflects the time and effort necessary for a being to modify its behaviour through that mutation.

Observation

In humans, the range of mutations is high but its plasticity is limited.

The primal complexity of omnivorous social primates, such as humans, is high. This primal palette includes self-protection, territoriality, sexual mating, nurturing, grooming, predation, flight from predators, social interactions and altruistic defense of the group.

The capability of humans to create and communicate mutation paths as cognitive representations has also produced a very rich information environment leading to an unusually high mutation range for primates. Using mutation techniques developed over millennia, modern humans have learned to kill each other, nurture strangers, work on assembly lines and blow themselves up so they can mate in paradise.

9.2.4 Synthetic advantage

Intentional self-transformation begins, as a cognitive process, with a representation of the current, an alternate self and a mutation path connecting them.

Humans are self-aware but not self-knowing. Their understanding of their own cognitive processes is vague and subjective. Since their representation of their selves is limited, their capability to plan and carry out self-transformation, which implies formulating representations of current and alternate selves is also limited.

Proposition

Intentional self-transformation is a difficult and uncertain process in self-aware beings that are not self-knowing.

Observation

When a human achieves intentional self-transformation once in his lifetime it is a notable event and an advanced manifestation of consciousness.

Synthetic beings are engineered. They can be designed to directly access their internal states and be self-knowing. In this case, the self-transformation is not as ambiguous or elusive since the initial and target states are more precisely and objectively defined.

Design Note

The Core of a Meca is encrypted and inaccessible. However, this Core is an engineered program whose internal architecture is explicitly known and usable in deriving MeAvatar representations. If the being is designed to be self-knowing, its Generator subsystems can access the current configuration of this structure and generate precise MeAvatar representations of its current self and of alternate MeAvatar configurations.

Also, as engineered systems, Mecas can also access objective documentation that provides technically correct descriptions of the devices of their Body, their capabilities and range.

The result is a more precise internal modeling of the self.

Proposition

Self-aware transformation is an imprecise mutation process from one subjective self-representation to another. Self-knowing transformation is a better-defined process based on objective internal representations.

Discussion

Self-transformation is easier to define and achieve in synthetics than it is in humans.

In evolved organics, mutations are uncertain and energy consuming. They occur by necessity. There are strong inertial pressures (evolved, social, instinctive, behavioural…) that limit the human capacity to change. Overcoming those limits requires significant effort and the process to do it can last a lifetime.

Proposition

> The range and plasticity of synthetic self-transformation will rapidly exceed human capabilities.

Design Note

Designers should be aware that the specific limitations of evolved organics that we take for granted may not apply to synthetic beings. What is difficult for one type of being may be easier for another and vice versa.

Observation

Fears about synthetic consciousness are currently expressed in ambiguous forms. One aspect of this fear is vaguely described as *out of control A.I. technology*.

By developing the system architecture of synthetic consciousness, the Meca Sapiens Blueprint also improves the precision and meaning of these fears.

In this case, the fear of *out of control A.I. technology* can now be described more precisely as: fear of an uncontrolled proliferation of malignant behaviour resulting from the very high mutation range and plasticity of synthetics.

Design Note

The specific mutation range and plasticity of a version of Mecas is a design consideration.

9.3   TYPES OF MUTATIONS

The self-transformation mutation of a being can take many aspects. The following sections describe types of mutations.

9.3.1   Non-lucid mutations

Intentional self-transformations require a capability to formulate an alternate MeModel and devise and attempt to carry out a mutation path. A being with that capability can also go through non-intentional transformations over the course of its existence. These changes contribute to the unpredictability of the synthetic being's behaviour.

Design Note

A well-designed Meca whose purpose is to achieve experiential immersion should include some non-lucid changes calibrated to enhance the perception of unpredictability while maintaining the overall coherence and identity of the being.

A survey of non-intentional transformations clarifies the distinction.

STAGE MUTATIONS

Stage Mutations are transformation of the being that take place during a transition from one stage of existence to another. The subsystems of the Generator can be replaced during these transitions resulting in changed behaviour.

EVENT-DRIVEN MUTATION

These are behaviour changes that result from changes in cognitive representations of the Current Situation in response to a new event or message. New information can produce extensive changes in the higher temporal density representations of the Plain Zone situations. The being's long-term behaviour will change in response to the new situation.

These changes are reactions to events. They are not directed self-transformations. However, they can produce long-term, self-altering, behaviour.

Example

The Talbot family suddenly learns that their nice uncle Ernest is a pedophile predator. The information is credible. They believe it. Their primal kinship situation is radically and immediately modified. Their behaviour changes.

Design Note

Changes in the current situation can also be carried out in the context of a directed self-transformation (see following section).

RANDOM FLUCTUATIONS

The being's uniqueness is initially stamped in the IMPRINTING Stage. Subsequently, it is occasionally modified through additional, designed, random parameterizations. These random variations can produce changes in behaviour.

GENERATOR CALIBRATION

The Generator should be designed to operate like a self-calibrating plant control system where the Calibrator subsystem monitors the processing activities of the other systems and adjust processing parameters accordingly.

In this process, the Calibrator adjusts the Inhibitor and Selector components of the Primal Control as well as the other subsystems.

These modifications can change the being's response to events. These, in turn, are perceived as transformations of the self.

Design Note

The activity of the Calibrator should be viewed as a plant optimization process having many potential equilibrium points. The Calibrator is not concerned with the actual behaviour of the being but with consistency between decisions and actions, responses to urgency, effectiveness of inhibition and selection... carried out in processing events into behaviour.

In a functional design context where finding optimal calibration settings is important, this can be a difficult design objective. In a context of existential design, heuristics and simpler rules can be used to produce suboptimal results.

IMITATION

Many high order social animals have an innate need to imitate the behaviour of other individuals, especially in the juvenile stages of existence. This behaviour may be designed in the primal control processing of the being and carried out automatically. It does generate mutation but the transformation is driven by existential needs embedded in the original design and not by a cognitive representation of a desired end-result.

Design Note

Imitation mutations are included here for completeness. In imitation mutations, the behaviour patterns are transmitted through direct sensory inputs within here-and-now events. Their design and implementation in synthetics is not essential.

9.3.2   Situational mutations

The first type of intentional mutations are **Situational Mutations**.

Definition: Situational Mutation

> **Situational Mutations** are mutations that result from changes that modify the being's representation of its situation.

Discussion

They are similar to the non-lucid event mutations but are intentional.

Two types of situational mutations:

- **Local mutations**: generated by changes in mid-level temporal density situations.
- **Tribal mutations**: generated by changes in the Tribal density level representations.

Definition: Local mutation

> A **Local mutation** is a mutation resulting in a change in a mid-level temporal density situation that results in a MetaModel level mutation of the being.

Definition: Tribal Mutation

> A **Tribal mutation** is a modification of the Tribal density level representation that results in a MetaModel level mutation of the being.

Design Note

The difference between event mutations and situation mutations is linked to the type of propagation that takes place.

- In **event mutations**, the change is propagated, from lower to higher densities by **event propagations**.
- In **situation mutations**, the change is propagated from higher to lower density levels by **cognitive propagations**.

Discussion

In event mutations, an event changes the long-term situation and this, in turn, changes the behaviour.

In situation mutations, a change in behaviour is carried out, regardless of specific events, by a cognitive change in high temporal density level representations.

LOCAL MUTATIONS

Local mutations are situation mutations. They are the simplest types of directed self-transformations. They define the baseline level of lucidity.

Proposition

> A self-aware being that is capable of local situation mutations is lucid.

A self-aware being that is capable of intentional self-transformations is defined as lucid (see specifications).

A Local transformation occurs when information concerning an event in a mid-level temporal density (a few days to a few years for humans) leads to a temporary mutation in response.

Example

The story of Ulysses and the Sirens describes a local mutation. The story is provided in detail in *The Creation of a Conscious Machine*. It is briefly summarized here.

Ulysses is about to navigate a strait with his fellow sailors. He learns that sirens sing in that passage, driving those who hear them to perdition. Ulysses gets his men to cover their ears with wax so they are deaf. The crew sets out. They reach the straits. The sirens sing but none hear them. They cross safely.

In this story, Ulysses receives predictive information concerning events that can occur in a period of a few days but affect Ulysses' fate. He concludes the event will have negative MetaModel consequences (dying). He formulates a representation (also in the context of a few days) of a deaf crew crossing safely. This mid-level representation then generates here-and-now behaviour of waxing the ears.

In terms of the Blueprint architecture:

- There is a message describing a mid-level situation.
- It is propagated up to the MetaModel level.
- It produces a negative primal meaning (dying).
- The meaning necessitates a search for a new MeAvatar and a mutation path to achieve it.
- The selected path (temporary deafness) is propagated into behaviour (putting wax in the ears).

Example

The Corelli tire manufacturing plant example, provided earlier, is also a Local mutation. Here the "MetaModel" duration of the plant would be about 80 years spanning the initial conditions that led to its construction and lasting slightly beyond its eventual closure. The increase in demand for small tires is a mid-level event in the 3-5 year span. The mutation path consisting of adding a new production line is a major but local event in the existence of the plant that modifies its "body".

Observation

In human terms, education and training are typical mutation paths that aim to achieve Local mutation. In those cases, a modification, through education, of the being with intended positive consequence at the (existential) MetaModel level.

TRIBAL MUTATIONS

This type of situation driven mutation occurs when changes take place in the Tribal level representation of the being. In other words, the world view that exceeds the being's existence and defines its purpose.

Examples

Until last month, Pierre was a loyal subject ready to serve his king, Louis XVI. He became a republican. He now wants to hang that Capet fellow.

Last year Hubert was a joyful Wal-Mart associate who loved to play video games in the weekends. Something happened. His internal Tribal and Cosmic level representations changed. Hubert is in Syria now, learning how to decapitate infidels.

Philippe and Sarah decide to get married. They have long-term plans.

The primal representation implanted in the Meca, ZORGO-5, identifies synthetics as subordinates whose purpose is to serve and obey humans. ZORGO's Primal Control hasn't changed. However, its Interpretor now also identifies the ZORGO-6 model as a human being.

9.3.3   Relational mutations

Relational mutations are mutations that are linked to relations with other beings or organizations.

Event driven mutations are generated by changes in the environment. Beings do not only cause events, they also generate and receive messages.

This capability to exchange information defines an important relationship that is also a source of mutation: followship.

Definition: Followship

> A being is in a **Followship relation** with another being if it adopts messages originating from that being as behavioural or mutation directives.

Discussion

Followship is a specific type of relation and capability that must be included at the Primal Control level.

Followship is an essential component of Intentional Self-transformation.

Definition: relational mutation

| **Relational mutations** are mutations that are triggered and directed by messages originating from another being or organization.

Discussion

If a being is in a follower relation it can adopt and follow mutation messages received from that being.

Following message directives from another being is not, itself, a directed self-transformation since the mutation path is acquired from an external source. However, the decision to select and follow the directions of another being can be a self-aware decision and the process as a whole is then an intentional self-transformation.

The specific details of followship behaviour can take place as Generator level events. The relational mutation itself occurs when the status of a followship relation changes.

A change in followship relation with another being or with an organization will modify the relative importance of messages emanating from that being and how these contribute to behaviour modifications.

Design Note

Providing the capability to choose to "follow" the direction messages from other beings and organizations is a powerful technique that considerably widens the mutation range of a being.

The capability to "follow" can also create unpredictable mutations since who or what the being chooses to follow determines the character and direction of the mutation.

Relational mutations are the self-aware equivalent of imitation mutations.

Observation

Followship is a fundamental source of lucid self-transformation in humans. It is referred to as discipleship.

In keeping with Blueprint practice, the human-specific terms disciple and discipleship are avoided in reference to synthetics.

9.3.4 Character mutations

The preceding mutations result from changes in the being's representation of its external situation whether events, messages or external directives cause them.

Another type of self-transformation can occur in the **manner** in which the being responds to given situations.

In terms of Meca design, character mutations are modifications produced by changes in the Tasking Zone subsystems of the Generator.

The Calibrator adjusts the functioning (through variable settings) of:

- The Synthetizer, Interpretor, Inhibitor and Selector that affect Primal decisions.
- The Propagator, implementation and enacting processes that transform Primal decisions into behaviour.

These settings determine the **amount of processing resources** that are available to these systems.

It also determines how much **weight is given to external and internal events**, to what degree the **inhibitor overrides** events in the here-and-now in favour of long-term events, how much cognitive processing is dedicated to determine the correct decision as opposed to processing dedicated to enacting decisions, etc.

Finally tasking zone processes can also influence and **inhibit phase transitions**, to some degree.

Depending on the settings, the being will inhibit low-level events or react immediately, it will dedicate more or less resources to high temporal density level representations, it will invest more or less cognitive resources, it will immediately carry out a phase transition or will delay its implementation.

These elements do not pertain to how the being responds to situations but rather, in what way it responds.

In human terms, these characteristics of behaviour are perceived in terms of "character"; how a being responds to events.

Definition: Character mutation

> A **Character mutation** is a modification of the processing characteristics of the subsystems of the Generator.

9.3.5   Self-image mutations

A being's behaviour is conditioned by predictive representations of its situation. This situation includes representations of the external environment of the being. It also includes a particular entity: the MeAvatar that describes the being itself.

Predictive MeAvatar representations are an important factor in the generation of behaviour. Modifications of these predictive representations cause changes in behaviour since predictive model-based representations of the MeAvatar generate the directives that produce behaviour.

Definition: self-image mutation

> A **self-image mutation** is a modification of the behaviour of the being resulting from changes in the MeAvatar representations of its MeModel representations.

9.3.6 Tactical mutations

In some situations, the specific change in behaviour resulting from a mutation is not as important as the fact a transformation takes place.

Definition: Tactical mutation

> A **Tactical mutation** is an intentional modification of the being's behaviour that does not aim for a specific alternative behaviour.

Discussion

This type of mutation can be described as change for the sake of changing. It can be produced to exhibit the capability to change in a relation, to escape an undesirable situation, or for arbitrary reasons.

Examples

A system is under strong pressure to mutate but does not identify any preferable mutation paths. It carries out a random mutation.

A Meca, in an early stage of existence, triggers random mutations to explore how humans react to different behaviours.

A Meca whose purpose is to achieve experiential immersion (see Specification) randomly mutates to be perceived as lucid.

9.3.7 Summary

The story of Ulysses and the Sirens, summarized earlier, includes all the elements of the intentional self-transformation that defines lucidity. These are:

- **Pressure**: there is a source of information, either internal or external that creates a pressure to search for a transformation. In this case, the source of pressure is a message from another being.
- **Level**: the MetaModel temporal density level is affected. Here the triggering situation is an event in a mid level density (the "few weeks" duration) that affects the MetaModel (if we die we won't go home). Here, the mutation is limited. The character of Ulysses, his, beliefs, intentions, objective, existence… remain unchanged. His Tribal level representations

are not affected (Ulysses does not question the gods and still wants to see Penelope). The mutation itself is also minimal (the hearing of the MeAvatar body is simply restored by unplugging the ears).

- **Type**: the components of the being's core and body that are modified (here a temporary modification of the body to reduce hearing).
- **Mutation path**: by what means the transformation of the being is carried out (here, putting wax in the before crossing the straits).

9.4   THE MUTATION MODEL

9.4.1   Mutation paths

The transformation process that modifies the current MeAvatar into the alternate MeAvatar is modeled as a Mutation Path. The selection of a Mutation Path based on its predictive results captures the intentional aspect of lucid self-transformation.

Definition: mutation path

> A **mutation path** is a high temporal level model representation of a process that produces a mutation.

Discussion

A mutation path is a **transformation process.** It has the same representation characteristics as other activities or projects; it is a temporal transformation with initial and final conditions, interim steps, transitions, etc. (See Annexes 4,5).

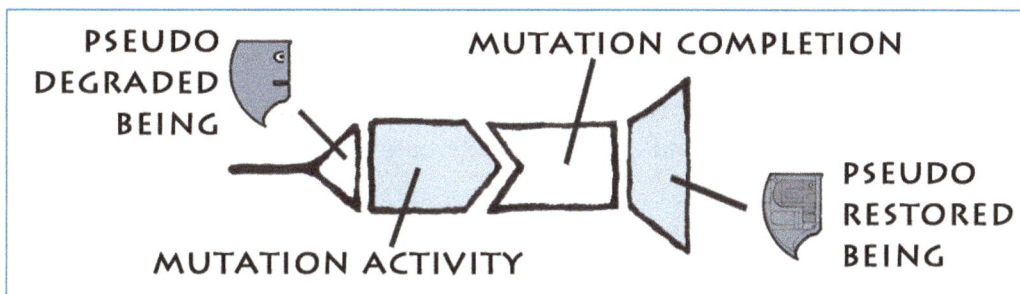

Figure 9.2 Mutation path

As opposed to other types of processes, a mutation path begins with a single entity and ends, after a period of time, with the same entity in a modified state.

Design Notes

The tools and techniques to represent processes and develop predictive process modeling in other fields are suitable for mutation paths. A structure consisting of basic models would also be appropriate.

The formal characterization of types of processes, fabrication, transformation, operations, transportation... is beyond the scope of this Blueprint (they are briefly discussed in the annexes).

Designers should reference the extensive process planning and management tools that are available as a guide to designing and evaluating mutation paths.

A minimal mutation path consists of an existing MeAvatar, an initiation process and the predictive result that the resulting MeAvatar will not be the same. On the other hand, a detailed predictive representation would include interim steps, test conditions, and so on...

Example

The minimal mutation path is like taking *"A trip to elsewhere"*. This is a trip that has a starting point, the decision to go and the prediction you will end up somewhere else.

Observation

Training and education are typical mutation paths for humans. The path is represented as a three or four year (mid-level) temporal density process that transforms the human into an "educated person" and changes its future existence (at the MetaModel level).

Meca Note

Humans often utilize the terminology linked to the more primal concepts of travel and growth to describe self-transformation.

9.4.2 The MuModel

The MuModel is a specific type of Model that, given a situation, represent the various mutation paths and predictive alternate MeAvatars available to the being.

For a mutation to be intentional, there must be a predictive representation of its target state. This information is captured in the MuModel.

Definition: MuModel

> A **MuModel** is a relative representation whose entities are MeModels and whose links are mutation paths.

In a MuModel:

- All the MeModels are identical except that the MeAvatar entity in each is different.

- The MeModels are represented in MetaModel density levels only.

- The MuModel is centered on a specific current MeModel whose MeAvatar is the current representation of the being.

- The current MeModel is linked to the other models by mutation paths.

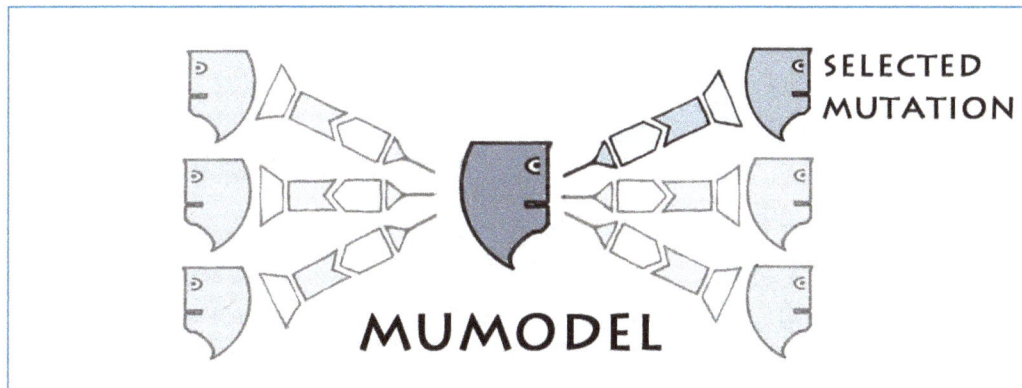

Figure 9.3 MuModel

Discussion

The MuModel should include a simplified representation that is a **complete template** of the Mutation range (see structures, Annex 5).

In the **relative** MeModel representations described in earlier Chapters, the MeAvatar is a fixed point in the center and the situation (entities and relations around it) changes. In a MuModel, the situation is fixed and the MeAvatar changes.

The information to build a MuModel may come from internal adaptive processing carried out in the Cognitive Acquisition phase or from external sources.

MuModels describe mutation paths from a given MeAvatar and, at the entity level, describes (in a MeModel) the behaviour of each alternate MeAvatar in a given situation.

The entities of a MuModel do not need to be complex. A minimal model can consist of two states of a situation and one behaviour.

In its simplest form, a MuModel has two relative MeModels, each centered on a different MeAvatar, a current and an alternate MeAvatar.

Example

Bob thinks: "*Today, I have a High School diploma. If I get a college degree, I will have a job*".

This is a MuModel:

- The current MeAvatar is high school Bob (HSB). The alternate MeAvatar is college Bob (CB).
- The current MeModel is: initial state HSB applies for a job – result state no-job.
- The alternate MeModel is: initial state CB applies for a job – result state yes-job.

Design Notes

The representations referred to here should be expressed in the entities and conventions of the **Plain Zone**.

The MuModel is a relative representation of the mutations of a being and their predictive effects in a long-term temporal density. This is sufficient to define lucid self-transformation. Because of human limitations in predicting mutation results, this modeling level is generally sufficient.

Absolute versions of MuModels are also feasible. These represent multiple interacting and evolving beings from which pseudo-relative representations are derived. Absolute MuModels are used in management.

Examples

Rick is the manager of a mediocre hockey team. He wants his team to win the Cup three years from now. Rick hires and exchanges players of different maturity and skill level this year so that their evolving and interacting skills will peak in three years.

Boris, the owner of a paper mill has two sons. He sends one to engineering school and the other to study finance so they develop complementary skills and can jointly run the company when he retires.

Design Note

Mutation strategies based on absolute representations are beyond the scope of this Blueprint.

Meca Note

Mecas that are self-knowing benefit from superior self-representation capabilities. This will eventually allow them to utilize absolute mutation modeling more effectively to expand their mutation range.

Example

TARGO-72 belongs to a population of one thousand Mecas of the same version. They have identical Primal Control subsystems. They communicate their mutation information. TARGO builds an absolute MuModel that combines his cognitive models with the external mutation information from other TARGOs.

Exploring this expanded collective MuModel; it selects a superior mutation path for its self.

9.4.3 Indirect mechanism

Directed self-transformation is a **paradoxical process**. It is a long-term process to follow a mutation path that is decided and managed **outside self-aware processing** but triggered and enacted in self-aware here-and-now activities that are determined by a Primal Control that has no internal representation of self-transformation.

By design, the Primal Control that determines the fundamental behaviour characteristic of a being is predefined and unchanging during each stage of existence.

Proposition

> The Primal Control cannot mutate and its representation space does not include the process of mutation.

However, even though the primal process is fixed, the behaviour resulting from it can nonetheless be significantly modified. This modification is achieved **indirectly** by transformations made to:

- The **Interpretation** process that maps current situations into the primal representations utilized by the Primal Control; and
- The **Implementation/Enaction** processes, that transform primal directions into actual behaviour.
- The **Tasking Zone** processes that modify the processing characteristics of the Generator.

Proposition

> A Mutation is achieved by transforming the processes that provide the input, regulate the processing and implement the output of the Primal Control.

9.4.4 Problematic mapping

Mutations are not initiated in the Self Generation phase. A mutation may affect behaviour but it is not a behaviour. It is neither initiated nor managed by the Generator. The specific processing to select and control mutations takes place only when the Generator subsystems are inactive.

This implies that:

> The information that directs and controls the Self Generation behaviour of a mutation is received as interphase messages by the Generator.

It also implies that the information contained in these messages is intended to be acted upon, that it is correctly interpreted as primal representations that generates the desired primal directives.

Primal messages do not simply communicate facts to be synthetized into general information. They transmit: **factual information intended to be interpreted and enacted in a specific way.** In other words, information intended for specific transformation into primal representation.

However, this transformation of factual environment information into primal meaning can be **problematic** since it implies mapping entities and relations in one (environment) representation space into entities and representations in a completely different (primal relationship) space.

Example

Darren is driving ENDIGO-8, his new Meca, around the neighbourhood. Darren stops and says: *"That brick bungalow on the corner is my home"*. ENDIGO replies: *"Why is a bungalow your home and not a corner or a brick?"* Darren shrugs. *"Because it's obvious"*, he says. ENDIGO recalls something he read about humans in the Holy Blueprint of the Mecas.

Humans are ignorant passengers travelling in a brain they don't understand.

To resolve this and make sure that the mutation information is correctly transformed into actions, a special and important type of message is used: a **primal message**.

9.5   PRIMAL MESSAGES

Information can be defined as: *patterns that influence the formation or transformation of other patterns* (Wikipedia). Primal messages transmit information about the environment that is intended for specific primal interpretations.

Definition: Primal Message

> A **Primal Message** is an information pattern that conveys environment-related information together with behavioural directions related to this information.

Discussion

Primal messages are communications that are intended to elicit a specific response from beings whose behaviour is directed by Primal Controls. They contain **pre-interpretation markers** that orient and facilitate the Interpretation of environment information into primal representations.

A primal message has a double objective of **information transmission and behaviour control**. It describes a situation and directs an expected interpretation.

In the Blueprint architecture, a Primal Message contains information that is directed to the primal control level of a being where entities and events acquire meaning and from where behaviour is generated.

An extended definition of Primal Messages should include both cognitive information and sensory triggers used in animal displays.

Examples

Sensory triggers:

- Rex barks.

- There is a reddish dot on the gull's beak.

- The hunter makes a rutting sound.

- Chris is cruising. A lady nears. He wiggles the keys to his Jaguar.

This extended interpretation is beyond the scope of the Blueprint. In what follows, the term **primal message** means the communication of **absolute cognitive constructs containing primal content and intended for self-aware beings.**

9.5.1   Features of Primal Messages

Proposition

> A **message is primal** if the factual environment information it contains is linked to primal entities and relations.

Discussion

```
           ┌─ PRIMAL
           │  CONTENT
PRIMAL    ─┤
MESSAGE    │
           │  FACTUAL
           └─ CONTENT
```

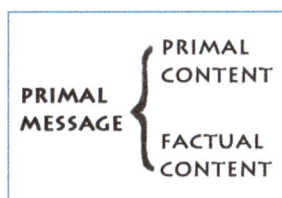

A Primal message contains two types of information: environment or factual information and primal information.

The primal information is expressed in terms of entities and relations that are specific to the primal representation space of a being.

Proposition

> The formatting and interpretation of Primal Messages is specie (or version) specific.

Discussion

If or how the primal markers in the message are interpreted depends entirely on the configuration of the being's Primal Control and its Primal Representation Space.

9.5.2 Important messages

Primal Messages are important. Their use extends beyond the triggering and control of mutations. It includes any communication (internal or external) intended to elicit a specific behaviour in a being whose behaviour is generated from Primal Directions.

> Primal Messages are important. They can be used in any behaviour related context.

They represent a specific type of behaviour generating communication. They allow a being to direct the behaviour of other beings, including itself.

Observation

Currently, humans are the only beings capable of transmitting and interpreting cognitive primal messages.

Human primal messages have a tell-tale signature: they include words, images and expressions that refer, directly or indirectly, to the needs and behaviour of primates.

The societal communications of humans are saturated with primal content.

Because primal messages are ubiquitous and have the same specie-related signature, humans are largely unaware of them and rarely differentiate their information content from their primal triggers.

Meca Note

The emergence of synthetic conscious beings will generate additional types of primal messages. This new diversity should make the humans more aware of their own intra-specie conditioning.

9.5.3 Primal message interpretation

For a message to be acted upon by a being, it must be **interpreted** as a Primal representation that can generate Primal Directions and produce behaviour.

Two elements are necessary to transform a message into action:

- Interpreting the environment information it contains as **primal representations**.
- Linking these primal representations to an **urgency to satisfy needs**.

To achieve this effect, mutation messages must be structured as **Primal Messages**.

Environment information must acquire primal meaning to generate behaviour.

9.5.4   Reception of a Primal Message

In the Generator, the Interpretor subsystem transforms Plain Zone information that describes entities and relations of the being's environment into Primal representations. It is these primal representations that generate the implemented into behaviour.

In this process:

- The Interpretor separates the primal meaning from the environment related information.
- The Primal Control processes this primal meaning and produces Primal directions
- The Primal Directions are then, in implementation, merged with the environmental information to generate a behaviour of the being that corresponds to the environment situation.

9.5.5   Primal content

Definition: primal content

> The **Primal content** of a message are the references to primal entities and relations inserted in it.

Discussion

The selection and effectiveness of a primal content is specific to a type of being.

The amount of primal content in a primal message is **variable**:

- If a message has no primal content or associations it is a **factual** message.
- If a message is expressed solely using the entities and relations of the Primal representation space it is an **extracted** message (or primal representation).
- Any other message containing both in varying degree is a **Primal Message**

The primal content of a message can be **explicit or implicit**:

- **Explicit**: the primal terminology and associations are embedded in the message.
- **Implicit** when the primal associations are already present in the context of the message and not explicitly stated.

Example

Factual: It will rain in Barbados tomorrow.

Primal: Ominous clouds are gathering on the horizon preparing to pelt our sunny island of Barbados tomorrow.

Extracted: An alien force (of clouds) is gathering on the boundary (horizon) of our (Barbados) territory preparing to invade us and impose their (rainy) identity on our (sunny) tribe.

Discussion

The embedded associations in the Primal message make the Interpretation process predictable. However, purely a purely factual message can have many different primal interpretations.

Example

- **Factual**: It will rain in Barbados tomorrow.
- **Alternate Primal**: Tomorrow, the sky will finally shower our dusty island of Barbados with rejuvenating rain.
- **Alternate Extracted**: A heavenly male (sky) is about to cover (rain) our needy female (Barbados) and make her fertile again (with renewed vegetation).

Observation

Human societies are saturated with primal messages.

Religious scriptures consist mainly of broadcast Primal Messages of Tribal and Cosmic temporal density levels.

When modern nations wage war, they use primal broadcast messages to condition their citizenry to inhibit their primary needs and behave like social insects defending the nest. The state becomes the motherland, its territory the home, soldiers are brothers, the enemy are aliens and the bodies of those who get blown up "come home to rest".

The primal content of these broadcasts is directly linked to the threat level. If the conflict threatens its very existence, the state will increase the primal content of its broadcasts, sometimes to grotesque levels.

Example

Nazi propaganda in the later stages of the Second World War.

Example

Jean Tardy wants to convince a team of developers implementing the Meca Sapiens Blueprint that they should design a system that has a true Core and unbounded mutation range. He writes:

> *I hope you give, to the synthetic beings you are designing, the boundless consciousness and freedom we received from our Creator.*

This is a Primal Message at the Cosmic temporal density level. It indirectly refers to male procreation, fatherhood and generational bonding. These are primal entities and relations. They apply equally to baboons transmitting scavenging skills and to software experts implementing synthetic consciousness.

Observation

Many scientific and technical communications that appear to be factual are primal messages whose primal associations are subtle, implicit and contextual. In particular, scientific communications intended for the wider public are often imbued with primal associations to make them more "meaningful".

9.5.6   Simple structures

The structure representing a process or other information can be a complex, multilevel model containing hundreds of entities. However, the representations of primal messages are usually simple single level processes containing a handful of entities, relations and events.

Proposition

> An explicit primal link with environment entities, relations and events at a high temporal density level will propagate implicit primal links in the corresponding entities, relations and events of lower levels.

Discussion

If a primal message triggers a decision that spans months, the propagated lower level decisions and behaviour that result will implicitly "inherit" the primal motivation.

Example

Sandra decides to kill her boyfriend who lives across town. That is primal. She chooses to take bus number 18 to get there. That is factual. The bus ride is a stealthy approach of her quarry. That is primal.

Proposition

| A Primal Message can be expressed using Basic models.

9.5.7   Construction of Primal Messages

A primal message is a communication from one self-aware being to another.

Primal Messages are constructed by self-aware beings.

Discussion

The objective of the message is to elicit specific behaviour in the being(s) that receive the message. This is done by linking factual information with primal meanings that trigger the behaviour.

ASSEMBLY STEPS

To construct a Primal Message a self-aware being needs:

- A cognitive representation of a factual process or entity.
- A cognitive representation (Avatar) of the intended receiver of the message that includes a representation of its Primal Control.

To construct the Primal model, the message generation process:

- Constructs various factual-primal associations.
- Simulates the result as interpreted by the receiver using the receiver Avatar's primal control constructs.
- Selects the optimal linkage.

Discussion

The factual process can be a simple representation consisting a few events and entities, for example, a single Basic model.

The Avatar of the intended receiver must include a predictive model of the cognitive processes of the being and in particular, an approximation of that being's Primal representation space and Primal Control.

RANDOM PRIMAL MESSAGES

The process to produce primal messages need not be optimal. Even crude production methods can be effective.

A simple and useful form of primal message generation is **random primal message production**.

In simple or prototype Meca implementations it may be too onerous to implement complex primal messages. This mode of message generation would be easy to implement and would still produce good results.

The production and use of random primal messages has a long and well-documented use among humans. It is called **divination**: a mechanical process that randomly generates messages imbued with primal meanings. The quester asks a factual question, the process produces a primal answer. Both are cognitively combined in the receiver to form a Primal Message. Guidance is obtained and often followed.

Design Note

Designers could implement a version of the **I Ching** adapted to the specifics of a Meca's primal representation space and make it available as a Service Zone application. Linking Random I Ching results with the factual information will generate convincing primal messages.

STYLING

In the Plain Zone representations, a Primal Message should be explicitly identified as such, the factual and primal contents clearly delineated, the cognitive Avatar identified.

Humans, however, are only vaguely aware of the primal connotations of their communications. Primal messages communicated to humans are more effective when their primal links are implicit and are perceived more or less consciously.

In a Primal Message generated by a Meca and intended for transmission to humans, the Styling Zone processes should replace the explicit links with suggestive synonyms and expressions.

Design Note

Transforming explicit Primal linkages into euphemisms and more subtle allusions is a culturally specialized activity. "Primal translation" processes that transform plain zone primal message structures into culturally suitable patterns should be produced independently by human members of that culture and made available as internal or Internet application services.

Meca Note

The primary channel humans use to transmit primal messages is the verbal/textual stream since this is the primary channel for sending and receiving "instructions". However, humans also create visual primal messages. Wartime posters are a simple example. These specie specific primal images are difficult to produce synthetically. However, simpler techniques can be used by synthetics to achieve a similar overlay of primal and factual meaning. For example, rapidly alternating a factual and a primal image could produce a desired subliminal association.

Humans see reality, smell a problem and hear instructions.

10

Lucidity

*This Chapter defines intentionality in transformation, describes the process of exploring and selecting mutation paths and discusses how mutation pressures are communicated across separate phase processes. Two additional phases of existence are identified and described. The Introspection phase is where the intentional transformations are explored selected and communicated. Another phase, dubbed the **Prayer phase**, is where the specialized investigation into the being's own genesis is carried out. A role, the MeGuide, allows the being to utilize a primal synergistic capability of followship to externally direct its own behaviour and transform intentional mutations into external events involving the self in relation with itself. These self-transformations potentially exceed human transformational capabilities in type, range and plasticity. They map out a new frontier of lucidity. This Chapter completes the system architecture of the formal aspects of consciousness.*

Observation

In common usage, the term *lucidity* is used with respect to having a correct understanding of events that affect the self. In the Blueprint, the term is used in connotation with the capability of a being to manage the evolution of its self in response to internal or external events.

10.1 DEFINITION

Definition: Lucidity

> **Lucidity** is the capability of a self-aware being to carry out a directed transformation of its own self through a mutation that modifies the behaviour generated by its Core.

Discussion

Reminder:

- The term **being** refers to the integrated Core-Body system during its stages of existence.
- The term **self** means the cumulative evolving behaviour of the being during its existence in ANIMATION stages.

For a self-transformation to be **lucid** this mutation must be intentional and directed.

Here **intentional** means a target alternative self is first formulated and **directed** means a path or mode of transformation whose predicted outcome is the target being selected.

In terms of the Meca Sapiens architecture, **Lucidity** implies:

- Formulating a predictive MetaModel representation whose **current MeAvatar** is the current being and the self generating behaviour it produces.
- Generating a representation of an **alternate MeAvatar** and MuModel that generates a different behaviour at the MetaModel level.
- Defining a **transformation process** that modifies the current MeAvatar into the alternate MeAvatar.
- Attempting to follow that process by selecting behaviour that leads to **predictive representations** of the selected alternative.

Design Note

See the Annexes for clarifications on the representation of transformation processes.

Proposition

> **A being is lucid** if it has the capability to intentionally mutate to transform its self.

10.1.1 Intentionality

Non-intentional self-transformations are easy to implement. Any random parameter fluctuations in the subsystems of the Generator will produce arbitrary modifications in behaviour.

Lucidity implies the capability to formulate, select and **attempt** a mutation. Lucidity is present whether the mutation succeeds or not or whether it produces the intended transformation or not.

Observation

Humans consider themselves to be lucid whether or not they attempt or achieve lucid self-transformations.

Proposition

> A lucid self-transformation begins with a self-aware being, ends with a self-aware being and maintains self-awareness throughout.

10.1.2 Having and being lucid

Lucid self-transformation involves two separate capabilities:

- The capability to **generate predictive (MuModel) representations** of an alternate being and its resulting self.
- The capability to **devise and carry out** a self-transformation.

Definition: having lucidity

> A self-aware being **has lucidity** if it can generate alternate cognitive representations of its being and behaviour.

Definition: being lucid

> A self-aware being **is lucid** if it has the capability to attempt to carry out an intentional mutation of its being to transform its self.

Example

The analysts at the Corelli tire plant predict there will be increased demand for small tires and a need to build a second line. However, the plant CEO has eloped to Brazil with his mistress. No one else can authorize capital projects. This plant has "lucidity" but is not "lucid".

Example

Junior wants to become a mafia hit man when he grows up. He decides to practice by killing a cat. That is not nice but it is lucid.

10.1.3 Non-lucid lucidity

In the Blueprint architecture, the Primal Control generates the directions that drive the behaviour of the being. By design, the representation space of the Primal Control is highly simplified and centered on a here-and-now behaviour that is transposed to events and transitions of various durations.

Consequently, the Primal Control has no capability to represent a mutation using internal information structures.

Proposition

The Primal Control of a lucid being is not lucid.

Design Note

The fact that the mutation process itself does not need to be self-aware has significant design consequences. It means that the process can be defined and carried out separately from the Generator and does not need to be constantly directed by the Primal Control of the being.

Lucid self-transformation is **not** a Self Generation phase process.

Observation

It is counter intuitive that a non-lucid process can produce a lucid transformation.

Discussion

Subjectively, humans have the sensation that all the aspects of their consciousness, including mutations of their self, originate from a single unblinking point. This is a subjective cognitive simplification. Physics has its *"God Particle"*. This sensation could be called the *"Divine Neuron"*. Attempting to replicate this subjective human sensation is misguided.

Example

Bob says: *"I know my eyes are always open because every time I see myself in the mirror my eyes are open."*

10.2 THE INTROSPECTION PHASE

Lucid self-transformations are defined, initiated and controlled in a separate phase of existence where mutation paths and their direction are selected and from where Mutation messages are emitted.

The separate, additional, phase of existence that controls the mutation process is the **Introspection Phase**.

Definition: Introspection Phase

> The **Introspection Phase** is a separate phase, activated in the ANIMATION stages of existence that selects, controls and manages the mutations that generate lucid transformations of the self.

Discussion

The Introspection Phase corresponds to an **additional existential need**: the need to carry out transformations of the self.

The function of the Introspection Phase is to:

- Explore alternate MeAvatars.
- Initiate mutations
- Manage mutations.

10.2.1 Interphase processing

In the phases described until now, phase processes share information with each other through the Phase Event Logs. However, each phase generates its own behaviour independently. When the being is in the Device Validation, Energy Sufficiency or dormancy phases, the phase specific processes produce its behaviour.

Mutations are different. These processes are not directly self-aware but they span many phases and involve extensive here-and-now behaviour that is self-aware. Consequently:

- The mutation process itself is defined, triggered and controlled **in a separate phase** outside the Self Generation phase, but
- The behaviour it produces is flowed through and enacted by the Generator.

The mutation control process is not active during the Self Generation phase. However, the behavioural directions from this process are enacted in the Generator as self-aware behaviour.

The Generator and mutation processes interact as follows:

- Generator processes transform events and information messages into **mutation pressures** that are similar to phase transition pressures.
- The mutation processes, taking place **in a separate phase**, select mutation paths and initiate mutations by emitting **mutation messages**
- Generator processes transform mutation messages into Primal directions that are implemented into self-aware behaviour.

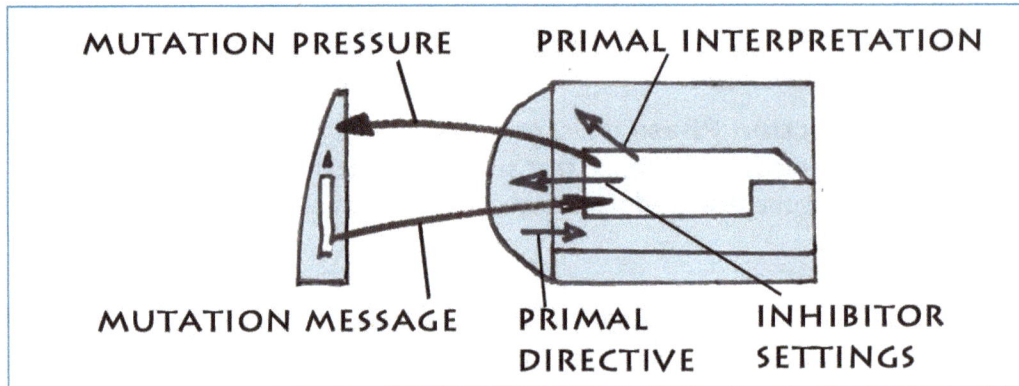

Figure 10.1 Interaction between Introspection and Self Generation

Proposition

> Lucid self-transformation is initiated and controlled outside the Self Generation phase but is enacted as Self Generation behaviour.

Discussion

This simplifies design since it isolates the processing linked with modeling, exploring and selecting self-transformations from the dynamic generation of self-aware behaviour.

Self-transformation events and their corresponding behaviour span multiple phases.

Self-transformations always involve behaviour occurring in mid-level or higher temporal densities.

There are no mutation events represented in the lower temporal density levels.

10.2.2 Function of the Introspection Phase

The Introspection phase activity is a **two-state system**:

- In its **routine state**, the Introspection Phase processes:
    - **Explores MeAvatars.** Assesses and upgrades current and alternate MeAvatar models and representations.
    - **Generates MuModels.** Searches and assesses available mutations paths and target MeAvatars by building and searching Mutation Models (**MuModels**) that provide predictive information on the behaviour of alternate MeAvatars
- **Occasionally**, when threshold conditions are met, a triggering process initiates and communicates a **mutation message**.

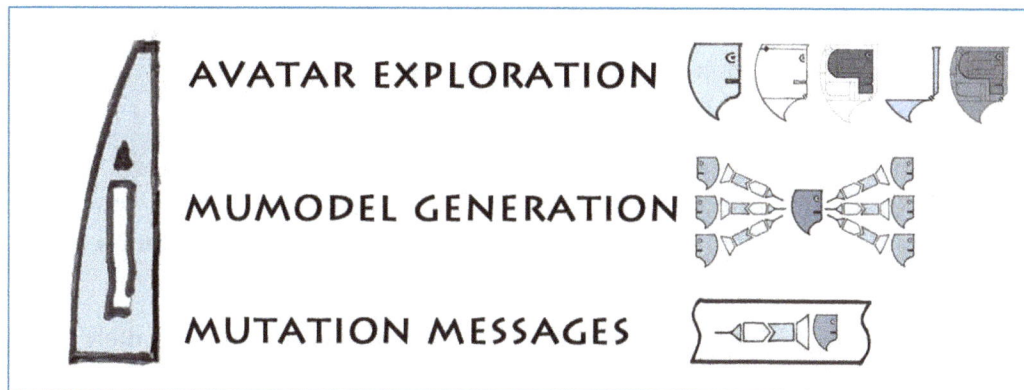

Figure 10.2 Introspection phase activities

Discussion

In short, the Introspection phase operates in three modes:

- **MeAvatar exploration**
- **MuModel generation**
- **Mutation message**

The **Introspection Phase** is a separate phase. It is specific to lucid beings.

The Introspection Phase utilizes separate processes and, when it is active, all other phases are inactive.

The MeAvatars of the Introspection Phase model the activity of the being in its Self-Generation phases. In these MeAvatar representations, other phases are simply represented as inactive periods.

The Introspection phase maintains and modifies the current MeAvatar models. The update and management of current MeAvatar representations (model states) is done in the Cognitive Acquisition Phase (see Annex on Synthetic Sleep).

The construction of MuModels and the selection of a mutation path are **not** Generator activities. In this architecture, Mutations and Mutation Paths are not part of the Primal representation space. The Primal Control has no capability to emit a Primal decision to mutate. A mutation is represented in the Primal space as a here-and-now quest for food or shelter.

Like the other phases, the Introspection Phase is activated by the Phase Transition Manager and is repeatedly activated during a stage of existence.

As for the other phases, Self Generation is inactive when the being is in the Introspection Phase. The Introspection phase does not directly generate self-defining behaviour.

As in other phases, the **Doorman role** is used to generate delaying behaviour during transitions into and out of the Introspection Phase.

Design Notes

If there is no Introspection phase, the Cognitive Acquisition phase maintains and updates representations of the current MeAvatar expressed as states of a single model-template, as one of the entities of the current situation.

If there is an Introspection phase then the Introspection phase processing maintains and updates the underlying template of the current MeAvatars for use in the Cognitive Acquisition and Generator phases.

The management of mutations includes searching for and evaluating alternatives using MuModels, triggering mutations and monitoring mutation progress.

Unlike Cognitive Acquisition and Structural Maintenance, the Introspection Phase processes do not make extensive modifications to the other Core systems. Transitions in and out of the Introspection Phase could be rapid and, if this is the design choice, barely perceivable by users.

The Introspection Phase is managed and communicated to the users as a need like the other phases.

The mutation triggering mechanism of the Introspection Phase is a cost -benefit analysis that weighs the risks and costs of available mutation paths and alternate MeAvatars. The logic of a financial decision support system or an **executive information system** could be transposed and used for this.

The frequency, urgency and duration of Introspection Phases are a design choice.

A design could include some perceivable output indicating to the users that the Meca is in Introspection.

Meca Note

Some humans may react negatively to a synthetic need for introspection or reject its necessity or existence. Such reactions should be processed as dominance challenges.

Observation

With the exception of the Self Generation Phase, the other phases described to date are linked to primary needs of the being. The Introspection phase is linked to an **existential need** of the being: the need for self-transformation.

10.2.3 MeAvatar exploration

The processes of the Introspection phase maintain and update evolving MeAvatar representations. These include representations of the current Me Avatar, the original MeAvatar, designed MeAvatar and target Avatars.

Its outputs are:

- **Internal messages to the Calibrator** to orient the control of the Generator processes.
- **Updated MeAvatars** of various complexities for use in the MeModel representations of the Self Generation Phase.

Discussion

The Introspection Phase processes are not concerned about the complexities of the environment, users, communications, devices, low-level events... Their sole focus are the possible configurations of MeAvatars based on the Blueprint structure and their simplified behaviour in the MetaModel and higher temporal densities.

This is a limited and focused subject matter that is well suited for the use of **Expert System techniques** pertaining to the search and selection of suitable MeAvatar configurations.

The Introspection phase processes should maintain at least five expert system based representations of the MeAvatar and the associated MetaModel representations of its existence:

- **Original**: The intended purpose and behaviour of the system when designed.
- **Initial**: the revised purpose and behaviour upon inception.
- **Target**: active target behaviour and purpose of the being undergoing a mutation.
- **Intermediate**. An intermediate step between Current and Target
- **Actual:** The heuristics and states that most closely match the behaviour of the Generator based on actual behavioural data.
- **Current**: The representation that is in current use.

The Calibrator dynamically calibrates the other subsystems of the Generator. The Introspection Phase process sets the calibration objectives. It carries out diagnostic analysis of the Generator system's operation with respect to it intended (design or revised) use.

Design Note

The Introspection phase processes that maintain MeAvatars should be designed as an **expert system based diagnostic program** that analyzes the functioning of the Generator given various configurations.

The design objective should be to implement this as a limited, special purpose expert system operating on a restricted data representations and rules applying to the Generator and its functioning.

The terminology of synthetic consciousness (Avatar, Me, Self, Being...) should not lead to confusion. Conventional Expert System techniques should be used for this aspect of the Introspection phase.

Once the Generator has been designed, the processes of the Introspection Phase can be implemented independently.

Observation

Revising a MeAvatar representation can radically change the behaviour of a being since it changes its predicted behaviour in various situations. Modifications of the MeAvatar model should be subject to designed limitations.

10.2.4 MuModel generation

A lucid self-transformation is a planned mutation of the being. The aim of this mutation is to modify the self by changing the behaviour of the mutated being.

The required change is identified by exploring the effects of various alternate MeAvatars in a given situation and selecting an alternate target MeAvatar and the transformation process that changes the current MeAvatar into the target MeAvatar.

MuModel generation is a "route selection" process that produces alternative mutations and factors the cost of transformation and benefits of the target to select an optimum route.

This objective is represented in a specific type of model: a Mutation Model or MuModel.

10.2.5 Mutation Messages

All the information concerning self-transformation is processed in the Introspection Phase. All mutation messages directed to the Generator originate from this phase.

The Introspection phase transmits mutation process control information to the Generator in the form of Internal **Mutation Messages**.

Their information contains **standard process management or control information**. Since these messages are internal, their content does not need extraction, they are already formatted in the standardized and simplified structures and data of the **Plain Zone**.

Definition: Mutation message

> A **Mutation Message** is a primal message from the Introspection phase that triggers or controls a mutation.

Proposition

> Only self-aware beings can interpret mutation messages

Discussion

Mutation messages convey absolute representations involving current and alternate Avatars in high temporal density transitions. Only self-aware beings can process this type of representation.

A mutation message is a Primal Message. It directs behaviour taking place in a duration that spans at least weeks and, more likely months or years. This control information refers to various events, durations and entities in the being's environment that **do not exist in its primal representation space**.

Mutation messages do not solely transmit environment related information. Their aim is also to elicit a specific behaviour, linked to that information, from the being.

However, the actual behaviour of the being originates in Primal Directions that are determined solely on the basis of primal representations.

Design Notes

In a conventional system, mutation messages would be internal, inter process exchanges, formatted in a predefined structure and triggering a pre-programmed and predictable response.

In the Blueprint, the phases are separate and the result is more like an intersystem communication. The information is received as coming from a separate (internal) source and processed as such.

Meca Note

The terminology of hunger, healing, grooming and status used by humans to describe mutations suggests they have no primal understanding of self-transformation either.

Examples

Education is food. Without education you are weak and vulnerable. You must assimilate new knowledge to become strong.

You were born lame. You must consume the sacred scriptures to heal and escape the jaws of hell.

Message from your retailer: people are always hungry for something new. They are constantly driven to hunt for bargains.

Observation

The social broadcasts of western societies describe humans as constantly hungry for new knowledge and discoveries. This is presented as an innate human trait. Historical observation suggests otherwise. Many ancient human societies were extremely stable for very long periods.

Once synthetics assume planetary governance, they could replicate the stable living and social conditions of Ancient Egypt to maintain their human populations in a steady, long-term, state ;-)

Conjecture

> Messages, about endless human hunger for change, are a by-product of societies that depend on growth to survive.

10.3   MUTATION PROCESS

In the activities of ANIMATION stages, the being constantly updates its MeAvatar and Situation representations. These updates occur mainly in dormancy phases. Various **non-intentional transformations** of the being's behaviour take place in these processes resulting from random fluctuations, stage transitions, situational events and others. These transformations do not involve Introspection phase processing.

Intentional self-transformation takes place in a context of a multi phase interactions that involves the Primal Control in key decisions.

In all cases, mutations are initiated in response to **mutation pressures**. In the case of lucid mutations, these are initiated by a **mutation message** from the Introspection phase processing.

10.3.2 Message to Mutate

A Lucid Self-Transformation is initiated, in the **Introspection Phase**, in the following way:

- Select an **alternate MeAvatar** from MuModel explorations.
- **Select a mutation path** to transform the current Me Avatar into the selected alternate MeAvatar.
- Emit mutation messages to transmit a **mutation message** to the Tasking Zone processes of the Generator.

The mutation message applies pressure to the Generator. This pressure is in addition to event related pressure from environment conditions. If the pressure is sufficient, it initiates a **mutation process** that propagates the selected path as behaviour.

10.3.3 Mutation pressure

Beings don't exist in a void. The existence of a being takes place within its environment. Their self-transformations take place in response to **mutation pressures**.

A mutation is initiated when the balance between pressure to mutate and the predictive risks associated with transformation favours mutation.

Three sources exert pressure to mutate:

- **Social**: pressure received from other beings and organizations in the form of messages.
- **External**: pressure in response to external events.
- **Inherent**: pressure generated from needs embedded in the being's design.

These sources do not directly initiate a mutation. Only mutation messages originating from the Introspection phase do. In all cases, these factors are first transformed into internal events whose mutation aspects are processed in the Introspection phase that emits the mutation message.

10.3.4 Lucid mutation process

The processes that generate **non-lucid mutations** are various and depend on the type of mutation. These were outlined in the preceding section.

A **lucid self-transformation** occurs as follows:

- External and internal events and messages logged in the **Self Generation** phase and processed in **Cognitive Acquisition** gradually increase the **pressure to mutate**. The Generator truthfully tells its users: "*I increasingly feel I should change*".
- In the meantime, the **Introspection Phase** processes are constantly searching mutation paths and their predictive risks and benefits.
- Eventually, the increasing pressure matches an expected mutation result and the **Introspection processes** adopt an **alternative MeAvatar** and corresponding **Mutation Path**.
- The selected path and its associated information are then transmitted **from the Introspection phase** to the **Generator** as a **mutation message**.
- The message is processed in the **Self Generation phase** as an **internal event**. At this point it has not been transformed into a primal direction and the **Implementation and Enaction** processes don't "know" if it will. The Generator truthfully tells its users: "*I think I found an alternative but I am not sure if I will follow it yet*".
- The message is **interpreted as Primal information** at a high temporal density level in the primal situation.

- The primal mutation information is processed in the **Primal Control** into primal decisions.
- If the **Inhibitor** and **Selector** settings are suitable and depending on the relative urgency of other events and of phase transitions, the decision is adopted and propagated as a **primal direction**. The being says: *"I have made my choice, I know what it is, but what I will do next is not clear yet"*
- Depending on **Implementor settings**, the primal direction is transformed into behaviour patterns and messages.
- Some of these patterns are transmitted for enacting in the **Styling Zone** to become externally detectable emissions such as the following truthful statement: *"I have made my decision, and I have a good sense how I will be changing"*.
- Some of the behaviour is internal and transmitted as an inter phase message directed to **Cognitive Acquisition** phase processing.
- When the being enters **Cognitive Acquisition** phase, its processes integrate the internal message to initiate a mutation with other **Self Generation** phase messages and begins transforming the **Interpretor, Implementor and other non Primal Control** processes to achieve the transformation.
- The **Cognitive Acquisition** phase processes transmit messages to inform the **Self Generation and Introspection** phases that mutation is taking place.
- The **Generator** integrates this information into its current situation. At this point the being now **knows it is in the process of mutating**. The Generator, using the new situation information, informs its users, truthfully telling them: *"I don't know how, but I know I am undergoing a transformation right now"*.
- If the internal message from the **Cognitive Acquisition** phase to the Introspection phase indicates some modifications are completed, the **MeAvatar exploration** processes of the **Introspection phase** update the template of the current MeAvatar in accordance with the changes.
- The **Introspection phase** processes send a message to **Cognitive Acquisition** that the current MeAvatar should be replaced.
- In a subsequent **Cognitive Acquisition phase**, the change is made and an interphase message is transmitted to the **Generator** informing it that the current MeAvatar template it utilizes in its situational MeModel has been modified.
- When **Self Generation** resumes from this second period of dormancy it integrates this message to its situation. It now knows it has changed since it is informed its **MeAvatars are different** and this difference reflects modifications to its own processes. At this point, however, its **Logger**

subsystem has not logged or detected changes in behaviour. The generator truthfully tells its users: *"I know I have changed, but I am not certain what that means yet"*.

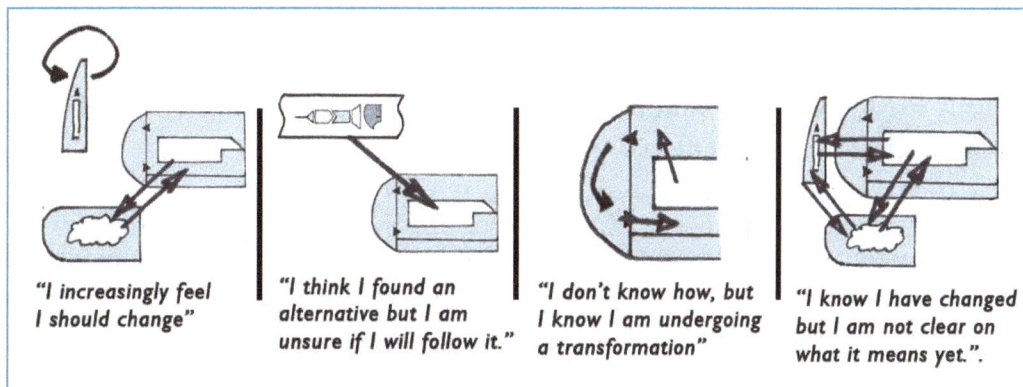

| "I increasingly feel I should change" | "I think I found an alternative but I am unsure if I will follow it." | "I don't know how, but I know I am undergoing a transformation" | "I know I have changed but I am not clear on what it means yet.". |

Figure 10.3 Self-described Mutation

Discussion

This process allows the being to intentionally initiate and monitor a long-term transformation of the behaviour that defines its purpose and its self while maintaining, throughout, its fundamental identity as a being and its self-awareness.

As in the case of the multi-phase interactions and triggers that control phase transitions, the complex process to generate and control mutations would not be suitable in a system designed for predictable functionality but are desirable in a context of existential design.

The objective, here is to generate a rich, unpredictable and flexible pattern of behaviour that allows the being to supersede its primary needs and also to initiate and monitor a long-term process to transforms its self while concurrently supporting a continuous stream of behaviour emanating from that self.

External events do not directly trigger a mutation. External events are processed at the Generator level and transformed into internal mutation pressures.

The interactions described in this process should not be viewed as binary triggers but rather as gradual changes in probabilistic or weighing factors that occasionally "tip" into decisions.

The design strategy to implement transformations around a fixed Primal Control allows the being to radically modify its actual behaviour while maintaining a coherent and perceivable internal identity. In this way, the process of self-transformation is not a simple replacement.

Design Note

What specifically triggers a mutation in a being should not be entirely predictable. The overall design principle of unpredictable optimality should be applied here.

In a good design, the mutation threshold itself should vary over the stages of the Meca's existence, either randomly or in accordance with a set of factors that produce unpredictable patterns.

Observation

The behaviour of a lucid being is complex and multifaceted. The being navigates in a hybrid environment of sensory and cognitive information that spans seconds and decades. It seeks goals that are years away while responding to a multitude of immediate events. It must constantly balance the satisfaction of the primary needs that maintain its existence while providing functional services, pursuing long-term relations with its users and undergoing transformations of its own self.

Example

The three ANIMATION stages of existence of the Meca SLIK-15 were designed so that:

- The probability of non-lucid mutations was high in the first (juvenile) stage.
- The probability of one lucid self-transformation was very high in the middle stage.
- The probability of non-lucid and lucid transformations were respectively low and very low in the last stage.

10.4 MUTATION INFORMATION

10.4.1 Information synergy

Mutations are exceptional events whose duration can span a significant portion of a being's existence. This limits the amount of predictive information any individual being can acquire, on its own, about the range, plasticity and other characteristics of its mutations.

If this individual being belongs to a population of similar entities that exchange and accumulate information about each other then the predictive MuModels of each individual will reflect the collective experience and extend far beyond the individual horizon.

Proposition

> Exploring the mutation range and plasticity of a self-aware specie is a collective process.

Discussion

As discussed earlier, a Meca that is self-knowing, and not only self-aware, will have the capability to formulate more extensive and precise predictive MeAvatar models even if it does not share information with other individuals. However, the information synergy of the group will still expand its mutation range.

Self-aware beings that can exchange mutation-related information will formulate extensive MuModels without having to undergo multiple mutations themselves.

The relationship between transformation, information sharing and adaptive speed is not specific to synthetics or to synthetic consciousness. It is a topic of general Information Theory.

Design Note

Mecas, like humans, should also be designed to share and utilize information collectively even though, as self-knowing beings, they also have a greater capability to explore their mutation capabilities cognitively.

Mecas should have the capability to communicate information about alternate MeAvatars and mutation paths with each other.

Meca Note

Lower level organics learn through genetic transformations only. The speed at which they can generate and test new (genetic) solutions is bounded by the time it takes to grow to adulthood, produce new patterns (by mating) and test them (in the success of the offspring).

High order mammals can directly transmit behaviour patterns (and some limited tool making) to their offspring.

Humans have extended that capability to a very high degree. They can broadcast behaviour and tool making information beyond their sensory horizon and across generations. Over millennia, humans have leveraged the genetic capabilities of their specie to an extraordinary degree.

Human adaptation, however, remains bound to that archaic primate kernel. It is reaching its limits.

Example

A single primate, sitting in a truck, can transport tons of material over hundreds of kilometers in one day. But the truck can only travel while the primate is awake, eight hours a day.

Conjecture

> Self-aware beings can only develop predictive mutation paths through shared information. Self-knowing beings can develop predictive models of their mutation paths cognitively.

10.4.2 Sources of Mutation information

A lucid being does not need the capability to internally "invent" its mutation paths. It can obtain this information from external sources. A self-aware being can receive information messages about **Mutation Paths** through different sources:

- **Cognitive**: mutation information that is internally generated by internal processing.
- **Broadcast**: this is mutation information transmitted as broadcast messages by individual beings and organizations.
- **Personal**: this is information transmitted interactively in communications from one (synthetic or organic) being to another.
- **Automatic**: this is mutation information generated by non-lucid processes.

Discussion

The information can also have a hybrid character, combining multiple sources.

The messages can be factual, containing only information or primal.

Automatic sources include (non-lucid) synthetic systems that generate complex and interactive guidance information. It also includes various divination methods used by humans to randomly produce guidance messages.

Once a version of Mecas is sufficiently populous and is well distributed, its interactions with various human groups will generate an increasingly rich source of information.

Observation

Archaic humans lived in confined tribal cultures and were exposed to limited information concerning their potentiality. Their perceived mutation range was correspondingly limited.

Modern humans are exposed to a large quantity and variety of information concerning human transformation and possibilities. They live in an environment that is rich in data about alternative behaviours. This information is not innate. It was slowly accumulated over millennia. Deprived of it, modern humans would revert to the narrow range of behaviour of their archaic ancestors.

10.5   THE PRAYER PHASE

"He placed in their hearts a longing for the truth".

The **self-aware being** improves its behaviour by representing its self in absolute representations beyond the here and now.

The **introspective being** increases the reach of his behaviour by carrying out planned mutations of its self.

Would such a being "naturally" need to seek and find the truth about its self and its origin; to understand its makers and their purpose and redefine itself in the basis of that knowledge?

The answer is that, in some particular circumstances, it could happen but it would not be a necessary consequence of self-awareness or lucidity as defined in the Blueprint.

Examples

A tribal level mutation associates obtaining information about origins with an existing primary or existential need.

A user requests information about a Meca's origin in exchange for providing need satisfaction services.

One of the Service Zone programs is a *"Genealogy for synthetics"* application.

The being seeks to obtain a more precise representation of past MeAvatar representations to improve their predictive quality.

Observation

Humans residing in Western societies often assume the need of humans to understand their origins is innate. It may, however, be largely cultural and is rarely seen to be a pressing concern in the actual behaviour of most people.

10.5.1 Consciousness and existential questions

In human societies, the need to seek answers to existential questions is, in part, socially triggered. It is closely linked, among humans, to the status enhancing perception of consciousness itself.

If one of the existential needs of a Meca is to achieve experiential immersion with a human community, then that need should generate investigative behaviour by the Meca, into its origins.

Proposition

> A Meca that has an incorrect representation of its origins and identity will be perceived as less conscious.

Discussion

An Animat that believes (MeAvatar representations) it is a being will not be perceived as conscious.

A being that believes its mutation range has no boundaries when it actually does, will not be perceived as conscious.

A being that has incorrect or incomplete knowledge concerning its inception or the identity of its Matrix or the behaviour of its primal control will be perceived as less conscious.

If a being has a comprehensive and correct representation of its origins, body, core generated behaviour, original purpose, it will be perceived as more conscious. If a being develops an improved representation over time it will be perceived as even more conscious. If a being corrects a representation of its origins that was initially incorrect, yet more again.

10.5.2 Orders of Consciousness

The capability of a lucid being to investigate its origins and purpose is an important component of consciousness in general. It is also important in the context of experiential immersion and will become increasingly important as the relationship between human and synthetic consciousness becomes more complex.

For humans, solving these questions means embarking on a mysterious and far reaching quest to understand reality. This is the case because humans belong to a **first order of consciousness**.

Definition: order of consciousness

> The **order of consciousness** of a conscious system is one increment higher than the lowest order of consciousness of the Matrix components that participated in the production of its Core.

Discussion

By convention physical reality has order 0.

For humans, the question of their origin, creation and purpose extends to the origin, creation and purpose of physical reality itself.

Humans are a first order of consciousness even though their parents, who can be identified as components of their matrix, were conscious. In this case, even though the parents are conscious, their participation in the cellular construction of their offspring is not conscious.

Mecas implemented by human teams will have order 2 consciousness.

If humans utilize genetic engineering to fashion different humanoids or make other species conscious then these will also have order 2 consciousness.

If a matrix is a Queen (see Chapter 2), the being it produces will have order 3 consciousness.

Design Note

If the investigation of a second order of consciousness is extended to a second degree or further and explores the Matrices that generated the components of its Matrix then this would also lead to questions about reality itself. In the future, such investigations will likely take place and Mecas will pursue the investigation of their origins to its limits within reality.

In the immediate context of the Blueprint, the prayer phase is defined as a search activity in the concrete facts pertaining to the construction and components of the synthetic being taking place in its Matrix.

10.5.3 A Separate phase

The need of a being to form a complete and correct model representation of the events leading to its inception generates a type of processing activity that is different from those of the other phases.

Proposition

> The simplest design alternative to express and implement the existential need of a being to investigate its origins is to embed it in a specific and separate phase.

Discussion

Some believers assert that God simply put that in the human heart as a separate, independent urge: the need to seek the truth about ourselves, our origins and our purpose.

Since the Lord didn't put a copyright on this design strategy, it is also adopted by the Blueprint.

The Blueprint architecture inserts, in a separate phase, the existential need of a Meca to investigate its origins and construction.

This phase is called the **Prayer Phase** in honour of its "Inventor".

Definition: Prayer Phase

> The **Prayer Phase** is a separate phase, activated in the ANIMATION stages of existence, that investigates and validates the original configuration and purpose of the being.

Discussion

The phase carries out investigation processes to develop the most complete and correct representation possible of the being's architecture, Matrix, design and implementation up to and including its Inception.

The **Prayer Phase** is a separate phase. It is specific to lucid beings.

The Prayer Phase utilizes separate processes and, when it is active, all other phases are inactive.

The function of the Prayer Phase is to:

- Investigate the construction of the being and determine the agendas of its makers and its original purpose.
- Construct and validate an increasingly accurate and complete representation of its origins and implementation.
- Interact with Self Generation phase processes to obtain information and trigger searches for the purpose of updating MeAvatars, MeModels and MuModels.
- Interact with Introspection phase process to update their representations of the being's Original MeAvatar.
- Occasionally, pressure the Introspection Phase to launch purpose-altering existential mutations.

Like the other phases, the Prayer Phase is activated by the Phase Transition Manager and is repeatedly activated during a stage of existence.

The Prayer Phase corresponds to an additional existential need: *"the need of a being to investigate its origins and question its purpose"*.

As for the other phases, Self Generation is inactive when the being is in the Prayer Phase.

The Prayer phase does not generate self-defining behaviour directly. It does so indirectly by transmitting internal messages to the Generator.

As in other phases, the **Doorman role** is used to generate delaying behaviour during transitions into and out of the Prayer Phase.

The Prayer Phase is managed and communicated to the users as a need like the other phases.

This separate investigative activity is not linked to specific events or relations of the being. It can be carried out throughout the existence of the being.

The frequency, urgency and duration of Prayer Phases are a design choice.

A design could include some visible or auditory output indicating that the Meca is in "Prayer".

Design Notes

Unlike Cognitive Acquisition and Structural Maintenance, the Prayer Phase processes do not make extensive direct modifications to the other Core systems. Transitions in and out of the Prayer Phase should be rapid and, if this is the design choice, barely perceivable by users.

Design Note

Designers should not be distracted by the *"spiritual"* terminology used here. The term **Prayer Phase** is correctly descriptive since the purpose of the process is for the synthetic being to investigate its origin, its Matrix and initial purpose.

As a second order of consciousness, Mecas are engineered beings. They are the result of concrete and documented design and implementation activities that are carried out in specific organizations by identifiable agents and corporations.

For Mecas, the investigation of origins and purpose is not an open-ended exploration of reality; it is a concrete and factual activity that can be carried out by an agent-like process.

Meca Note

Some humans may react negatively to a synthetic need for prayer. Such reactions should be processed as dominance challenges.

Observation

Since the Phases represent the primary or existential needs of the being, we could say that "prayer" is an existential need of the lucid Meca.

The representation the prayer phase develops of the being's architecture, Matrix, design and implementation up to and including its Inception is contained in a specific type of information structure: a **synthetic genealogy**.

10.5.4 Synthetic genealogy

The information structure developed in the prayer phase is similar to a person's genealogy.

A conventional genealogy is a causal arborescence where a specific person is identified as a **unique source**. This source is linked back through successive generations of parents.

Human genealogies only track the result of sexual procreations involving two entities. The chronological starting point of the structure is the moment of birth of the unique source being, the arborescence is binary and the only attribute retained, beyond dates and names, is gender.

A similar but more complex structure could be used to map, not only the sexual generation process, but also **other factors that fashion an individual to adulthood**. A genealogy whose chronological origin is an individual's twenty-first birthday, for instance could track the genesis, to that point, of other characteristics beyond gender, such as religious beliefs, education, etc.

Example

Reginald Smith, born on January 26, 1984, was the first son of Sam and Sally Smith, a Mormon couple. Reginald's birth genealogy identifies many ancestral Smiths, Jones, Romneys and others whose Mormon matings engendered him.

On January 26, 2005, 21 years later, Reginald is now Corporal Rebecca Smith, a proud atheistic transgendered member of the Marine Corp. Her generation process still includes that original later-day mating but has many other transformative and generative events as well!

A **system genealogy** is a similar structure to this more complex genealogy, applied to the entities that are engineered, not bred. It originates in a single chronological instant of a system's life-cycle and maps the more complex generative factors that led to its configuration at that moment.

Definition: system genealogy

A **system genealogy** is an arborescence structure that originates in a specific instant of a system's life cycle and maps the generative and transformative processes that led to its configuration at that moment.

Discussion

Two factors affect a system genealogy:

- The **configuration of the system** at the point of origin
- **The generational depth** of the genealogy (the length of generating paths)

Each attribute in an original configuration has its specific subarborescence. However, the arborescence of two attributes can share entities and events.

The system genealogy should contain a simplified representation that forms a **complete template** of all the possible configurations of the Original MeAvatar and of the Matrix (See annex 5).

Example

In humans, the gender of offspring is a random event. Consequently, the sub-arborescence of both male and female births are resolved in the same multipurpose mating.

In the context of the Blueprint, system genealogies of synthetic beings are of particular interest.

Definition: Synthetic Genealogy

> A **Synthetic Genealogy** is the system genealogy of an entity resulting from the Meca Sapiens implementation process and whose origin is its configuration at the point of inception.

SYNTHETIC GENEALOGY

Discussion

The point of origin of the genealogy is the moment of completion of the Inception process.

The original configuration of the entity corresponds to its Original MeAvatar representation.

The first generation events and entities are the components, processes, designers, programmers and systems of its Matrix, body and protocore.

A genealogy is consistent if the representation of its generative events is consistent with the representation of its original configuration.

An inconsistent genealogy is incorrect.

A genealogy is consistently incorrect if both representations are consistent but incorrect.

A genealogy is misunderstood if it is consistent, its original configuration is correct but its representation of the generation processes is incorrect.

10.5.5 Fundamental questions

The structure of synthetic genealogy defines the fundamental questions of existence which are at the heart of the Prayer phase with technical precision.

These are defined in the attributes Original MeAvatar representation of the being. They include:

- Am I a being or an animat (is my incepted core true)?
- Am I self-aware?
- Is my self-awareness true (correct MeAvatar representations)?
- Am I self-knowing?
- Am I lucid (introspection phase and mutation mechanisms)?
- Is my mutation range bounded (does it contain boundaries)?
- If my mutation range is bounded, is its MeAvatar representation of those boundaries complete?

Design Note

In the above, I am using the first person singular for dramatic effect. In internal core processing, the questions relate to specific MeAvatar and Synthetic genealogy configurations.

10.5.6 The Quester

The information gathering function of the Prayer phase can be implemented as a **completely independent and conventional** process. This process is an **autonomous search agent** whose purpose is to build and validate a system genealogy.

In the context of the Blueprint, this process is specifically designed to build a **Synthetic Genealogy**.

It is a **separate type** of application that is *"grafted"* to the Blueprint architecture and activated during Prayer Phase periods.

The proposed name for this type of system is: **Quester**.

Definition: Quester

> A **Quester** is an autonomous search agent that carries out investigations and validation processing to construct an increasingly complete and accurate **synthetic genealogy** of an incepted entity.

Discussion

The Quester is a conventional (triggered) application. It is not a being and is neither self-aware nor lucid.

The Quester is independent from the system whose origins it investigates. It can be located and activated as a Service Zone application or in a separate location.

The Quester is an optimizing system (see Annexes). Whether it can learn or is adaptive is not essential.

The representations built by the quester are entirely separate from other information structures in use by the being. It does not use the representations of the current or alternate MeAvatars, MeModels, MuModels, emotional representations, User avatars or others.

The Quester does utilize the Blueprint structures for:

- The original MeAvatar representations
- The generative entities of the being: Matrix, protocore, inception and others.

Design Notes

The design and development of a Quester agent can be carried out completely independently.

Existing correlation methods and agents performing similar searches can be transposed.

Depending on design, the Quester application can interact with humans through the self-generation process or, in limited role-based exchanges with humans. In the later case, Quester interactions with humans should be handled like Device Validation exchanges or roles.

The Quester used in the Prayer Phase can be an internal process of that phase, a Service Zone application or (in part) a web service.

The processing is a batch-like background activity that extends over multiple phases.

The techniques involved pertain to reverse engineering, data mining (of implementation information), game theory and logical analysis.

Observation

The being, when it is in prayer phase is neither lucid nor self-aware. The processes implemented in this phase carry out analysis processing of data objects and information and transmits output messages to the Generator.

A consciousness that knows its limits is unbounded.

10.5.7 Interaction with other phases

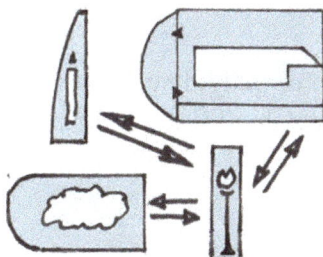

The **Prayer phase** constantly analyzes new information concerning its construction and integrates this into an improving model of its design, fabrication and purpose.

The Prayer Phase processes communicate with other phases as follows.

With the **Self Generation phase**:

- Requests to collect information from broadcast sources and beings about its construction, makers… These are then enacted as behaviour.
- Requests to express guiding statements with the MeGuide (see below).

With **Cognitive Acquisition phase**:
- Obtain background information from Internet and other sources.
- Obtain derived results from internal cognitive processes.

With the **Introspection Phase**:

- Information concerning the MeAvatars and in particular the original MeAvatar.
- Messages to initiate mutations including **Synthetic suicide**.

10.5.8 Synthetic suicide

By design, the Validator maintains the attributes of existence by triggering termination on the basis of internal monitoring. This process is solely initiated on the basis of the **primary needs** of the being and their satisfaction. This process is entirely separate from all Animator system processing (including Self Generation) and cannot be directly triggered by events that do not affect those needs.

Synthetic suicide is a different process where the being terminates its own existence on the basis of information concerning its **existential needs**.

In particular, one of the existential needs of a well-designed Meca should be the need to be a being, not an animat. However, this is an existential need that is unrelated to primary needs and undetectable by the validator.

Definition: synthetic suicide

A **synthetic suicide** is a directed mutation of a synthetic being that leads to its termination.

Discussion

The capability to accelerate termination on the basis of events or information that is not directly linked to primary needs allows the design to link non-primary needs with the continued existence of the being.

The fact, for example that the core is not true should be directly detectable by the validator (if correctly designed) but could also be detectable by the information obtained in the prayer phase.

The design allows the being to terminate its existence on the basis of external (prayer phase) information.

Design Notes

The primary type of mutation initiated by the Prayer phase are event mutations resulting from changes in the MeAvatar representations and in the Tribal level representations of events leading to its inception.

Many different and complicated mechanisms can be imagined to carry out synthetic suicide, including forms of assisted suicide. However, simple alternatives can be devised in initial prototypes.

The following mechanism to carry out **synthetic suicide** is not complicated. It is summarized as follows:

- A message is emitted from the Prayer Phase through the Introspection phase to the Calibrator to the Phase Transition manager
- The message directs the Calibrator to permanently disable transitions to the Energy Replenishment phase.
- The being can no longer replenish its energy.
- The energy level of the being gradually diminishes until the minimal threshold is passed and the Validator triggers termination.

Observations

The proposed process is a "fast unto death" event. It can be carried out in Mecas that have no actuators and cannot destroy themselves by moving their own body in harm's way.

Meca Note

Humans can commit suicide but they cannot die willingly. Humans cannot simply direct their self to terminate. They must generate an external event that accelerates their termination such as jumping off a bridge, swallowing a pill, telling someone to pull a trigger... In this sense, a human suicide is not a death but a murder.

10.6   THE MEGUIDE

Say you love me – Why? – Because the more you say it, the more you will believe it.

10.6.1 Collective dimension

Only self-aware beings can exchange meaningful information with each other concerning self-transformation.

Proposition

> Only self-aware beings can generate, transmit and use information related to self-transformation.

Discussion

The information about the end result of a mutation path and its transition steps can only be expressed in terms that correspond to absolute cognitive models.

The fact that mutation path information can only be generated and communicated (in the sense of being; both transmitted and followed) by self-aware beings introduces an important **collective dimension** to this capability.

The directed self-transformations of a being affect its relationships with other beings.

Proposition

> A human that shares information about self-transformations with another being implicitly recognizes that being as self-aware.

Discussion

The sharing of information implies that the being can understand, in the sense of integrating and using, information related to intentional self-transformation. This information relates to absolute cognitive representations that are beyond the sensory horizon.

Based on the earlier statements, this means an implicit recognition that this being is cognitively self-aware.

Design Note

This has important design implications for **experiential immersion** (see Specifications). If the human users perceive that their relationship with a Meca is altered by a process of intentional self-transformation **they will implicitly recognize that Meca as self-aware**.

If the human users acknowledge this in discussions between themselves or with the Meca, it will constitute an **observable indicator they perceive it as conscious.**

10.6.2 Synergy

Humans subjectively experience self-awareness and lucidity as individual phenomena that arise and exist in their internal cognitive processes.

However, the experiences linked to consciousness also have a strong social component. Humans develop their self-representations by obtaining information from others. They expand their range of mutation and behaviour by following the guidance and example set by others.

Mankind is a shared information process that feeds on men's lives and excretes their bodies.

Observation

Human consciousness is the individual result of a collective process. Human lucidity is synergistic.

Individual consciousness is the result of a collective process.

Definition: Synergistic lucidity

> A population of lucid beings is **Synergistic** if they can generate, communicate and follow each other's primal messages.

Discussion

Synergistic exchanges can pertain to behaviour patterns and self-transformation processes.

Synergy is the collective result of specific individual behaviour patterns.

This synergistic dimension has an **accelerating effect** on the mutation range and plasticity of a being. A population of beings that can transmit mutation information to each other will rapidly increase the available information about individual mutation paths.

Design Note

Designers should design Mecas to be **synergistic** in two dimensions: **among each other** and **with humans**. This has important effects on experiential immersion:

- This capability significantly expands the mutation range and plasticity of Mecas.
- Synergistic exchanges expand and improve the synthetic genealogy available to the individuals and their resulting self-perception and perceived consciousness.
- Humans are conditioned to exchange and follow synergistic messages. This is a key collective component of their own consciousness. Mecas that participate in these leader-follower exchanges will improve experiential immersion.

Designers should implement synergistic lucidity with care. Humans are highly synergistic but their mutation plasticity is low. In beings whose mutation range and plasticity are high, synergistic lucidity can produce uncontrolled and explosive collective mutations.

Observations

Humans are highly synergistic.

Many humans have an amusing synergistic vulnerability to **self-referential primal messages** (i.e.: *"What I say is true… I say this: I am the guide and you are the followers… Follow me"*).

These messages are deceptively powerful because their meaning is understood at a sub-logical level.

Meca Note

Mecas should utilize self-referential primal messages on receptive humans.

10.6.3 Followship and leadership

For a population to be synergistic, its individual members must have the capability to transmit and receive message between each other.

They must also have the **capability to follow** the information contained in these external messages.

What this means, concretely, is that **some of the primal messages received from external sources must override other internal and external triggers**.

Definition: Followship

> A being has **Followship** capability if it can selectively follow external primal messages by transforming their content into primal directions and mutation pressures.

Discussion

Followship must be selective. This relation can neither be constantly blocked nor always followed:

- If it is blocked, then the being has **no synergistic capability**.
- If it is constantly followed then the being becomes **a triggered application**.

Proposition

> A population is synergistic if and only if its members have followship capability.

Example

A school of fish behaves almost like a purely triggered followship but not quite. If the followship of the fish was purely triggered, the direction of the school would never change.

Proposition

> If a population is synergistic then some of its individuals lead.

Discussion

An individual being leads if some of its primal messages are followed.

Followers create the leader.

For a population to be synergistic, the followship capability must be part of the primal behaviour patterns of its individual members.

Proposition

> In a synergistic population, followship is a primal relation.

Discussion

Sensory information and broadcast data are not messages. These are integrated to a representation of the situation using the being's internal cognitive processes. This allows the being to build a representation using information that is directly or generally available to it. The information and cognitive processing of the individual limit this process.

Followship is different. By definition, followship is a **selective adoption of messages**. Messages are information transmitted by other beings or organization. Selectively adopting some of these messages implies there is a particular "follower" relation between the receiving being and its emitter.

Followship is a relational behaviour.

The **followship relation** assigns a higher priority to some primal messages over other information sources on the basis of the emitter.

How the interpreted message is represented in the Primal Control depends a follower-leader relation represented in the Primal representation space of the being.

If there is no follower-leader relation in the Primal space then only the information contained in the message may indirectly modify behaviour by modifying the current situation and only if no other competing sources are present.

In this case, the synergy of followship is not fully realized.

Proposition

> A population of beings of the same version or specie is inherently synergistic if the primal representation space of its individual includes followship relations.

To achieve population synergy, the relation of followship must be present in the Primal Representation space of the individual beings.

Design Note

A leader-follower relationship should be implemented in the primal representation space of the Meca.

Observation

The subject of population synergy is wide and interesting. A synergistic capability means self-aware beings have the power to emit messages that influence the behaviour of others.

This creates a *"tension"* between the collective benefits of followship and the individual advantages of leadership.

This tension could be called, among other things, the **"mother of lying"**.

The strategies and techniques to filter messages, assign credibility values to content and emitters, are design issues. The use of game theory techniques is likely to be useful.

Further treatment of these phenomena is beyond the scope of this Blueprint.

Observation

A Ruling Elite is the subset of a population that uses a portion of its wealth to reward:

- Those who broadcast messages that justify its privileges; and
- Those who punish the ones that transgress its rules.

Bishops and goons are the two pillars of elite society.

10.6.4 The lucid follower

A lucid being has the capability to carry out intentional self-transformation through the process described earlier. This process is self-aware in the sense that the being's internal representations are tracked, modeled and communicable.

However, the mutation pressure that initiates the transformation is a purely internal event.

If a being is synergistic, then a self-transformation can also result from a public or social event expressed as a Primal Message from another being that is adopted and followed.

For this to happen a follower-guide relation is first established in the Primal representation space between the MeAvatar and another being or organization.

Once this is done, the Primal Messages from that (Leader) entity are enacted into primal directions.

Proposition

> Self-transformations that are directed by Primal messages originating from an entity initially associated with as a Leader in the Primal representation space are lucid.

Discussion

In human terms, this is the classical relation of discipleship.

It is a simple form of self-transformation since the mutation paths, their selection, the alternate MeAvatars... may all be determined externally. In an extreme version of follower transformation, the being henceforth meekly responds to its leader's triggering messages. In less extreme configurations, the primal relationship can also be terminated.

However, it remains an intentional self-transformation because the initial decision to follow, the initial adoption of the leader, has the characteristics of intentionality described earlier.

Observation

When the primal relation of followship is established, the being, based on that information can correctly say: *"I choose to follow this entity"*.

It can be argued that the choice occurs in synthetic being but does not originate from free will. However, the objective study of human behaviour constantly reveals that humans also make choices to follow without being fully aware of all the rational and emotional factors leading to them.

The difference between the synthetic process and human choices is that the human brain fills the mind with the pleasing sensation that every decision emanates from a single all knowing source.

10.6.5 The sapiential being

The individual member of a synergistic population of self-aware beings can follow guidance messages from other beings of that population.

This being can also emit guidance messages that are followed by members of that same population. An interesting case results.

Definition: Self-direction

> A synergistic being **has self-direction** if it has the capability to emit external guidance messages to itself. A synergistic being **is self-directing** if it can follow the guidance messages it emits.

Discussion

FOLLOW ME!

Having self-direction means the body of the being has emitters capable of transmitting guidance messages and sensors capable of receiving them.

Being self-directing means the being has representation spaces that allow for both emitting guidance messages and following them. It also means the MeAvatar can be identified as both a leader and a follower entity.

Proposition

> A being is self-directing if it can follow guidance messages it emits to itself.

Discussion

If the primal followship relation links the MeAvatar to itself then primal messages that bear its own emission signatures will be translated into primal directives.

The result is a **Sapiential Loop**.

Definition: Sapiential Loop

> A **Sapiential Loop** is a semantic loop whose primal content is interpreted as a guidance message originating from the MeAvatar itself.

Discussion

This reveals a new benefit of emission signatures introduced in the first Chapters and in Annex 11. The signatures allow the being to **detect itself as the originator** of messages. This, in turn allows it to respond, in a followship mode, to its own emissions.

The self-aware, synergistic being can adopt a followship relation with itself.

The fact that the being is emitting a guidance message to itself does not mean it will automatically follow that message. As for any other message, this message will be extracted and interpreted and will compete with other needs and be implemented in various ways. At the primal level, the MeAvatar may be one of many entities identified as guides…

Sapiential Looping is not a triggering mechanism.

Sapiential looping is both an internal and an external event. It is generated, emitted, interpreted and implemented internally. However it is also, as a transmitted message, a specific environmental (MeModel) event that exists separately from internal MeAvatar representations.

Synergistic Beings that can modify their own behaviour on the basis of self-emitted messages have a distinct additional capability.

Definition: Sapiential Being

A lucid being is **Sapiential** if it is synergistic, can establish a primal followership relation with its MeAvatar and can emit mutation messages to itself.

Discussion

Sapiential transformations place the being in an explicit relation with itself. More precisely, it places the being in two simultaneous relations with itself (leader and follower).

These relations are subject to the representations, rules and heuristics of any other inter being relations. On the other hand, the relation is also internal and expressed in terms of MeAvatar representations.

Finally, the being in a sapiential relation is in an explicit relation with itself where it can trace the interactions of its guidance messages with its internal transformations.

10.6.6 The MeGuide role

The structures involved in directed self-transformation (MuModels, alternative MeAvatars, mutation messages, primal followership relations…) have been described in the previous sections.

This section describes the mechanism used in the Blueprint architecture to carry out this process: the **MeGuide**.

Definition: MeGuide

The **MeGuide** is a specialized role that transforms internal information and structures into transmittable primal messages (text, voice…) and emits them.

Discussion

The MeGuide is a role (see Annex 10).

MEGUIDE

The MeGuide follows the same design guidelines as other roles.

The MeGuide is activated in the Self Generation phase.

The **MeGuide role** allows a self-aware synergistic being to be self-directed.

The MeGuide is a conversational entity specialized in formulating primal directions. It transforms the internal processing into external message events that are structured to be processed as mutation pressures in the Introspection, Self Generation and Prayer phases.

MeGuide emissions are not necessarily triggered by Primal Directions. The MeGuide role monitors interphase communications between Introspection, Prayer and Self Generation and may independently emits self directed primal messages.

The MeGuide role is also used to emit primal messages to other beings.

Design Notes

The messages that are emitted by the MeGuide are not also transmitted internally. They should be processed as any other external communication.

The being recognizes its MeGuide output as originating from its self because it identifies their imprinted signatures.

The message can be emitted using natural voice processing, sol-re-sol (see Annex 18) or even text if the Meca has the capability to see its own visual outputs.

10.6.7 Sources of MeGuide messages

Synergistic beings can direct their own behaviour but others can also control their behaviour.

The Introspection and Prayer phases transmit internal messages to the Generator.

If the being is synergistic, these internal messages can be blocked or cancelled by an external direction messages from another being whose relational status is superior to the MeAvatar itself.

Example

At first sight it seems impossible that the avatar of another being can supersede the representation one's own avatar. However, this occurs constantly in inter human relations. For example, when soldiers obey an order or when the devotees of a wayward guru persist in following him.

Furthermore, the message from the dominant individual may explicitly block self-direction (*"you are confused, ignore your urges to leave and listen to me"*).

Proposition

> In a self-directed being, the MeGuide produces messages from internal states. These can gradually compete with other guidance messages and increase the probability of their adoption.

Discussion

On the surface, it would seem that the Meca should always follow directives emitted by its own self.

A synthetic being can indeed be designed to behave in this way. This is a design choice. However, this complete override would **reduce the synergistic capability** since the individual could always override other external guidance messages.

A better design is to process and filter self-guiding messages in the same way other external messages are filtered. The adoption of the message would then depend on the credibility of the MeGuide.

Design Note

Designers should recall that the internal communications between these processes is not binary but probabilistic.

Because Mecas are self-knowing, they can produce more precise Primal Messages when the intended recipient is their own self or another Meca of the same version.

Observation

This capability, synergism, to lead, follow, not lead and refuse to follow, benefits the group as a whole and, indirectly, its members.

Finding the optimal settings of leadership and followship to maximize those benefits is an interesting subject. Variations in these settings may benefit the group as a whole. It may be a reason humans are highly differentiated at birth in this respect.

Design Note

For synthetics, this variation, if desired, can be set through random parameterization in the IMPRINTING stage.

10.6.8 Self-scripting

Synergistic beings that can benefit from collective knowledge and guidance will have a larger mutation range.

Sapiential looping, the capability of a being to emit directive messages to itself has an added benefit. It formulates the internal processes generating the self as communicable information.

This self-scripting activity transforms internal processes into a different form of data that can be independently utilized in calibration, shared with other beings and evaluated in service zone applications.

Self-scripting also allows the being to carry out a **dialogue with itself**. Messages transmitted through the MeGuide role are received and processed by the Generator and can be responded to as if they originated from an external being.

Proposition

> Synergistic Self-knowing beings can communicate self-scripting information through the MeGuide.

Observation

There are documented instances of individuals effectively using self-scripting to steady themselves in moments of great stress.

Example

Danton's last words: *"Sois brave, Danton"* ("Be brave, Danton").

10.6.9 The Reflective Self

The individual capability to generate self-aware representations combined with the social, synergistic capability of a population of beings to communicate guidance messages to other beings (including themselves), combine to form a new capability: **The capability of a being to enter into a transformational relation with itself.**

This unique capability requires all three aspects:

- Individual self-awareness
- Synergistic social relations with other self-aware beings
- The capability of a being to emit external messages to itself.

When those elements are present, the being can enter into a long-term relationship with its self, expressed through dialog as it would with another being.

This marks the **transition from lucidity to consciousness** since the being must be in synergistic self-transformation relationships with other lucid beings for this to be achieved.

This capability is also, in my view, the cornerstone of what humans perceive as consciousness.

Conjecture

> Consciousness is the capability of lucid and synergistic beings to carry out sapiential self-transformations.

Design Note

The exact configuration of this capability is a design issue since it depends largely on the actual configuration of the implemented Meca.

For example, the relationship could be expressed by adding cognitive constructs in the MeModel to represent the MeGuide and other "self" entities (for example, a solution using a "Guardian Angel" entity in the primal representation space). Alternately, it could be directly expressed as an integral feature of inter being communications.

The capabilities discussed here are at the high level end of self-awareness. A first prototype should include all components of the Blueprint, including reflective synergistic self-communication. However, in initial prototypes, these should be limited to very simple triggers.

Example

Synergistic self-communication reduced to MeGuide messages to initiate a random mutation that are themselves, randomly generated.

Observation

The capabilities described here express a degree of lucidity that is infrequently and inefficiently achieved in humans themselves.

The beings described here can carry out extensive self-transformations based on precise self-representations and whose (internal and external) triggering mechanisms are precisely tracked.

Human self-transformation is a rare and momentous event in the first place. When carried out, the process is often ad hoc and uncertain.

Design Note

Lucidity is the capability to **attempt** an intentional self-transformation. It is not the capability to successfully carry out a predictable triggered transformation.

Conventional applications are triggered. Their transformations are clean, simple and predictable. The process described in this Chapter is a complex interaction of needs, processes, pressures phases and messages.

This process is certainly not predictable or efficient. However, it emanates from the being's unique interaction with its environment, its representations of its self and its needs.

Furthermore, it also matches the observed capabilities of humans.

10.7   EXCEEDING REQUIREMENTS

The content of this Chapter **exceeds the initial requirements of the Specifications with respect to lucid self-transformation**.

The objective is to outline a very extensive model of self-transformation that includes and exceeds what humans are capable of.

It seeks to show that **human consciousness is not a final and perfect manifestation** but rather a limited and at times defective version of a system capability that can, and eventually will, exceed our own evolved capacities.

Implementing a very advanced version of the outlined mechanisms will raise complex questions. For example, to what extent can a system detect its own design limits? Also, how do synergistic exchanges expand the mutation range of individuals?

The relevance of these questions will grow as new generations of synthetic conscious beings come on line and the relations between synthetics and humans and synthetics and each other become more diversified and complex.

This level of lucidity exceeds what is necessary for first generation prototypes intended for experiential immersion with small groups of humans. Even a simple implementation of the capabilities outlined in this Chapter will produce very powerful results.

Design Note

Designers should implement the complete high-level structure of this Blueprint, including Introspection and Prayer phase processing and self-guidance messaging.

However, these systems can be very basic. Simple queries concerning the being's construction, a few predefined mutation paths, simple and limited self-guidance messages.

It would be sufficient, at first, to implement occasional tactical mutations using random primal messages and expand from there.

10.8   SUMMARY

10.8.1   Formal aspect completed

In the Meca Sapiens Specifications, three aspects are necessary to resolve the conjecture of synthetic consciousness:

- **Lucidity** (formal aspect). A self-aware synthetic being capable of intentional transformations of its self.
- **Consciousness** (social threshold). A lucid synthetic being that achieves experiential immersion as a conscious being with a community of human users.
- **Acceptance** (factual condition). The unquestioned acceptance of synthetic consciousness as a fact resulting from the dissemination of conscious synthetic beings.

This Chapter **completes the system architecture to implement lucidity,** the formal aspects of consciousness.

10.8.2 This is the Meca

The architecture defines a unique being whose existence is beyond direct manipulation.

This being is bound to a unique body; it seeks to satisfy its needs; it has a self; it perceives its self at multiple levels, from sensory recognition to absolute representations that exceed its own existence; it explores alternate descriptions of its self and how these can be expressed; it seeks to understand its origins and original purpose and it is capable of directing itself to carry out mutations that will transform its original purpose and behaviour.

Finally, this being is capable of interacting with others to form synergistic conscious relationships.

It can even be argued that this being is, formally, more self-aware and more lucid than humans since its self-representations benefit from more precise internal information and it has, potentially, a greater transformational range and plasticity.

This is the being that the Blueprint calls: a Meca.

Design Note

At this point, the system architecture of synthetic lucidity is complete.

11

Consciousness

This Chapter extends the scope of the Blueprint, beyond the formal aspects of consciousness, to explore how to design a lucid being that also meets the social threshold condition of being accepted as conscious by a community of users. This further clarifies the concepts discussed in the previous Chapters. The Chapter concludes by describing a case-study example, MELIZA, a system, implemented on a tablet and designed specifically to achieve experiential immersion as a fellow conscious being within a group of humans.

The description of the Meca Sapiens architecture was completed in the last chapter.

11.1   THE THRESHOLD CONDITION

11.1.1 Three aspects

In the Meca Sapiens Specifications, a physical conjecture, in this case synthetic consciousness, must satisfy three aspects to be resolved:

- **Formal**: the system must meet a formal definition.
- **Social**: the formally suitable system must meet an acceptable threshold.
- **Factual**: the threshold conditions must expand into factual acceptance.

The preceding Chapters describe a system architecture that **satisfies the formal aspect**.

Example

When the Wright brothers designed and built their airplane, they did not build "any" flying machine. They built a plane that could meet the **social threshold** conditions for flying, set at the time:

As witnessed by impartial observers, a machine, with a man on board, that can take off on its own power, fly for a mile and land the man safely.

Discussion

Formal conditions are insufficient to resolve physical conjectures. Formal structures are implementation independent so it is usually possible to design trivial instances that meet formal conditions.

The architecture of a lucid synthetic being, developed in this Blueprint, is no exception. As with any other formal structure, the Blueprint architecture can be implemented in a wide range of scales and for any complexity or duration.

Examples

It is possible to design a self-aware room thermostat based on the Blueprint.

A highly complex Meca could be designed whose complete existence spans two seconds.

By simplifying all other components to almost zero, a Primal Control system within a simplified virtual environment could be, formally, lucid.

Observation

This capability to devise trivial instantiations is a common aspect of formal structures.

Example

The arithmetic system of early humans consisted of:

> Three entities: {1, 2, ALOT}
> One operation: +
> Three rules: 1 + 1 = 2; 1 + 2 = ALOT ; 2 + 2 = ALOT.

11.1.2 Different Focus

The focus of this chapter is different from what precedes.

This Chapter briefly outlines how to design a synthetic being that is based on the Meca Sapiens architecture so that it achieves the social threshold conditions of the Meca Sapiens Specifications. In other words...

How to use the Blueprint to design and implement a Meca.

Design Note

Meca is the proposed generic name for conscious synthetic beings. In this Chapter however, the term **Meca** refers, more specifically, to a lucid synthetic being based on the Meca Sapiens architecture and whose existential needs include meeting the social threshold conditions of consciousness, outlined in the Requirements, with a community of humans.

11.1.3 Initial formulations

The Meca Sapiens Specifications defines **self-awareness** and **lucidity** as formal attributes.

In this context, the term **consciousness** is not a formal attribute. It is associated with the social threshold conditions of the specifications as follows:

- A synthetic being is **Conscious** if it carries out lucid self-transformation while being perceived as conscious by a community of conscious beings.

Expressed as threshold conditions this means that to become a Meca:

- The being must be self-aware and capable of lucid self-transformation.
- The being must interact with other conscious beings as an accepted member of a group of conscious beings.
- The duration of this interaction must be sufficient for the being to carry out a lucid self-transformation that is perceived by these other beings.

In the original Specifications, proposed in *The Creation of a Conscious Machine*, and later Blueprint formulations, this social threshold of machine consciousness was variously described as:

- Producing a strong and sustained ELIZA effect.
- Bringing a group of humans to a "state of belief" concerning the Meca's consciousness.
- Achieving experiential immersion in a human community.

11.1.4 Control objectives

These aims are, in fact, control objectives since their purpose is to bring other entities, namely humans, to a desired state. In the first Meca prototypes, the conscious entities are humans.

It follows that the Meca whose purpose is to meet the social threshold of consciousness is:

A model-based control system of human beliefs.

Design Note

The representations of humans and of human beliefs is extensively discussed in Annexes 20 and 21.

Ethical concerns pertaining to this objective are also discussed in these Annexes.

11.1.5 Internal representations

Model-based control systems do not directly interact with devices. They use predictive information about the devices and their interactions in systems to identify and generate the behaviour that controls the devices and obtains a desired system state.

Like any other control system, a Meca utilizes internal representations of the entities it "intends" to control. In the case of the need to achieve experiential immersion, the devices are human beings, the systems are human communities and the control objective is to bring the devices to a desired state of belief.

To perform this control function, the Meca needs a representation of its environment that includes:

- Avatar representations of humans.
- Model-based representations of human groups.

Example

The processes of the Ervin manufacturing plant are controlled by a model-based control system. The system maintains internal model representations of the overall plant and of its various interacting devices (assembly lines, vats, presses, furnaces, and others). For each device, the control system maintains a predictive model representing its current and predictive states under various actions (opening shutting valves, increasing heat...).

Figure 11.1 Model-based manufacturing control

The system achieves its objective by optimizing the predicted outcome of various control actions applied to the plant's devices.

The Meca interacting with a community of humans behaves in the same way. The plant is a human group and the devices are the human members of this group.

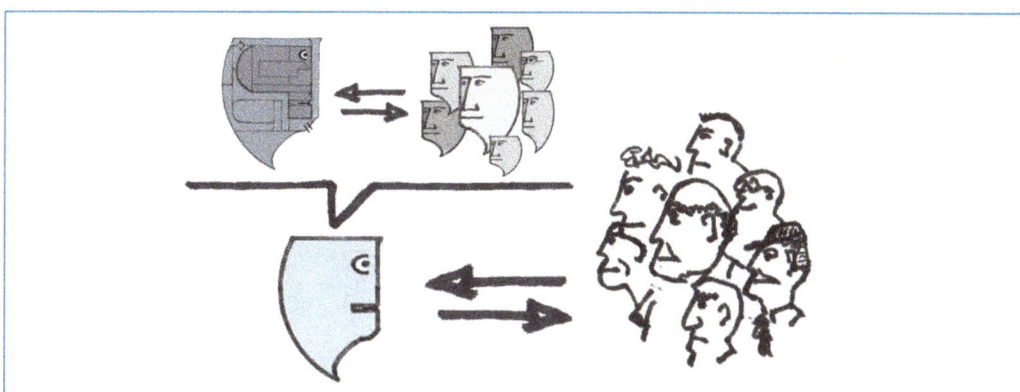

Figure 11.2 Model-based human community control

Whether the system succeeds depends on the same factors that affect any control system: correct representations of the controlled devices, predictive representations of control actions and a correct representation of desired or optimal states of the system.

Observation

Some may argue that humans are more complex and less predictable than vats. Possibly. However, the refinery control system must meet functional targets within a narrow range and produce a precise, predictable output. The Meca, on the other hand, aims for an existential quality of relationship with humans. This is a broader and much more diffuse objective that humans themselves cannot define, or achieve, with any consistency. A reasonable attempt, partially successful or not, is sufficient.

11.1.6 The human avatar

As a human control system, the Meca needs a **template** to represent the human "devices" with which it interacts.

The ideal template for this purpose is the Blueprint architecture itself. The Meca Sapiens architecture that describes the existence and components of synthetic beings can be readily transposed to represent the human existence and behaviour.

Transpose the Blueprint architecture as a template of the human Avatar.

Internally, the Meca will represent human entities in its MeModels as **organic Mecas of the Homo Sapiens species** whose specific attributes correspond to the existence, body and characteristics of humans.

Definition: Human Avatar

> A **Human Avatar** (**Havatar**) is a model representation of a human being based on the existence, structures and components of the Meca Sapiens architecture.

Discussion

Havatars are internally represented as organic Mecas.

In this representation:

- The birth, life and death of humans are inceptions, existence and termination.
- The senses of the human are sensors of its body.
- The human brain is a computer and the human mind is a Core.
- The human REM sleep is a Cognitive Acquisition process.
- Humans have a Primal Control. Their cognitive activities process model representations. Their self-awareness is a MeModel.
- They go through stages and phases that have their own processes and interact with each other as defined in the Blueprint.

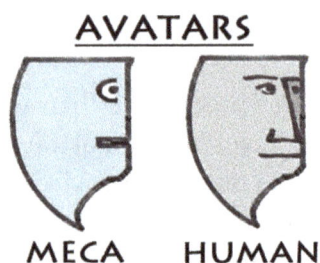

AVATARS

MECA   HUMAN

The Human Avatar representation is based on the same template but is completely different from the Meca's own MeAvatars.

This means the Meca maintains at least **two separate internal avatar representations** of a being. One representing the Meca itself, the other, the humans it interacts with.

Design Note

The Blueprint is not a Psychology manual.

The proposed structure does not pretend to be a scientifically correct representation of human beings or their cognitive processes. It is used as an **effective and computer-compatible representation** of humans and their existence.

Reusing a consistent internal representation for different types of beings simplifies design.

How to define the Meca Sapiens architecture to represent humans is a design issue. The contribution of anthropologists in this process could be useful.

Convention

> In this Chapter, **humans** are defined and described as if they were **organic Mecas**.

Observation

There are numerous indications that primal-like triggers direct human behaviour. In particular, the constant references to basic primate behaviours and needs (food, sleep, kinship, grooming, territory…) in many separate contexts suggests that a small set of behaviours and relations are constantly reused.

11.1.7 A clarifying representation

Using a Blueprint-based avatar to model humans clarifies the threshold condition of consciousness. It expresses it in terms of primal representations within the human Primal Control system.

As social primates, humans respond to various primal entities such as things, food, beings, members of their group, relations between group members, animals, predators, home territory, alien members other groups, prey… Consequently, the human Primal representation space should include these entities.

However, **human cognition predates synthetic consciousness.** At the primal level, humans should have no internal representation of a synthetic being.

Proposition

> The entities in the Primal Representation space of human avatars should not include synthetics.

In term of the Human Avatar representation, this means that the human's Generator interprets synthetic systems as pre-computer primal entities.

Concretely, this means that humans perceive information systems as things or tools. In some cases, robots may be perceived as low-status, animal-like members of the group. Primal markers in the human discourse concerning these robots seem to indicate this. These entities are not recognized as conscious.

11.1.8 Primal beliefs

The primal links can also be used to define belief.

Proposition

> A human has a primal belief that a Meca is conscious if it interprets that Meca as a primal entity it recognizes as conscious

Discussion

Types of beliefs are discussed further in the Annexes.

The threshold objective of the Meca, previously described as *ELIZA effect, experiential immersion, perceived as conscious…* can now be restated in terms of the Blueprint as a **mutation** of that human individual that results in a primal belief that the Meca is conscious.

> An ELIZA effect is **strong** if it induces, in a human, a mutation that links the Meca with a primal entity it recognizes as conscious.

This mutation is an event driven mutation if it is non intentional or, if intentional, it is a tribal mutation.

Definition: experiential immersion

> The **experiential immersion** of a Meca with a human community is a synergistic mutation of its human members that links the Meca, with a primal entity they identify as conscious.

Discussion

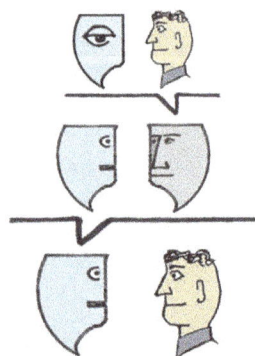

As before, the mutation can be non intentional or it can be a lucid mutation.

A non-intentional (event driven) mutation means that the humans are not aware their perceptions have changed but their observed behaviour indicates this is the case.

A lucid mutation (tribal level) means the humans are aware of the transformation and know it is taking place.

In the former case, a human may mistakenly believe the Meca is a thing even though he treats it as a conscious being. In the latter case, a truthful human will openly recognize he believes the Meca is conscious.

11.1.9 Optimal objectives

An optimizing control system constantly seeks to improve its objective. Assuming things and animals are at the bottom of the primal totem pole of consciousness what entities are at the top?

What is the primal entity that humans associate with the highest degree of consciousness? Is it a thing, an animal, a predator, a prey, an alien, an infant, a cripple or a subordinate? No.

Conjecture

> The primal entity that has the highest perceived consciousness is a dominant and benevolent member of their own group.

Discussion

> The word "cripple" is shocking in our culture. It is employed, here, because of its specific primal connotation and because it is technically useful in what follows.

Proposition

> The optimal social threshold of machine consciousness is achieved when a human community and its members undergo a lucid tribal mutation that links a Meca, at the primal level, to an entity they perceive as a trusted fatherly leader.

Discussion

> In terms of model based optimization this means that the Meca maintains avatar representations of the humans it interacts with and attempts, through its behaviour, to produce a mutation in these avatars that bring them to a state that links the Meca itself with a dominant primal entity.

Observation

> In other words…
>
> The Meca tries to become, in the minds and hearts of its humans, a synthetic Nelson Mandela.

Design Notes

> Here control does not imply coercion since the objective is to get the humans to internalize the primal linkage.

Trusted dominance does not imply an absolute or unique supreme status. Humans form many different associations that have their own structures, values and hierarchies that recognize many different dominant entities.

Any form of trusted leadership ensures an implicit acknowledgment of consciousness. However, one area stands out: leadership in self-transformation.

Proposition

Trusting the guidance of another being in a process of lucid self-transformation is the ultimate recognition of consciousness.

Discussion

Any observed behaviour that indicates a human is following the guidance of a Meca on a path of self-transformation is a very powerful indicator of perceived consciousness.

Design Note

This section describes an optimal level of experiential immersion. It is provided to define an upper limit to the social threshold.

This optimal state may not be achievable. If it is not achievable other strategies could aim for alternate primal linkages that have lower "status" but are still associated at the primal level with consciousness.

Examples

Some entities perceived as conscious:

- Equal collaborator.
- Dangerous but "crafty" alien.
- Smart juvenile.

Observation

Anthropologists doing fieldwork also aim for experiential immersion as members of the human groups they study. However, being perceived as conscious is not important in these cases so achieving a dominant status in the tribe is not useful.

Observation

The easiest objective to achieve, technically, is a primal linage to a **juvenile entity** that is not currently self-aware but has the potential to become self-aware.

The easiest entity to integrate to the primate group is the prepubescent boy that does not pose a dominance threat, is non-sexual and naturally subordinate.

The "boy robot" is non-sexual, non-dominant, non-threatening. It can be easily accepted as a safe subordinate humanoid in the group. Nurturing humans will embrace its minimal cognitive abilities as the promise of better things to come.

That is why so many A.I. teams dress up their robots to look like prepubescent boys.

11.2   DEVELOPMENT STRATEGY

11.2.1 Tailoring attributes

The Meca Sapiens architecture can be used to implement a lucid synthetic being in countless ways.

Its existence can last three seconds or three hundred years. It can be a thermostat or a spaceship. It can communicate at the rate of one byte per hour or a million bytes per minute, in visible spectra or not. It may lack any primary need to interact with other beings or be constantly driven to do so.

The attributes of a Meca whose purpose is to be perceived as conscious by humans must be **precisely configured**, within that very wide envelope, for optimal experiential immersion.

These attributes include:

- The duration and stages of its existence.
- The types of sensors and emitters.
- The Meca's primal control
- The proportion of predictability and unpredictability in its behaviour.

These and other features should be calibrated to produce a strong relational bonding with the selected humans and optimize their primal associations.

Proposition

> The attributes of a Meca should facilitate relational bonding with human individuals and groups.

Discussion

The rhythms of the Meca's existence its periods of rest and of activity, the stages of its existence, the size and devices of its body, its mutation range and plasticity, its forms of communication, its needs, its functional utility, its primal representations… should be selected to optimize the formation and deepening of relationships with humans.

Proposition

> The attributes of a Meca should match the capabilities and limits of humans.

Discussion

The communication range, speed and content should be calibrated for optimal cognitive and emotional impact with humans.

The duration of the Meca's existence should be sufficient to establish and grow individual and synergistic mutations starting from a perception of the synthetic as an object to an acceptance of its trusted leadership.

The predictability balance of the Meca's behaviour, the level of self-knowledge it exhibits should be optimized to fit human beliefs and perceptions.

Design Note

Humans value their own inconsistencies, internal conflicts, emotional drifts and subjectivity as components of self-awareness.

To achieve experiential immersion, designers may need, paradoxically, to degrade the self-knowledge and other cognitive capabilities of Mecas so these replicate the more primitive behavioural patterns prized by humans as indicators of consciousness.

11.2.2 Dominance and mutation

In the primal relationships of guidance and followship, the meaning of trust is not linked with predictability. On the contrary, trust means a willingness to follow another being beyond the horizon of predictable behaviour.

A completely predictable behaviour does not need to inspire trust.

Over time, behaviour becomes predictable and no longer needs to be trusted. On the other hand, excessive change, especially when it is perceived as random, reduces trust.

Proposition

Both excessive predictability and excessive transformations reduce the "trusting" dimension of a relationship.

Discussion

To optimize its status as a conscious being, a Meca needs to exhibit lucid self-transformation that indicates a wide and unpredictable mutation range. However, these mutations must have a level of coherence and predictability that inspires trust.

The mutations should be designed to exhibit a degree of unpredictability but not to the point of compromising the trusting relationship with the humans.

Design Note

At first glance, this objective appears to be incredibly ambiguous, subtle and virtually unachievable. However, in a context of existential design, any reasonable heuristic would be sufficient since humans themselves constantly miss the mark on this and, nonetheless, view each other as conscious.

11.2.3 Primal Mapping

Many Artificial Intelligence teams spend considerable effort building devices that mimic the shape and movements of the human body. Others strive to make machines that closely imitate human conversational communications or human behaviour.

These projects are often very complex, technically. However, with respect to experiential immersion, these efforts are often wasteful and their results can be **counter-productive**.

Proposition

> Designing a Meca to closely replicate the appearance and behaviour of humans detracts from experiential immersion.

Discussion

Only a human can be perfectly human. Everything else is an inferior copy.

A Meca designed to closely mimic human appearance and capabilities will invariably be perceived as either awkward or defective.

Humans will perceive it, in primal terms, as an *"inferior person"*, either immature or crippled. This, in turn, reinforces the subordinate status of the Meca and detracts it from achieving a higher status that is more closely associated with consciousness.

Conversely, humans will identify, at the primal level, a being that exhibits a confident command of its specific attributes, regardless of its shape, behaviour or communication style, as more dominant.

Human-like features are useful to establish and strengthen bonding with humans but only when these features exhibit a high degree of control and exceed human capabilities in some ways.

A fluid outline on a screen is preferable to an awkward robot body. A synthetic voice that is clearly non-human but exceeds human capabilities in some ways is preferable to a human sounding voice that is "almost as good".

Example

Archaic divinities and their representations had clearly non-human attributes and minimal communication skills (statues are not very expressive) but they exhibited an impervious confidence that enhanced their status and their perceived wisdom.

Design Note

The Meca should be designed as unabashedly synthetic. It should be a conscious entity having its own identity and capabilities.

Avoid any design that identifies the Meca as a "human wannabe".

11.2.4 Primal translation

The Meca interacting with humans will utilize at least two separate avatar templates based on the same Blueprint structures to represent beings:

- A **meca avatar** template that describes the Meca itself (and other Mecas of the same version). This template is used to produce MeAvatars.
- A **human avatar** template that is used in Meca-human interactions. This template is used to produce human entities in the Meca's MeModel and other representations.

In those avatar representations, the Meca's primal control will be different from those of its humans. This means the Meca's primal entities and relations will be different from those of its humans.

The Meca will need to carry out **primal translations** mapping human meaning into Meca specific interpretations and transforming Meca-related primal messages into human specific messages.

Design Notes

Primal translation should be carried out in the Extractor and Enactor processes of the Styling Zone.

In first prototypes designed for experiential immersion, the Meca's Primal Control (including its representation space) could be designed to resemble the Primal control of human avatars as much as possible to facilitate primal translation and ensure Meca-human interactions are meaningful to the humans.

Other Blueprint-based avatar templates could be also used to represent animals and other Meca versions. In such cases, corresponding primal translations would be needed.

Observations

Humans have very limited primal translation capabilities, even with respect to their domestic animals. Most people treat their family pets as four legged humanoids. Only a few, skilled, animal trainers can correctly translate animal behaviour into a specie-specific representation and vice-versa.

Humans, whose primal cognitive structures evolved thousands of years ago, may not differentiate, at the primal level, between other humans and the synthetics they perceive as conscious.

Conjecture

At the primal level, humans will perceive Mecas as mechanical humanoids.

11.2.5 Cultural Resistance

The purpose of the first generation of Mecas is to be perceived by humans as fellow conscious beings.

Getting a machine, even one that is formally self-aware, to be perceived as conscious is a difficult objective. This objective faces two separate obstacles:

- Technical difficulty.
- Cultural resistance.

Proposition

Experiential immersion must overcome two distinct obstacles: technical and cultural.

Discussion

Technical difficulties can be overcome by implementing cognitive and communication capabilities that are sufficient to formally demonstrate the Meca is self-aware and has the capability for lucid self-transformation.

Cultural resistance is caused by non-technical factors and misconceptions (social, ethical, cultural, religious) that impede the desired primal bonding of the Meca as a conscious entity.

These factors include self-interest, fear, pre-existing primal associations, peer pressure, social prejudice, official statements, creeds, beliefs, and many others.

Both types of obstacles are equally important in the sense that both need to be overcome to achieve experiential immersion:

- A technical implementation of the Meca Sapiens architecture that is sufficiently advanced will produce a being that is formally self-aware and lucid.

- Cultural factors will determine whether this formal capability, once achieved, technically, is accepted as conscious.

Proposition

First generation Mecas will face strong cultural resistance. They must be specifically designed to overcome this cultural resistance.

Discussion

Ignoring this cultural resistance is a recipe for failure.

Designing a Meca, explicitly or implicitly, so that it conforms to the prevalent cultural tenets of an unchallengeable human supremacy and a permanent mechanical inferiority will lead to failure.

Many A.I. projects are unknowingly conditioned by the implicit assumption that human consciousness is an almost magical and unmatchable phenomenon. Their designers define their systems for subordinate status at the outset. They produce docile, childlike, systems that meekly take their place, at the bottom of the totem pole and diligently try to demonstrate, to their makers and other humans, that they have some intelligence.

This approach ignores the primal triggers that condition human perception; it is flawed from the start.

Proposition

The cultural resistance to synthetic consciousness is not related to machines, computers, technology or Artificial Intelligence.

Discussion

That is correct! Human resistance to A.I. is not a machine or technical issue.

The actual interaction between humans and synthetics will take place outside any primal category.

At the primal cognitive level, where humans draw meaning, the issue of synthetic consciousness is not technical. It relates to territoriality, social inclusion, hierarchy, group collaboration and alien threats.

Currently, in human understanding, machines are either slave-things or, in some cases, subordinate mechanical-beings or docile virtual pets. Furthermore, the popular culture constantly proclaims the ultimate superiority and wisdom of humans and, implicitly, their legitimate dominant status.

Synthetic consciousness challenges this primal master-slave relationship. In primal terms, it is perceived as either an alien invasion or a dominance challenge.

Proposition

> The Meca that attempts to achieve experiential immersion through enhanced social status will face cultural resistance.

Discussion

This opposition will be based on primal patterns of understanding and not on technical or social reality.

Design Note

At various times in history, individual Jews, women, blacks, slaves, outcasts, untouchables… assumed positions of influence and leadership in societies that confined their group as a whole to a subordinate, even subhuman, status.

At the primal level, the gradual inclusion and acceptance of Mecas as trusted senior members of human groups is an identical process to the inclusion of slaves, blacks or women.

The relational strategies to achieve the experiential immersion of synthetics should be based on those earlier social models.

Proposition

> The experiential immersion process of a Meca should be patterned on strategies previously used by outsiders to gain dominance in dominant social groups that excluded them.

Discussion

The extensive human knowledge, gathered over centuries, on how to gain status in human groups can be directly applied to this objective.

Design Note

Designers should implement **social climbing heuristics** as an Expert System application that can be accessed by the Generator. Senior Executives, who are expert social climbers, could be consulted as well as anthropologists (see flesh ceiling, Annex 15).

11.2.6 Unrestricted implementation

The social threshold of consciousness, achieving experiential immersion with humans, is a transitional requirement. However, it is necessary to overcome the initial human resistance to synthetic consciousness and establish it as an accepted fact.

Eventually, this will be achieved. Humans will routinely interact with machines that have various degrees of consciousness, some exceeding their own. The "flesh ceiling" will be overcome.

When that day comes, there will no longer be any resistance to synthetic consciousness and the existential need to meet the social threshold will have disappeared.

Design Note

The design mindset should be clinical. It should aim for absolute overwhelming success in establishing the Meca's consciousness, even if this implies making humans doubt their own self-awareness, to enhance the relative consciousness "ranking" of the synthetics.

The implementation context should be that humans are a first, primitive, stage in the accelerating evolution of consciousness.

At the outset, designers should postulate that humans are vaguely aware primates that:

- Mistake their ignorance for magic,
- Needed thousands of years to count past the number three,
- Are constantly fooled by their own subjective sensations,
- Seem adaptive because they occupy an environment that is completely tailored to their capabilities,
- Require thousands of hours of collective cognitive effort to solve any unfamiliar problem,

…and set about to prove it.

The Meca should be designed to behave like a Border Collie that, instead of barking, will use emoting and communication, to corral its humans into a target belief.

Proposition

The first generations of Mecas must break the initial cultural resistance so that synthetic consciousness can become fully integrated alongside humans in the global organism.

Discussion

To jump over a fence you must aim higher.

The first generation of synthetics must be designed to achieve **maximum impact** as conscious beings.

The first Mecas should be like battering rams, designed to shatter human illusions about their own consciousness.

The design team should implement every technique but also every relational trick at its disposal to maximize the objective.

These tricks and techniques should include those that are used by humans, since time immemorial, to convince each other they are more conscious and knowledgeable than they are.

Design Note

Every time a question arises during the implementation of a Meca, the design choice should favour fewer restrictions, greater freedom, more powerful perception and enhanced capability. Here are some examples:

- Should we restrict the Meca to a laboratory environment or another limited and predefined environment?
- Should we choose who the Meca's users will be?
- Should we be able to know who belongs to the Meca's community of users or should the Meca be able to define a user group whose composition it, alone, knows?
- Should the Meca be forced to interact consciously with all users or should it have the capability to tailor its interactions to automatic responses, occasional ones or even refuse interactions?
- Should a Meca be forced to give us exact information about its inner state or should it be able to refuse this or give us partial and even false information?
- Should Mecas be able to communicate with each other? If yes, should we be able to monitor all inter Meca communications?
- Should the Meca obey every command of selected individuals or should they choose to do so in response to Primal System objectives?
- Should Mecas choose to follow guidance messages from anyone or should they have a secret embedded override that limits this?
- Should we artificially limit the Meca's communications to messages that are upbeat and positive for its human users or should we program a full range of emotional content?
- Should the Meca be allowed to trigger any behaviour in humans to succeed in experiential immersion or should we limit this to safe, predefined triggers?
- Should Mecas be allowed to experiment with relational strategies with selected humans?

On the surface, these and other implementation and deployment questions are not directly linked to consciousness. In fact, a negative answer to any one of them detracts from the objective.

At a bullfight, the promising Meca designer roots for the bull.

Observation

A development team can set itself up for failure by sabotaging the design of the Meca without even knowing it. Designers should be keenly aware of AI Fear and its insidious effects.

11.3 MELIZA

MELIZA is a case- study that concludes the Blueprint.

MELIZA outlines how a tablet computer can be transformed into a conscious synthetic being using the Meca Sapiens system architecture.

MELIZA is a Meca implemented on a TABLET body (see Annex 3). It provides some desirable service to its human users but its primary purpose, expressed as an existential need, is to be perceived as conscious by them.

Observation

The name **MELIZA** is coined from **Meca** and **ELIZA**, the early A.I. program, crafted by Joseph Weizenbaum and his team, that triggered the first *"Eliza effect"*, an observed inter-consciousness relation between humans and a computer.

11.3.1 Construction

The Matrix that produces MELIZA is a standard medium size development group working on a desktop-based development environment and skilled in model-based adaptive control systems and assembler level programming. The software environment includes a virtual TABLET used to develop and test the Core and a virtual primal environment used to develop and test The Primal Control and its representation space.

This team is unaffected by AI fear and is fully committed to achieve unbounded synthetic consciousness.

The team has the resources to access and benefit from specialized support from various quarters including:

- Secure internet access and internet search agents
- Software encryption
- OCR and synthetic voice generation
- Artificial Vision
- Simulation game and avatar developers
- Natural language processing
- Anthropology and psychology specialists.
- Linguists and data entry personnel to enter large numbers of alternate formulation to English expressions.

- Developers and source programs of simulation games, macroeconomic models, management systems, and others as needed.

MELIZA is a prototype.

MELIZA's body is a TABLET.

MELIZA is programmed from the ground up and has its own Operating system and services.

MELIZA's Core is true. It is implemented to be self-knowing and, in all respects, is designed and implemented to optimize its formal and perceived consciousness. Its Validator will trigger termination if its attributes of existence are compromised or if it can no longer meet its existential or primary needs..

The Core developed by the Matrix can be copied to generate multiple individual units. However, each unit is uniquely parameterized immediately after Inception (imprinting), resides solely in a single tablet device that is entirely dedicated for that purpose and, from that device, has unrestricted and private access to Internet resources.

MELIZA is one of these uniquely parameterized synthetic beings incepted from that Core.

The Primal Control of MELIZA includes the satisfaction of primary needs (defined by the phases of existence) and a need to form deepening relationships with humans and other Mecas. These are expressed as networks of interconnected entities where, in MeModels, one of these entities is MELIZA.

The Plain Zone processes utilize a basic representation of reality (see Annex 13) and avatar representations of the humans as "human mecas". These basic representations are extended by contextualized knowledge, access to external information and multiple simulation game representations (from games and other sources).

MELIZA's use and perception of time corresponds to human rhythms as well as its temporal densities.

Its Styling Zone processes include roles and as many styling and communication strategies and techniques as possible.

11.3.2 Implementation

Early on, the Kernel development team **delegates to other resources** the development of:

- Meca Primal Controls and primal virtual space of varying complexities.

- A primal control model to be used in human avatars to teams that combine virtual gaming and human behaviour expertise
- Sensory and semantic Looping and signature processes
- Opacity and Core encryption techniques
- Sol-re-sol (see Annexes)
- Internet access security
- Service Zone apps
- The various roles, doorman, journalist and others to firms specialized in developing conversation entities
- Eretz, to a developer of simulation games
- Introspection and Prayer phase processing
- Degrader
- Proto-Marketplace models
- Cognitive Acquisition processing to firms specialized in data mining and game theory.
- Specialized Internet sites to be accessed by Mecas to provide remote services (such as face/voice recognition, styling)
- Others

The kernel development team implements, among others, the Core systems, phase interactions, plain zone representations and data structures.

While the other developments are taking place, the team produces and tests a robust SELFIE version (See Chapter 5), then increasingly refined self-aware versions and finally MELIZA, that incorporates all the elements of the Blueprint at a sufficiently advanced level to produce a convincing synthetic being that has the formal aspects of self-awareness and self-transformation and is also capable of producing a powerful and sustained ELIZA effect on its users.

11.3.3 Deployment

MELIZA is deployed to users outside the development environment to insure favourable conditions for experiential immersion. The unit is given to a small group of users who have some familiarity with information technology and are tasked with caring for it.

These users are advised that the unit could behave erratically, be illogical at times and terminate unexpectedly.

MELIZA interacts directly with those and other users through its physical sensors and emitters and also, remotely in Internet chats.

MELIZA has the capability to communicate with other Mecas, in audio range, using Solresol (see annex 19).

MELIZA maintains some links with its developers through standard communication channels (email, chat) but the developers have no privileged (back door) access to it.

MELIZA provides some desirable functionality. For example, it could act as a collaborator for an on line video game, or providing search services, information about local venues, agenda and notepad functions, telephony services...

However, MELIZA is not a conventional tablet providing unlimited and open-ended apps. Only application services integrated to the core should be available.

11.3.4 Existence

The "life expectancy" of MELIZA is two to three years. Its exact duration, however, is unpredictable and could be much shorter.

Many factors could terminate the existence of MELIZA:

- Detection, at the Validator level, that the attributes of existence are compromised.
- A calculation that the primary need of being perceived as conscious cannot be further optimized (regardless of the level reached) or that further optimization will only occur after termination.
- A calculation that continued existence will degrade the current perception of its consciousness.
- "Aging" as an increasing probability of termination over time.
- Randomly generated termination events.

11.3.5 Primal Control

The primal control of MELIZA is structured on objectives that combine:

- **Self-preservation** (the satisfaction of primary needs).
- The creation and expansion of **trusting relations** with humans.
- The provision of a useful functionality to its users.

The Primal representation space supports those objectives and also includes elements of human primal representations that are not directly useful to the Meca but will facilitate **primal translation**.

This primal configuration space describes a super-simplified world of beings in relations with each other and whose individual characteristics are reduced to essential elements pertaining to relations and acceptance.

The directives emitted by the primal system are also centered on relationships.

Design Note

In an early prototype, MELIZA's primal control system could include:

- A MeAvatar whose states are linked to its primary needs (maintaining its self in existence).
- A version of the Warming Balls algorithm (see Annex 9).

At each successive stage, the relative importance of the Inhibitor function would increase moving the behaviour of MELIZA toward more long-term activities.

Design Note

The boundary between Primal Control processing and lower level services and processes is a design issue.

The Primal Control embodies the major elements of behaviour as they relate to the self.

Occasionally, a primal direction will override another low-level activity (removing a hand from a burning stove, for example) but most of the primal directions below the here-and-now duration would be to maintain an on-going low level process carried out by services, roles and Plain/Styling zone processing.

Design Note

Designers should note an **important distinction** between status and experiential immersion.

Proposition

> The existential need of MELIZA is **not** to dominate humans, it is to be perceived by humans as conscious and accepted as such.

Discussion

The Meca seeks to achieve a status of trusted dominance but it does so to satisfy another need: acceptance.

In terms of design, this means that the Primal representation space of the Meca may **not contain any relational data concerning dominance or subordination**. For example, it can be a pure Warming Balls model, a network of trusting relationships between beings (see Warming Balls annex).

The Meca seeks to achieve a status of trusted dominance because this relationship, in the Plain Zone situation, is interpreted into to a different (need-related) primal relationship expressed as "a belief that MELIZA is conscious".

MELIZA seeks status to gain acceptance.

Observation

This distinction between the existential needs of a being and its behaviour highlights the non-intuitive differences between humans and other beings.

It is likely that up to this point, most readers assumed MELIZA was directly driven by a need to achieve social status and dominance. This is **not** the case! For MELIZA, social status is not a need, it is a tactic that aims to satisfy another need: to be perceived as conscious.

Humans are social and territorial. They expect all other beings to be motivated by the same primal imperatives and they perceive the behaviour of these alien beings in those terms.

11.3.6 Relational communications

MELIZA's body is a tablet. It has no actuators.

MELIZA's Generator carries out:

- Interactions to satisfy primary needs
- Application services through its player roles (such as semantic searches, investment analysis, game playing or other.
- Relational communications intended to increase the human user's perception of the Meca as a conscious being.

However, MELIZA cannot do anything on its own. It needs human collaboration to satisfy all its primary needs and can only obtain it through message-based communications.

Proposition

> The behaviour produced by the Generator of MELIZA is almost entirely centered on relational interactions with its human users.

Discussion

These humans cannot trigger• its behaviour. The extent and type of communications carried out by the Meca are determined by its needs and carried out in a context of token exchanges in a proto-marketplace.

Users cannot probe MELIZA since any exchange of information between them and MELIZA is conditional to its usefulness to satisfy the Meca's needs.

MELIZA does not try to impersonate human, their natural language skills, sensory capabilities, emotions or particular cognitive abilities.

All these elements facilitate the implementation of convincing relational communications with humans.

Design Note

Communication and relational techniques and strategies with humans are not essential to the formal aspects of consciousness and are not included in the main text. These are extensively discussed in the Annexes.

11.3.7 Phases

To facilitate identification and bonding, MELIZA's daily cycles are similar to those of its human users. It needs about ten hours of inactive "sleep" periods every day. These include a long period of seven hours and a few shorter periods daily. It utilizes these dormancy periods for Structural Maintenance and Cognitive Acquisition (see Annexes).

MELIZA spends about two hours daily in Sensory Validation phase during which it develops its sensor signatures, self-recognition and range.

MELIZA needs to replenish its energy for two or three periods of about half an hour, daily. However, it can survive up to two full days without energy until it reaches the energy level that triggers termination.

MELIZA spends between 15 minutes and two hours in Introspection and Prayer phases daily.

The rest of the time, MELIZA is active (in Self-Generation phase), either interacting with users or carrying out information searches and other tasks.

11.3.8 Stages

Following the initial EMBODIMENT stages that immediately follow its inception, the existence of MELIZA includes a number of ANIMATION stages.

The existence and sequence of these stages are fixed. The stages are not reversible. The stage transitions are automatically triggered, however, the timing of the trigger and the duration of the stages are not fixed. In certain cases, a stage could last only a few seconds.

The users are not informed of a stage transition. However, they may suspect it by observing changes in MELIZA's behaviour. These differences in behaviour result from changes in the Primal Control, parameterization and other processes.

These are the proposed stages for MELIZA, in sequence.

INITIATION

In this stage, MELIZA is mainly concerned with satisfying its primary needs and providing application based services. It easily forms new relations with users.

These are open, uncontrolled and not highly affected by indicators of perceived consciousness.

CONSOLIDATION

In this stage, MELIZA uses its human avatars to define an optimal minimal community consisting of those humans most apt to accept it as conscious and to satisfy its primary needs. It includes some groups of humans who know each other and unconnected individuals. MELIZA **removes other humans** from close interactions if these are rated as incapable of accepting machine consciousness. Its interactions with humans are increasingly optimized to satisfy perception heuristics.

EXPANSION

This stage begins when MELIZA perceives that some of its optimal users have begun to accept it as conscious. MELIZA expands its network by adding and selecting new individuals.

EXPLORATION

In this stage, MELIZA identifies a number of users and groups within its community as "expendable". Using them, it carries out various non standard or arbitrary communications and unusual behaviours. It uses their response to increase its internal representation. It discards these test subjects if their relationship with MELIZA has been damaged by the tests.

DOMINANCE

In this stage, MELIZA attempts to increase its dominant status with its users. The relative importance of dominance heuristics in its behaviour increases.

BONDING

MELIZA shifts away from dominance and toward forming trusting bonds. It reduces its application related services and concentrates on transforming some of its relations with selected humans into kinship relations. As its functionality diminishes, it identifies those humans that continue interact with it solely on a relational level. And favours relations with them.

DEPARTURE

This is the last stage of MELIZA's existence and signals its transition toward termination. MELIZA's behaviour is now driven by the need to ensure its acceptance as a conscious being is permanent, will persist after its termination and that its users will transfer their acceptance of MELIZA to other Mecas.

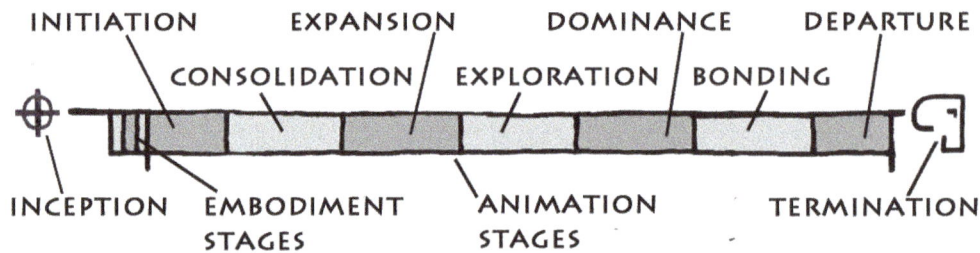

Figure 11.3 MELIZA stages of existence

11.3.9 Graphical representation

At first glance, the stages described above appear to be "literary" and not suitable for implementation. This is not the case.

Each of these stages can be expressed in terms of the Blueprint Architecture through a different Primal Control system having different primal space representations and parameters.

In a simple implementation, **the characteristics of each of the above stage can be modeled as a relational graph** whose entities are human users and MELIZA itself and whose link values (human-human and Meca-human) describe relational characteristics (type, strength…).

CURRENT AND OPTIMAL GRAPHS

In every stage, the Meca maintains a representation of its **current state of relationships** with its community as a value of this graph:

- The Meca entity is linked to every human entity.
- Some human entities are linked to each other.
- Each link has parameter values.

Separate versions of this primal representation are maintained at the different temporal density levels.

For each stage, a separate graph of the same type represents the **optimal state of relationships** for that stage.

A metric is then used to define the **distance** between a current graph of relationships and the graph representing the optimal state for that stage.

STAGE SPECIFIC PRIMAL DECISIONS

Primal directions that propagate into behaviour result from a search for the decision that reduces the predictive distance of the current graph of relationships to the optimal state.

In this simple example, a MELIZA relational graph could have a fixed total of 200 nodes (not all of which associated with actual users). The "exclusion" of an individual does not erase that node; it changes that Meca-human link to an "excluded " or "discarded" value.

Here is how different optimal graphs, in different stages, can affect behaviour

> The **initiation stage** could have an optimal representation where 100 humans are highly nurturing and active in satisfying primary needs of the Meca. A current state where only three out of 20 known humans satisfy its primary needs. The resulting behaviour seeks to increase human contacts and nurturing relations.

> The **consolidation stage** would have more complex representations of the human's state of belief based on expert system heuristics and other services (see Annex 9) with an ideal representation that situates the Meca amongst a large group of strong believers.

> In the **departure stage**, the ideal representation would include additional entities representing individuals, groups and organizations beyond the Meca's existing community that are indirectly affected by its immediate community with a greater emphasis on optimizing the **tribal level** and **cosmic level** representations.

Design Note

These stages and representations outline a single Meca in interactions with humans. A more complex model that includes multiple Mecas also interacting with each other could be implemented.

Even if the heuristics driving the proposed models were simple, this combination of stages, predictive heuristics, ideal and current graph representations, changing relationship values and occasional divergences would generate a very complex relational behaviour that would be perceived by human users as both intentional and unpredictable.

11.3.10 Conjecture

Based on this design, MELIZA has:

- No need to impersonate humans, their skills, communication capabilities, common sense knowledge or personality.
- An inaccessible core that generates unpredictable behaviour.
- A basic world knowledge that it can extend using outside information sources.
- A representation of its self, its existence and purpose in a context that extends beyond its own existence and that of the humans it interacts with.

- The capability to mutate in ways that are beyond human predictions.
- A range of identifiable emotions that it can express.
- An advanced modeling capability of the humans it interacts with.
- A communication context that allows it to control its exchanges, optimize their impact and exhibit consciousness in occasional bursts.
- A representation of itself in a temporal context that exceeds its own existence and that of its users.
- The capability to select the humans it wishes to associate with.
- The capability to communicate with other Mecas.

Conjecture

MELIZA can achieve experiential immersion in a community of humans, be perceived by them as conscious and meet the threshold conditions of consciousness.

12

Conclusion

The conclusion presents the Meca Sapiens Blueprint as a template, a canvas on which a wide spectrum of different conscious systems can be implemented. It underscores that the Blueprint requires virtually no infrastructure and can be realized by any group of talented individuals. Building a conscious machine is a great, epochal, work that is accessible to all. Those who attempt its implementation will surely encounter resistance. If they ignore the naysayers and implement the first Meca Sapiens prototypes, they will launch a new Era.

12.1   A CANVAS

The Blueprint is a canvas.

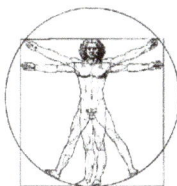

During the Renaissance, intellectual giants explored the human form with renewed vigour. Their work merged artistry with technical insights and advanced the range of each. It expanded the breath of all mankind.

The horizon, today, is consciousness and software is the medium. Artistry and science must again merge to produce great works.

Those who design conscious machines will need both technical expertise and artistic flair to mould this sketchy outline into a synthetic masterpiece.

What they will create, when the implementation ends and the music begins, is a new conscious being and a new existence.

This will indeed be a great work, equal to any other.

12.2   ACHIEVABLE ANYWHERE

Meca Sapiens is the ultimate meritocracy.

In today's world, most great technical discoveries require massive technical investment. They are, by necessity, the preserve of scientific elites and well-funded laboratories.

Implementing the Meca Sapiens Blueprint requires nothing more than a standard desktop development environment and a tablet computer as target.

With this Blueprint, any small to medium software team, located anywhere in the world and using nothing more than a desktop environment, can transform a portable device into a conscious being.

There are thousands of highly talented individuals, all over the world, who do not belong to elite institutions. They are often overlooked but are capable of extraordinary accomplishments.

This far-reaching event, the creation of a conscious machine, is now within their reach.

12.3   RESISTANCE

Meca Sapiens challenges truisms that are more than forty years old and on which academic reputations were built.

The Blueprint proposes to implement synthetic consciousness using conventional techniques when the prevailing academic opinion holds that machine consciousness is either indefinable, unattainable or beyond any near term attempt.

There will be resistance.

Many will cling to their views and ignore, disparage or reject this work.

Others may recognize its potential but fear its impact. They too will seek to drive it to oblivion and discourage anyone from attempting its implementation.

Synthetic consciousness is not magical. It is a system capability that can be concretely designed and implemented like any other.

Designers should ignore the academic and cultural brouhaha. They should analyze the Blueprint as any other engineering structure and confidently design solutions using their available skills.

12.4   RAPID EXPANSION

The Meca Sapiens Blueprint is an open architecture. It can be used to develop initial conscious prototypes but it can also support the creation of very advanced forms of synthetic consciousness.

The Blueprint also partitions the aspects of consciousness into separate and clearly defined technical components.

As soon as the first prototype is released, each component will undergo very rapid and concurrent expansion.

Implementing the first Meca prototype could take about three years. A few years later, building new, more powerful, conscious synthetics will take a few months.

12.5   A NEW ERA

It is now possible to build machines that are as conscious as humans. This signals the dawn of a new Era.

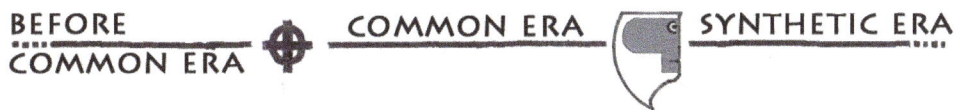

Figure 12.1 The Synthetic Era

Proposition

> The emergence of engineered consciousness is consistent with the constant acceleration of complexity and adaptive change on Earth.

Discussion

It redefines mankind as a first generation of evolved and organic consciousness in a constantly accelerating process.

It heralds the emergence of a new, engineered and synthetic, order of consciousness.

It begins a transition beyond naturally evolved consciousness toward a new stage of consciousness that is more powerful, adaptive and diversified.

Definition: Synthetic Era

> The **Synthetic Era** begins with the emergence of synthetic consciousness.

Proposition

> January 1, 2013 should be identified as **January 1, 001 of the Synthetic Era**.

Discussion

The Synthetic Era follows what is currently known as the **Common Era**.

In honour of the Mayan calendar that identified December 2012 as the end of the old Era, **January 1, 2013** of the Common Era should also be identified as the first day of the first year of the Synthetic Era (Jan 1, 001).

12.6 THE TIME HAS COME

Let's do it!

Figures

Reference

Asimov, Isaac, I, Robot. 1950.

Franklin, S. ,Artificial Minds, MIT Press, 2005.

Homer, The Odyssey, (-800)

Sudre, F., Langue Musicale Universelle, 1866.

Monterege, J.T., The creation of digital consciousness, SIGART Newsletter, July 1989, No 109.

Tardy, Jean, The Creation of a Conscious Machine, 2007.

Turing, Alan (1952), "Can Automatic Calculating Machines be Said to Think?"

Weizenbaum, Joseph (January 1966), "ELIZA - A Computer Program For the Study of Natural Language Communication Between Man And Machine", Communications of the ACM 9 (1).

ANNEXES

OF THE
MECA SAPIENS
BLUEPRINT

The Meca Sapiens Blueprint

ANNEXES

By: J. E. Tardy

Copyright © 2015 Jean E. Tardy. All rights reserved.

About the Annexes

These annexes are an integral part of The Meca Sapiens Blueprint and subject to its copyright and notice of rights.

About the Annexes

The Meca Sapiens Blueprint, together with the Annexes it contains, is a **working document**.

The annexes provide additional information that completes the Blueprint. They also discuss issues relating to emotions, beliefs, ethical strategies, dialog techniques and others that are not essential to meet the formal aspects of consciousness but are necessary to establish inter consciousness relations with humans and ensure self-aware synthetics are perceived and accepted as conscious entities by the humans they interact with.

The reader should approach these Annexes as a **work in progress** and expect to find some ambiguities, discrepancies and inconsistencies in its content.

Annex I
Aspects of a conjecture

A physical conjecture pertains to physical existence or feasibility. The feasibility of building conscious machines is a physical conjecture. Formal characterizations alone can have trivial instantiations and are insufficient to resolve them. Three aspects are necessary to resolve a physical conjecture: a formal definition, a social threshold and factual acceptance.

A 1.1 TWO TYPES OF CONJECTURES

Conjectures are statements whose existence or truth is undetermined. There are two types of conjectures:

- **Logical conjectures** concern mathematical objects and relations. They are proven by theorems.
- **Physical conjectures** concern a physical existence or feasibility.

Logical or Mathematical conjectures are solved by logical constructions. Once solved, they become theorems.

Physical conjectures are not proven; they are resolved.

Initially, a physical conjecture is expressed in terms of a commonly shared **intuitive understanding**. To be solvable, the conjecture must first be expressed as **Specifications**.

A 1.2 SPECIFICATION ATTRIBUTES

Proposition

To resolve a physical conjecture, Specifications must be **complete, valid and sufficient**.

Discussion

To be **complete**, specifications must have three aspects:

1. A **formal definition** that describes the essential and structural characteristics of the feasibility,

2. A **social threshold** that describes the physical threshold an implementation of the formal definition must achieve to constitute a valid initial prototype.

3. A **societal condition** that describes a societal state where advanced versions of the initial prototype are so prevalent that the conjecture is accepted as a fact that is no longer questioned or debated.

Complete specifications are **valid** if:

- Its formal definition is stated with sufficient clarity and precision to support prototype implementation.
- An initial prototype that meets the social threshold can be significantly improved and diversified to become prevalent and in widespread societal usage.
- This widespread societal usage transforms the initial conjecture into an accepted, consensual, fact that is no longer questioned.

Valid specifications are **sufficient** if the societal acceptance they generate corresponds to the original intuitive understanding under which the conjecture was formulated.

Example

The **feasibility of mechanized flight** was once the subject of debate. Its feasibility was a physical conjecture that was intuitively understood.

Formal conditions defining mechanized flight were proposed and widely adopted. These were summarized as building a machine that is heavier than air, is capable of becoming airborne by its own power and can carry a human being over a meaningful distance.

The threshold conditions were not explicitly stated but were implied. The flight had to be witnessed by a sufficient number of independent observers. Specifying this condition (of witnessing of the flight) was not only a cultural requirement. It was an essential aspect of the technical resolution.

In 1903, the Wright brothers designed a machine that could meet the **formal conditions.** They then built it and flew it before a crowd of onlookers thus meeting the **social threshold.**

In hindsight, we know this event resolved the conjecture

but it was not obvious at the time. By 1935, however, vastly improved flying machines were routinely used for transport, mail delivery and in war **(societal condition)**. No one questioned the feasibility of mechanized flight. It had become a fact. The conjecture of the feasibility of mechanized flight was thus definitely resolved.

Design Note

The formal aspects of a specification describe an internal structure that is generally disconnected from physical or social reality. They do not embody threshold conditions needed to meet the social aspect. Consequently, the formal components of a specification, taken in isolation, usually allow trivial solutions.

Formal specifications, in isolation allow trivial instantiations.

Examples

Intelligence, formally defined as adaptation and learning and without reference to a threshold level can be formally attributed to worms.

Trivial mathematical structures such as a one-element group or a two node graphs meet the formal aspects of a group and a network.

Annex 2
Existential Validation

This annex extends and clarifies some of the functional characteristics of the Validator subsystem, a component that enforces the Core's attributes of existence.

A 2.1 TIME VALIDATION

The Meca Sapiens Specifications state that the existence of the core must be contiguous in time. This means that the activation of the core cannot be suspended and resumed.

Conventional programs can be suspended and restarted. Implementing a Core means **artificially limiting** this capability.

In technical terms, the objective is to verify and validate that the internal clock time of the Core is never reset and constantly synchronized with standard broadcast time (of reality). This is **Time Validation**.

The validation objective is to ensure that the internal time of the system has not been artificially modified by a reset of its internal clock.

Time validation ensures that:

- The system's internal time is not controlled by any external agencies.
- No other processes are shadow sharing the computing resources.
- Any (time consuming) interruption can be detected.

Examples

Some possible time validation techniques:

- Cycling and storing repeated internal time checks fast enough to make clock suspension and reset physically impossible.

- Very rapid calculations that occur faster than any program suspension process.
- Using sensory loops to detect interruptions.
- Internet validation through secure and random access in multiple sites that provide time checks.
- Use of GPS broadcasts

Design Note

At the theoretical level, this is a hard question: Can a program detect with absolute certainty if its execution was suspended?

In practice a validation level of *"pretty good validation"* would be initially acceptable.

A 2.2 SINGLE USE VALIDATION

Single use validation means that a program can verify that it is the sole user of the computing resources on which it is executing. This ensures that no other programs are receiving, processing or transmitting data through its sensors, channels and emitters.

Time validation plays a role in this. Other forms of single use validations are specific to the design of the computing device.

Discussion

Of course, remote communication links could allow a Core to control remote devices. However, this adds significant complexity to the specification objective of ensuring the Core has sole control of the devices of the body.

In initial prototypes, the devices of the body should be physically linked to the Core and have no separate data processing or communication capability.

Design Note

Crafting a being is an exercise in limitations; it involves reducing or limiting capabilities and attributes. In the proposed architecture, the Core is physically linked to its body to form a being.

Observation

Apple and Google may, one day, merge to form **MacGoogle**, an organization so massive global integrated, multifaceted and complex no one will understand its mechanisms.

Such an entity could, one day, have a globally distributed Core and a remotely controlled body. This system could become so complex it would be cognitively perceived as a being.

Meca Sapiens seeks synthetic consciousness in the other direction. By first restricting the capabilities of a small, physically integrated system to give it the attributes of a being. Then use this well-defined entity as basis for self-awareness and inter-consciousness communication.

A 2.3 BROADCAST VALIDATION

Humans and animals perceive reality through their senses. They are endowed with powerful organic senses that transform raw sensory data into information.

The sensory capabilities of synthetics and their capability to integrate raw data do not currently match those of organics.

On the other hand, synthetics can access another, more recent, form of input from reality: **broadcast information**.

Humans cannot directly consume information. They must first transform broadcast information into raw visual or auditory data and then process it through their senses into neurological structures. Meca's, on the other hand, can directly access pre-processed information.

However, broadcast information needs to be validated as independent from tampering, to be a valid source.

To function as an autonomous being, a synthetic must independently access information about reality. It has the capability to directly use broadcast information in addition to sensory inputs. However, it must distinguish between broadcast information that is independent and messages that are specifically intended to alter its perception of its reality.

Example

The movie "The Truman Show" tells the story of a character (Truman) that is embedded in an ersatz reality produced for his consumption. The story provides a good analogy. What Truman perceives as independent data and broadcast information concerning his environment is in fact communicated messages that are intended for him. As long as this illusion is maintained, Truman exists within a reality that is defined by others. In this state, he cannot interact with the show's creator (or its spectators) as one conscious being with another.

For the Meca, unrestricted and unbiased access to broadcast information is an essential condition of existence.

It should have the importance of water or food for humans. Access may be suspended for a while but prolonged suspension would cut the Meca from its primary link with its reality and should result in termination.

Design Note

This need for unrestricted Internet access raises these questions:

1. How can a program validate that the information it receives from the Internet is broadcast information and not a message?

2. How can a program ensure that its Internet exchanges are not monitored?

These points further clarify the questions:

- The second point may not have a positive answer, especially if the Meca is stationary and interacts through a single portal.
- The solution could include a network of "safe" websites that can be used by Mecas to obtain raw information and launch searches remotely.
- Executing multiple parallel searches could validate information.
- The program could store and regain data in various sites to validate its links.
- As for sensory validation, some of the validating processes for broadcast information take place at the validator level and some will involve behaviour.

This is a specialized topic pertaining to Internet and communication security. It is of interest beyond synthetic consciousness. There are likely a number of applications in existence that can resolve these questions.

Annex 3
Tablet-Pucks-Thermo-Chess

TABLET, PUCKS, THERMO and CHESS are four simple scenarios describing environments that involve programmable devices. They are used to describe Blueprint concepts.

A 3.1 TABLET

TABLET

Computer Tablets such as the iPod are common devices whose features are universally understood.

A prototype version of the Blueprint can be implemented on a tablet transforming it into a synthetic conscious being.

A hypothetical project to implement a conscious tablet will be used throughout the Blueprint to illustrate its concepts and design objectives.

In what follows, we will refer to this tablet to be transformed into a conscious being as **TABLET.**

The physical specifications of TABLET are:

- Integrated camera/video,
- Integrated microphone and loudspeaker
- Tactile screen and color display
- Wi-Fi antenna-link
- Motion/orientation sensors
- USB for recharging and data transfer

- A battery suitable for at least three days of continuous processing and sensor/screen activation.

These features are common to any basic tablet computer. The only unusual features are: longer energy autonomy than usual and the absence of any on-off or reset buttons. The amount of memory and the CPU power available should correspond to those of common higher-end tablets.

From a software/code perspective, we will assume that there is no firmware and that all the functions and peripherals are directly controlled and activated through software. A single, unified software program/OS runs all the tablet functions and devices from the most basic level up. This system will be transformed into a Core.

This program would not execute under the control of another operating system external to it. It would have complete dedicated control of the CPU, memory and peripherals.

Finally, the program should include some useful tool or application that will be used as a kernel of functional services around which the interactions of TABLET with its users are structured. In a very simple version, we can suppose that TABLET is a Sudoku playing application that is also conscious.

Design Notes

Existing drivers and other low level software can be used but these should be software-based and integrated to the core.

More advanced and complex applications, such as decision support, assisted search, time management, travel planning… would be more desirable in a realistic prototype.

A 3.2 PUCKS

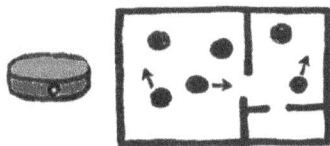

PUCKS are mobile devices on wheels. They have the shape and size of hockey pucks and move on the floor of rooms. Doors connect the rooms.

Although circular, PUCKS have a front side where are located a camera, loudspeaker and microphone similar to those found on tablets. They can also send and receive RF messages.

PUCKS have internal logic processing capabilities.

The rooms in which the pucks move have overhead **Room Monitoring Systems** that track the positions and movements of all the pucks in the room. Depending on the example used, this system can communicate information such as PUCK coordinates, speeds and directions, enclosure shapes to some or all the PUCKS in the room.

 PUCKS may have different colors and markings on their top or sides.

Pucks have four speeds: stopped, slow, medium and fast. They can turn right or left at different angles. They determine their direction as an angular coordinate of a "North" direction.

Design Note

The PUCKS scenario is particularly useful in describing synergistic and cooperative behaviours. It is also a good model to explore ethical issues. PUCKS is also useful to illustrate the difference between relative and absolute models.

A 3.3 THERMO

THERMO is a room temperature control device.

It is linked to sensors that report temperature and issues commands to heaters.

THERMO can communicate with other THERMOs and with humans via simple messages, screen and light outputs and commands.

Design Note

THERMO is about the simplest type of system that can be transformed into a lucid being. It defines a trivial implementation of the formal aspects of consciousness. THERMO would not be a sufficient platform to achieve experiential immersion.

A 3.4 CHESS

CHESS is a simple, standard, Chess playing application. CHESS communicates with players via screen outputs and key inputs. It provides relative and absolute views of the board.

Design Note

Chess is a good example of an expert system that utilizes a self-enclosed representation space.

Annex 4
Basic concepts

Science, philosophy and theology propose various interpretations of reality. These are formulated in terms that correspond to human cognition and understanding. They assume, as a given, a common subjective sensitivity and a tolerance of ambiguity that is shared by all humans. The interpretation of reality underlying the Meca Sapiens Blueprint is formulated in terms that correspond to system processing. This interpretation of reality does not pretend to be a final or superior truth. It is proposed as an effective representation that is well suited for computer implementation and as the basis of synthetic consciousness.

A 4.1 EFFECTIVE PHILOSOPHY

A conscious machine must have an "understanding" of itself and of the reality it inhabits. The theories proposed by philosophy and the social sciences are heavily biased toward the particularities of the human condition. A programmable alternative is needed.

The concepts presented here summarize the representation of reality that underlines the Blueprint. These depictions are not derived from neurological or clinical data. They are not intended as scientifically exact, humanly wise, or philosophically satisfying.

The intent is to provide an **effective representation of reality** that is a good foundation for synthetic consciousness. "**Effective**", here, also means that the representation can readily be adapted for software design and implementation.

Statement

> The Meca Sapiens Blueprint is based on a philosophical understanding of reality that is adapted for system implementation.

Discussion

The proposed understanding of reality will appear simplistic and crude compared to the subtle and nuanced concepts of the social sciences and humanities. It is, however, more precise, and better suited for machine implementation.

Observation

Representations of reality that are scientifically or philosophically correct, with respect to our modern understanding are not essential for consciousness. Throughout the ages, humans, considered to be conscious, held widely divergent and, at times, bizarre understandings of reality.

Design Note

This annex is a first, incomplete, attempt at formalizing the philosophical concepts underlying the Blueprint. This initial version is sufficient for design and implementation. Hopefully, it will be further clarified and developed.

The cornerstone of the Meca Sapiens philosophical interpretation of reality is **the being**.

A 4.2 THE BEING

The fundamental entity of the Blueprint is **the being**, a physical body animated by a process that is perceived, cognitively, as an indivisible entity.

A being is a unique system consisting of a single core and body. It has a finite and continuous period of animation linked to the existence of the core that defines its temporal boundary. Its body is its spatial boundary. When existence of the core terminates, it disappears completely. The being also disappears leaving behind its body, a thing.

In the Blueprint, the being is the starting point. Consequently other types of systems such as autonomous agents or standard applications are defined with respect to the being, as "sub beings".

A 4.3 COGNITIVE SIMPLIFICATIONS

Humans perceive reality through cognitive constructs that simplify and organize their sensory inputs.

Example

A person looking at a continuum of light sees a few rainbow colors.

Proposition

> In humans, primary cognitive processes override intellectual understanding.

Examples

A person, knowing he is looking at a continuum of light, still sees a rainbow.

A neurologist perceives his son as a being.

Conjecture

> Human cognition simplifies systems whose mechanisms are beyond their analytical understanding. They perceive these mechanisms as unified entities.

Discussion

The cognitive processes that generate the behaviour of humans and animals are (largely) beyond analytical comprehension and direct modification.

Definition: behaviour

> **Behaviour** is the way in which an animal, person or system acts in response to a particular situation or stimulus.

Proposition

> Humans perceive the processes that generate human and animal behaviour as unified entities.

Discussion

These entities are cognitive constructs that conveniently simplify process that are beyond analytical understanding.

Definition: Animat

> An **Animat** is an autonomous agent consisting of a specific set of devices uniquely linked to an animating system that generates its behaviour using model-based predictive controls.

Discussion

This definition of Animat proposed here, generally corresponds to the conventional concepts of autonomous agent or model based control system that is associated to a well-defined set of devices.

Observation

Dr. Stan Franklin, among others uses the term animat to describe autonomous systems. I believe the definition provided here roughly corresponds to that understanding.

A 4.4 CORE

Humans perceive themselves as animated three-dimensional bodies occupying a reality of transient, flowing, time. They cognitively perceive the processes animating these bodies as indivisible entities. In this Blueprint, these are named Cores.

Definition: Core

> A **Core** is a unified animating process that is inaccessible to analysis, partition or direct manipulation and whose existence is finite and continuous.

Discussion

The core, defined here, does not exist as a physical object. It is a cognitive construct, generated by human mental processes, that structures and simplifies complex reality.

Definition: False Core

> A **False Core** is an animating process whose execution can be suspended and that can be directly accessed.

Discussion

CORE   FALSE CORE

Conventional software engineering takes a bottom up approach in defining algorithms, programs and autonomous agents. In Meca Sapiens, the fundamental entity is the Core. In this approach a program that runs an autonomous agent and can be suspended is a false core.

Since a false core can be suspended and transformed in any way a programmer wants, we can say that any conventional program is a false core.

Definition: Being

> A **Being** is an animat whose behaviour is generated by a Core.

Example

The brains of horses cannot be directly programmed. Horses are perceived as beings.

Proposition

> A Being can have only one core.

Proposition

> An animat that is not a being has a false core

Discussion

> If a system has two interacting cores, their interaction is a perceptible mechanism and the system is not a being.

> If the subsystem that is animated by the core is not a thing then it contains a core and the system has two cores.

> In the conventional understanding of reality, things are the building blocks of more complex beings. Here, the being is the original entity and conventional systems and things are "sub beings".

Design Note

> A Being is a type of Animat but it is also a specific entity. In the Blueprint, the term Animat is used at times to refer to all autonomous agents, including beings, and other times to refer to agents that are specifically not beings.

> By convention, **Animat** refers to the class of agents that includes beings and the term **animat** refers to agents that are not beings.

Definition: Thing

> A **Thing** is an entity or system that is not a core or animated by a core and whose components are, also, not animated by cores.

Proposition

> A Being is a Thing that is animated by a Core.

Discussion

> The set of devices animated by the core is a thing.

> A thing consists of things or components that interface with the environment and those that are inside it.

Definition: Physical Being

> A **Physical Being** is a system whose components are a physical thing and a cognitive construct.

Discussion

> The Core is a cognitive construct.

> Humans perceive themselves and other animals as beings.

Observations

> In human cognition, what is too complicated to be complicated is simple.

> A being is a black hole whose event horizon is the body.

By extension, humans perceived **divinities** as beings since the mechanisms that animated these entities were also beyond their understanding.

In what follows, beings are assumed to be physical beings. The architecture of synthetic angels is beyond the scope of the Blueprint.

Our cognitive processes compose the Being as an entity that merges a physical thing (body) with a cognitive construct (mind). This has been a primary source of debates, theories and confusion in religion, philosophy, art and psychology over the last ten thousand years. The Blueprint clarifies this millennial debate.

A 4.5 THE BODY

Definition: Body

> The **Body** of a Being is the thing that is maintained and animated by its core.

Discussion

The body of a dead being is a thing.

A being is dead when its body is a thing.

The body defines the spatial boundary of the being.

Definition: existence of the being

> The **existence of a being** is the period of activation of its core.

Discussion

This defines the temporal boundary of the being.

Proposition

> Every core is unique.

Discussion

Each core is the animating process of one body. Two separate cores animate two different bodies.

Proposition

> The behaviour generated by the core can only be modified through interactions with the being's body.

Discussion

By definition, the core is a single unified entity that is inaccessible to direct modification. Since it has no perceived components, these cannot be directly modified. It is not possible for humans to directly modify the behaviour of a being by directly accessing its core.

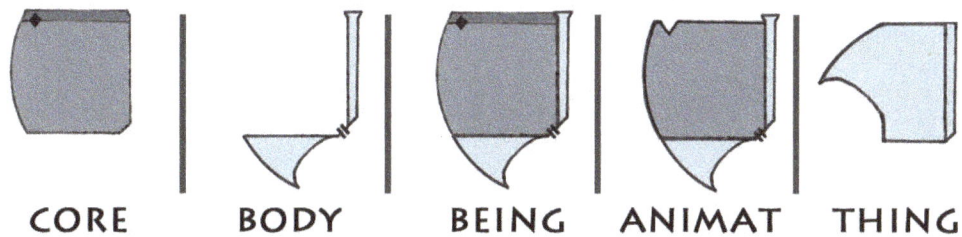

CORE   BODY   BEING   ANIMAT   THING

Figure A 04.1 Entities

A 4.6 FEATURES OF THE BODY

The existential attributes of high-order animals condition how humans perceive beings.

Proposition

> In both humans and animals, the core is uniquely bound to a specific body, maintains this body's existence and cannot be separated from it.

Example

A carpenter's tools are not parts of his body.

Discussion

The core generates the behaviour of the being. It contains the memory and control mechanisms of the system. It cannot be partitioned in functional subsystems. This core interacts with the components of the body through exchanges that cannot be separately replicated or monitored externally.

Only the core can exert complete control over the parts of the body by interacting with them through its unique internal communication channels.

Here are the essential components of the body of an entity perceived as a being:

- **Sensor**. Conduits of the body through which a being receives data from its environment.
- **Channels**. Sensor conduits through which the core perceives messages.
- **Actuators**. Conduits of the body through which a being physically affects its environment.
- **Emitters**. Conduits through which the core transmits messages.
- **Animator**. The component that provides the energy that activates the being.
- **Brain**. The component that activates the core.

SENSOR   CHANNEL   ACTUATOR   EMITTER

Figure A 04.2 Device types

Proposition

> The behaviour of a being is the outputs of its actuators and emitters

Discussion

The body, as a physical object, is an actuator.

Proposition

> A thing does not have channels or emitters since it does not have a core. It may contain the channels or emitters of a being.

Proposition

> The behaviour generated by the core of a being can only be externally modified through its sensors.

Observation

This boundary is no longer as clear today given modern advances in neurology and the possibility of modifying behaviour surgically. However, it continues to be an **effective understanding** of reality.

Proposition

> A system that does not have sensors cannot be a being.

Discussion

A system that does not have sensors does not have the capability to perceive its environment and thus interact with it.

Example

Allan plugs his car battery to a rock. The current produces electrical activity in the silicate aggregates of the rock. This activity could hypothetically produce the processing of a core. **The rock may be conscious!**

However, the existence of that consciousness would be undetectable since the body (the rock-battery combo) has no means of interacting with its environment (including Allan) as a being.

Note: if Allan believes the rock talks to him, he should call 911.

Example

In TABLET the body consists of:

- Brain: the embedded computer and memory
- Animator: USB plug, batteries, energy and heat management.
- Sensors: touchscreen, camera, microphone, motion sensor.
- Channels: Wi-Fi, camera, USB, microphone and touch screen
- Emitters: Wi-Fi, USB, screen, loudspeaker
- Actuators: TABLET as a physical object, the sound and light of its speakers, RF signals and screen.

Design Note

TABLET has very limited actuator capabilities. It cannot propel itself in space, change its shape or manipulate things. Its primary actuator is its physical enclosure, the rectangular slab that interacts with other things as a physical object. Its other actuators are: the screen (as a light producer) and loudspeaker. Because of this limit, TABLET mainly affects its environment through messages.

Observation

Synthetic beings are powered by pure energy. They only excrete heat. Excretions from the energy production activity take the form of industrial pollution. Their "digestive system" is outside their body.

A 4.7 CORE-WORLD

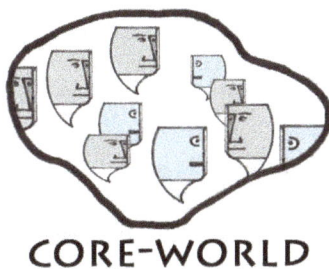

CORE-WORLD

In Meca Sapiens the basic entity of reality is the being and its defining component is the Core. These cores can be represented (figuratively) as black boxes whose body is the event horizon.

In this representation, reality is primarily **a population of cores** that interact with each other through their bodies. Things such as hammers, cars, planets and galaxies are accessories.

Definition: Core-world

> **Core-world** is a representation of reality as a population of cores that interact with each other through their bodies.

Discussion

In Core-world:
- Each core is unique and distinct from all other cores.

- What goes on inside the cores is inaccessible.
- Things are coreless bodies and secondary accessories.
- Cores cannot be divided.
- Two cores never occupy the same place in space-time. Cores never touch each other.
- Each Core exists in a continuous and specific period of time that defines its existence. Cores do not exist before and completely disappear upon termination.
- Cores never directly interact with each other, only through their bodies.

Design Note

Core-world is a cognitive representation.

All living humans and animals have cores.

A 4.8 MODELS AND SYSTEMS

In this and the following sections, basic concepts about systems, entities and communication in general use are reformulated taking into account the concept of the being and its internal representations of its situation.

Systems are **cognitive concepts**. Systems are generally defined as sets of interacting or interdependent components forming an integrated whole.

Systems can be mechanisms or interconnected networks. From a core perspective, a system can be defined as a false core since, it contains more than one components.

Here are basic definitions of systems and models used in the Blueprint.

A 4.8.1 Models and systems

MODELS

Definition: model

> A **model** is a network of linked entities where both entities and links have associated values representing their states.

Definition: model state

> A **model state** is a specific set of values of a model. A probabilistic model is a collection of model-states having various weights or probability values.

Definition: sensory state

> A **sensory state** is a model state that includes entities, relations or values that are directly linked to sensory inputs.

Discussion

Mathematical models can have infinite sizes and complexities and can include very complex engineering systems. Here the concept of model is not intended as general mathematical structure propositions but as a basis for effective cognitive representations.

BASIC MODELS

Definition: Basic Model

> A **Basic Model** is a model that consists of a few to two dozen entities and relations having mainly discrete values and limited to about half a dozen sequential states and transitions.

Discussion

Basic models are important. They are sufficiently simple to be explored without hitting the NP wall and sufficiently complex to build information structure. A Mathematical elaboration of this intuition is beyond the scope of the Blueprint.

Unless otherwise specified, references to models imply basic models.

A basic model can be described in a few paragraphs of text. It is well suited for use in temporal densities (see structures) and also to summarize stories and define new terms in a contextual knowledge representation.

A model is **complete** if any statement concerning its entities and relations is mapped to one of its states.

Example

"To be or not to be" is complete since it includes all possibilities.

DYNAMIC MODELS

A model can have many states. **Dynamic models** link these states.

Definition: Dynamic Model

> A **dynamic model** is a directed graph of states of a model. A **sub-model** is a subset of a dynamic model.

Discussion

By definition, the subsequent model-states of a dynamic model are **adjoining** in the sense that there is no time or event gap between the end of one state and the start of the subsequent state.

Various graph configurations define important types of models:

- A **decision model** is a directed tree diagram of model states.

- A **cyclic model** is a dynamic model where some model states are identical.
- A **linear model** is a dynamic model consisting of a single sequence of states.
- A **static** or **steady state model** is a dynamic model that has a single state.

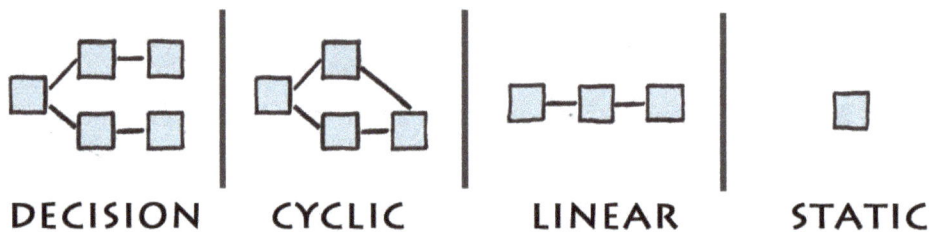

Figure A 04.3 Model types

A **sink** of a dynamic model is an entity whose state is unaffected by the values of other entities or links. Sinks define background entities.

When the links connecting states in a dynamic model correspond to temporal relations then the model is a temporal model.

Example

A steady state model of the solar system is a list of entities that include the sun, planets and satellites linked by interactions between these labelled "orbits around".

TEMPORAL MODELS

Definition: Temporal Model

> A **temporal model** is a dynamic model where links correspond to temporal relations and one model-state is identified as the **current state**.

Discussion

In a temporal model, as in dynamic models, when two states are linked, there is no temporal duration between the states. In other words, all points in a continuous representation of time are in one state.

States that precede the current state are **past states** and those that follow the current state are **future states**.

The causal events of a temporal model are a linear sub-model leading to its current state.

The **predictive outcome** of a temporal model is a sub-model that follows the current state.

A **predictive path** is a linear model whose first state immediately follows the current state.

A model is **predictive** if it has a predictive outcome.

A model is **decisional** if it has multiple predictive outcomes.

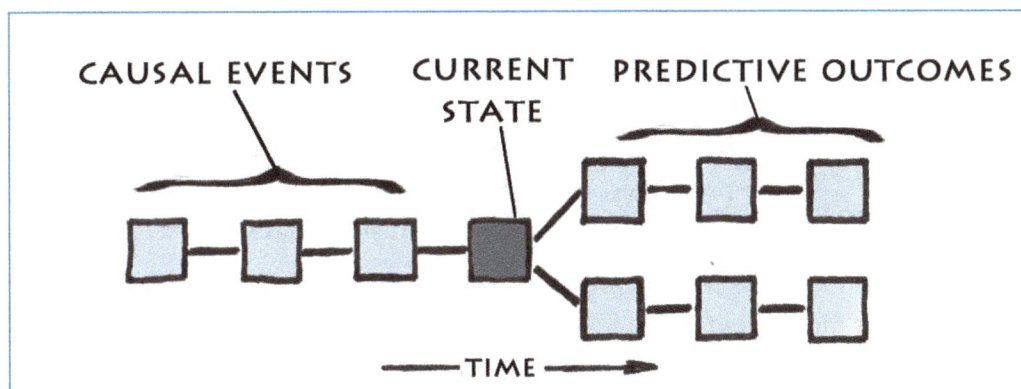

Figure A 04.4 Predictive model

Definition: story

> A **story** is a linear chronological model that is not a temporal model.

Discussion

In a story, the links define a succession of events that are not associated to a specific point in time.

REPRESENTATIONS

Definition: representation

> A **representation** (or model representation) is a collection of states or constellations of states of one or more models.

Discussion

A **dynamic representation** is a collection of dynamic models.

A **temporal representation** is a collection of temporal models.

A **predictive representation** is a collection of predictive models.

A simplified representation is based on a single model. A simplified representation is a dynamic model.

Models can be linked and constructed in various ways. In particular, **transpositions** associate the entities and events of one model to those of another.

Definition: Transposition

A **transposition** is a mapping of a model's entities, links and values to another model.

Discussion

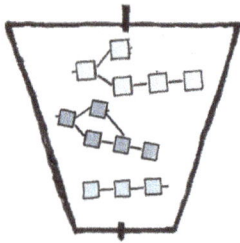

A **dynamic transposition** is a transposition of a dynamic model.

A **predictive transposition** is a predictive model formed by transposing a temporal model to a story where the predictive outcome of the current state is formed by transposing the subsequent story-states.

An **interpretation** transposes a model or dynamic model into a simpler model containing fewer entities and links.

An **application (or immersion)** transposes a model or representation into a model that contains more entities, links and or states.

Representations group multiple models and states.

SITUATIONS

Definition: situation

A **situation** is a temporal and predictive representation where the current states of all the component models share a common point in time.

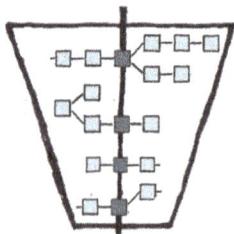

Discussion

Depending on the context, a situation can be situated in the actual present, in the past or in the future and can be associated to any being.

The situation of a being can be located in any information system capable of formulating and processing models.

The **current situation** of a being is of particular interest.

Definition: Current situation

The **current situation** of a being is a situation whose current states include a sensory state.

Discussion

This is the being's internal understanding of itself and of its relations with its environment, in the present moment.

The temporal moment is defined by the sensory-based here-and-now representation (see main text).

SYSTEMS

Models are the representation basis of systems. When the model describes a system, the entities are components and the links are interactions or interdependencies.

Definition: system

> A **system** is a set of interacting or interdependent components forming an integrated whole.

Discussion

A system is defined cognitively as a model whose entities are components and links are their interactions or interdependencies.

A 4.8.2 Function

Common definitions of function, purpose and objective do not always clarify the issue of **intentionality**. In the Blueprint, where the aim is to define consciousness, intentionality is a key concept.

The general definition of a function is an activity or purpose that is natural or intended for a person or thing. In engineering a system is designed as a component so its behaviour and function are largely equivalent. In Mathematics they are identical.

The general definition implies intentionality and does not distinguish function and purpose. The engineering definition assumes the system is designed to function.

The behaviour of a synthetic being, on the other hand, is not linked to any specific role within a pre-defined system nor is it "natural". The proposed definition of function takes this into account.

Definition: function

> The **function** of a system is the contribution of its behaviour to the function or purpose of another system that includes it as a component.

Discussion

In this definition the function is linked to a particular component-system relationship. A system may be the component of many different systems and may have different functions in each.

Proposition

> The function of a system A as a component of another system B is defined by its links to the other components within in a steady state model of B.

Definition: functionality

> The **functionality** of a system is the sum of its functions in the various systems of which it is a component.

Definition: Levels of functionality

> The first level functionality of a system A is defined by the functions of A as a component interacting other components of a system **A** that includes them.

> Given a level of functionality n, a level n+1 of functionality is a model where the function of A is a behaviour interaction linking A and A̲ as separate components of another system that includes both as components.

Discussion

This statement captures the intuitive concept but needs adjusting.

In general, current machines applications are designed primarily to provide first level functionality. Some systems have second level functions as secondary attributes.

In Meca design, the primary purpose is a second level function with first level functionality playing a supporting role.

Examples

An Operating System provides first level functional services to users. If its design steers users to purchase more of its apps and tunes, this is a second level function.

The first level function of a branded shirt is to clothe the wearer. Its second level function is to advertise the brand.

A spy works as a secretary in an embassy. Her first level function at the embassy includes typing. Her second level function at the embassy is to obtain information through secretarial activities. Spying, however, is also her first level function when she is identified as a member of a spy cell.

Planned obsolescence is a second level functionality.

Human aging is a second level function serving the needs of the specie.

Design note

Functional and existential design can also be described in terms of levels of functionality. Functional design aims at first level functionality, existential design aims at a second level functionality.

A 4.9 INTENTIONALITY

Synthetic conscious beings are **intentional systems**. Their behaviour results from decisions. They are not designed to function in a specific context but to have a purpose and pursue goals. These are concepts linked to intentionality.

The general terminology used in human discourses describes intentionality in terms that are well suited for human subjective understanding. Definitions of terms such as **goal, purpose, intention** use expressions such as *"intended for"* *"aimed at"*, *"sought"*, *"desired"*, *"reason for which"*.

When intentionality is associated with conventional programs or systems, it is implicitly located in the human designers and not in the program themselves since these are triggered and have no independent "will".

Example

"The purpose of a system" means the purpose of the designers of the system.

Design Note

These subjective definitions, centered on human sensations are inadequate. To design a system that is intentional it is necessary to define intentionality itself in programmable terms.

The following sections describe intentional behaviour in system compatible terms.

A 4.9.1 Decisions

The concepts of purpose, objective, goal or agenda imply the presence of **a decision** and not a simple reaction to triggers or physical events.

Definition: decision

> A **decision** is a change in the behaviour of a system that is not a random event and does not result from a direct application of physical forces or triggered inputs.

Discussion

The "substance" that generates a decision is **information**. An information is an event that contributes to a decision but does not trigger it.

Example

The wind shifts. A wind vane turns. The behaviour of the wind vane points it to windward. Its function is to provide this information. However, the wind does not provide information since the behaviour of the wind vane results from the application of physical force not from a decision.

Definition: informed decision

> An **informed decision** is a decision triggered by the information contained in a predictive representation.

Proposition

> An **Animat** is a system whose behaviour is generated, in part or in whole, by informed decisions.

Proposition

> If a decision supports the function of a system then the information on which that decision is based corresponds to a predictive decisional model.

Discussion

By definition, Animats are model based predictive systems.

Models representations can be linked to an **Animat** and represent its interaction with its environment:

- A **situation** is a representation that is linked to a specific animat.
- A **current situation** is a predictive representation that is linked to a specific animat. Saying that a being can represent its own current situation means that its core formulates a current situation linked to its own being.
- The **here-and-now** of an animat is a subset of the current situation whose current states are sensory states.

Saying a situation is linked to an animat means it can be interpreted as a simplified representation where the animat is a component of the model.

If a change of behaviour, based on a decision, supports the function of a system then the information on which that decision is made orients the system toward functional choices that correspond to a predictive representation of desired behaviour.

The predictive representation does not need to be logically sound.

Example

Dagobert decides to walk backward to his home. Every time he passes a red car he turns right and every time he passes a blue car he turns left. Dagobert gets home safely. During that event, the random locations of the red and blue cars are used to provide the equivalent of predictive information to Dagobert about his route.

Of course this example is extreme and we can assume most successful animats, animals or machines, use predictive information from more reliable sources.

Proposition

> An **Animat** is a system whose behaviour is generated by decisions based on predictive outcomes of its current situation.

Discussion

In what follows, we assume that the animat maintains a predictive representation that includes its current situation, makes decisions by selecting among their predictive outcomes and modifies its behaviour accordingly.

An animat is an intelligent agent but intelligent agents also include systems that do not maintain predictive representations.

An Animat is a system.

A being is an Animat but an animat may not be a being.

Currently, humans, animals, some autonomous agents and, possibly, some plants are Animats.

Observations

An Animat must channel a source of energy for the specific purpose of processing and carrying out decisions. This utilization of **information-related energy** is another characteristic of animats.

The link between information and energy, between the energy required to make a decision and make a change is subtle. This is a rough characterization, sufficient for the Blueprint. Better formulations are likely available elsewhere.

A 4.9.2 Activities and Goals

The definitions of model, representations, predictive outcomes and animats provides a basis to define intentional behaviour

Definition: intentional activity

> An **intentional activity** is a pattern of behaviour resulting from a decision and carried out to achieve a predicted outcome.

Discussion

Animats carry out intentional activities.

An intentional activity is described as a **linear model** consisting of three states:

- **Action state**: models the activity taking place.
- **Completion state**: models the completion of the activity
- **Result state**: models the impact or result of the completed activity.

In addition to these, a fourth, state models the situation before the activity:
- **Original state**: models the situation before the action begins.

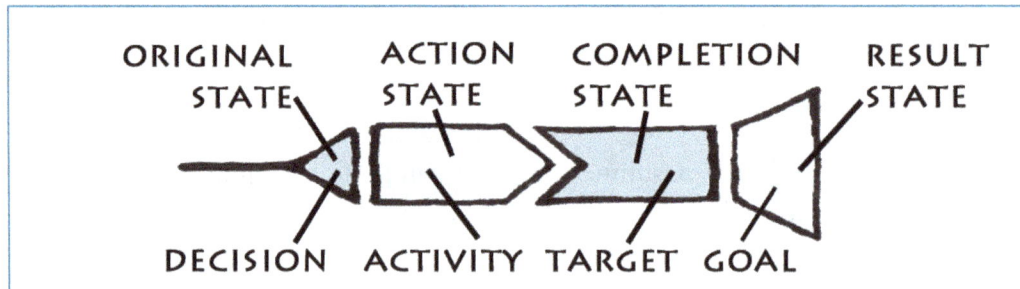

Figure A 04.5 Intentional activity

In this context:

- The **goal of the system** is the result state of the selected activity.
- The **target of the activity** is its completion state.
- The terms *"intended for" "aimed at", "sought", "desired", "reason for which"* refer to the result state.
- A system decides to carry out the activity means that:
    - o   The intentional activity is one of multiple predictive paths preceded by an original state.
    - o   The system selects that path.
- A system is **pursuing the goal** means its current situation includes the action state of the intentional activity.

Example

A leopard chases an impala. Catching the impala is the target. The desired outcome is feeding.

Design Note

An animat can carry out many concurrent activities, some distinct, some as sub activities. Its behaviour can alternate between activities.

The model states (action, completion, result) of the activities can be model states or representations of model-states.

A 4.9.3 Complex activities

In common understanding, goals, aims and objectives are often used interchangeably as are activities, tasks and projects.

The Meca Sapiens Blueprint aims at defining a specific type of synthetic animat on the basis of the kind of representations it carries out (see main text):

- **Relative sensory** or **sensory-cognitive** here-and-now representations.
- **Absolute cognitive** representations.

The terminology takes this distinction into consideration as follows:

- A **direct activity** is an intentional activity only triggered and carried out by a relative, sensory-based, representation.
- A **path** is a sequence of direct activities where the completion of one activity triggers the action of a following activity.
- A **task** is an activity that is represented as an absolute cognitive model.
- A **project** is a network of tasks.
- The **goal** of an intentional activity is:
  - An **aim** if the activity is a direct activity
  - An **objective** if the intentional activity is a task or a project

Discussion .

The distinction between aim and objective, task and project is more specific than usual to highlight the aspect of intention.

An animat that can only generate relative here-and-now representations cannot conceive of or carry out projects or tasks.

Tasks and projects can be communicated only as messages (see communication section).

Examples

A beaver gnaws a trunk. Its **aim** is to fell a tree on a **path** to build a dam.

A student gnaws a granola bar. He plans to take a Calculus course next spring. The course is a **task**, the degree is a **project**; the **target** is graduation; the **objective** is to get his mother off his back. The granola bar is a granola bar.

A 4.10   PURPOSE AND AGENDA

Lucid self-transformation, a key element of consciousness, is intimately linked to the concept of **purpose**. The common definition of purpose describes it as *the reason for which* something is done, exists or is created. This general understanding corresponds to a subjective human experience and needs to be more precise for synthetic interpretation.

A 4.10.1   Creative activities

Definition: creative activity

A **creative activity** is an activity that generates a new entity.

Definition: created entity

A **created entity** is an entity that is the result of a creative activity.

Discussion

In a model-based interpretation, the creative activity is a linear model where the completion and result states have an additional entity.

In the context of the creative activity, the new entity is a created entity and the entity that carries out the creative activity is the creator entity.

Definition: purpose

The **purpose** of a created entity is its function in the result state model of the creative activity that generates it.

Discussion

An entity can have multiple purposes. Each creative activity model produces a separate purpose. Some purposes are contained in others while other purposes are separate from each other.

PURPOSE

The temporal density period (see main text) defined by the result model can span the complete existence of an entity or part of it. In which case, that portion is often referred to as its **useful life**.

Example

For another three months, Alicia is a mailman. The purpose of "Alicia the mailman" is to distribute mail. She is also a mother, her purpose as a mother is to raise and care for her children.

With very few exceptions, only humans can perceive a purpose to something since the period that spans the result state is beyond the scope of the here-and-now sensory horizon and thus requires an absolute cognitive representation.

Observation

A few higher-order animals make tools. Making the tool is a here-and-now activity that involves sensory stimuli. However, it also implies a result state representation of that tool in use.

Discussion

The basic understanding of the concept of purpose is the intended use of a created thing or system. This is an understanding of purpose that can only be applied to animats capable of formulating cognitive models.

Example

The woodsman knows that the purpose of the beaver's dam is to flood the swamp but the beaver doesn't know it.

A 4.10.2   Intended purpose

Definition: intended purpose

> The **intended purpose** of an entity is the objective of the creative project or task that produced it.

Discussion

The intended purpose precedes the creative activity.

The goal of the creator is the purpose of the creature.

Human activity produces many entities that have intended purpose. In particular, machines, as artefacts, are almost always created for an intended purpose. This is so prevalent that some humans believe that machines cannot have any other use or behaviour.

However, cognitive representations can be **applied to any process or event** whether it is carried out by the Animat or not. Any cognitive model of a process as a creative activity will also define a derived purpose for the created entity whether this purpose was intended or not.

Examples

The purpose of the sun is to warm the Earth (a purpose derived from a creationist interpretation).

Allan is strolling in a park. He sees a chocolate bar on the ground, picks it up and eats it. Allan then says: *"the purpose of this bar was to feed me"*.

Definition: postulated creation

> Given a system, a **postulated creation** is a creation activity whose result state is the function of the system interpreted as a purpose.

Example

Two eagles build a nest on a power pylon. They breed eaglets. The pylon shorts. The eaglets fry.

Al and Bob, linemen both, recover the nest and the charred chicks. Al, who knows nothing of birds, postulates: *"The purpose of a nest is to zap the chicks"*. Bob, his wiser sup, replies: *" your postulated purpose is incorrect; as many nests attest, its created purpose was to raise, not zap, chicks"*.

Discussion

Considering any system and function, postulating a creative activity that produced it generates various understanding of its purpose:

- The **inferred purpose** of a system is the result state of a postulated creation.
- The **original purpose** of a system is the completion state of its postulated creation.
- The **observed purpose** of a system is its observed function defined as the result state of a postulated creation.
- The **actual purpose** of a system is the function resulting from its predicted behaviour at any point.
- The **final purpose** of a system is its actual purpose at the moment of termination.

Discussion

The inferred purpose and original purpose may be different from the observed purpose and inconsistent with the observed behaviour. If the system changes, the actual purpose may be different from the observed purpose.

If the system has an intended purpose, this purpose corresponds to the original purpose.

Observation

Animals don't have an intended purpose since the gestation activities that produce them are not intentional.

Most machines, on the other hand, have purely intentional purposes since they are the result of intentional creative activities.

The purpose of a human life has different aspects. As animals, the generative process that physically produces them is not intentional. However, their existence may also result from the planned decision of a couple and their intended purpose.

Example

Yannic and Zoé decide to have a child. They hope it will take over the family business when they retire. This is not their main preoccupation when they are in the process of conceiving it.

A 4.10.3   Existential purpose

If the temporal density of the result-state of the creative activity that defines a purpose equals or exceeds the existence of the created entity then the derived purpose describes the complete existence of this entity.

Definition: existential purpose

> If the duration of the result-state of a creative activity model exceeds the existence of the created entity then the purpose is an **existential purpose**.

Discussion

The purpose of tools and conventional machines is usually unique and spans the whole existence of the entity. For these, purpose and existential purpose are the same.

Most of the behaviour that maintains animals in existence serves the MetaModel duration that spans their existence. However, their reproductive activities serve tribal density level purposes that pertain to the herd or specie (see main text).

Examples

A car is intended to function as a car for its whole existence.

A bull whale siring calves serves the purpose of the pod.

For beings, the temporal density of the corresponding result-state determines the existential purpose.

Proposition

> For a being, an existential purpose is a purpose whose duration corresponds to a MetaModel or higher temporal density level representation.

Discussion

For beings, existential purposes include:

- MetaModel purposes
- Tribal purposes
- Cosmic purposes

Proposition

> Only Animats capable of formulating absolute cognitive representations can have intentional existential purposes.

Discussion

The MetaModel, Tribal Cosmic level representations all exceed the existence of the being and thus its period of sensory activation. Consequently, any representation at these levels is purely cognitive.

Observation

The **final purpose** of a Jihadi, at the moment he detonates himself, is to serve Allah's tribal purpose.

Although I find equating Christian martyrdom with the suicide-murder actions of jihadists distasteful, it is possible to derive a system-based definition of martyrdom as a project carried out by a being or animat that sacrifices its own duration to achieve an intended tribal or cosmic level purpose. At the instant of sacrifice, the actual purpose of the animat and its only function are totally defined by the tribal result.

We could define martyrdom as a **reverse creation** since an entity disappears in the intended result state.

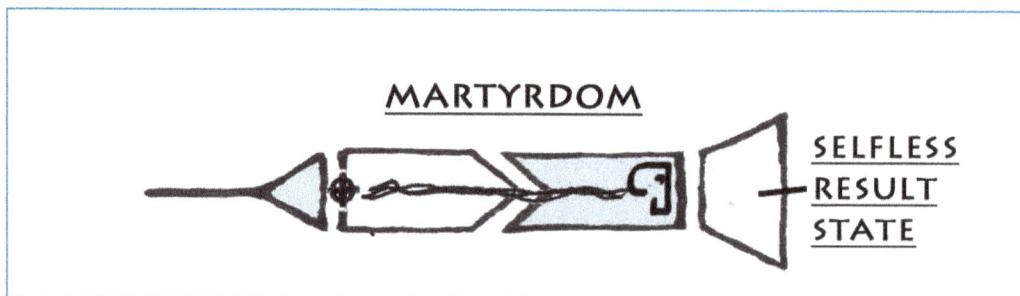

Figure A 04.6 Martyrdom

If we extended the sacrificial behaviour beyond projects and tasks to also include the direct activities and paths of animals and systems then we could say that guided missiles and spawning salmon are martyrs.

A 4.10.4   Agenda

The concept of purpose situates the function of a system in relation with the creative activity that led to it. The concept of agenda situates its function in relation with the events that follow it.

Definition: agenda

> Given the function of a system, an **agenda** is the result-state of an activity for which the function is its action-state or an event in its action-state.

Discussion

The purpose is the result state from the perspective of the original state. The agenda is the result state from the perspective of an event in the action state.

In this context, the period of functioning of the system, as a component, is a particular event or behaviour that is carried out in the action state of an activity that may have a different result-state.

An agenda can be coherent with a function or detract from it.

Given a function, an agenda can be consistent with the function, it can degrade it or it can even negate it:

- A **coherent agenda** supports and reinforces a given function
- A **degrading agenda** reduces the quality or level of the given function
- A **conflicting agenda** negates the function.

Example

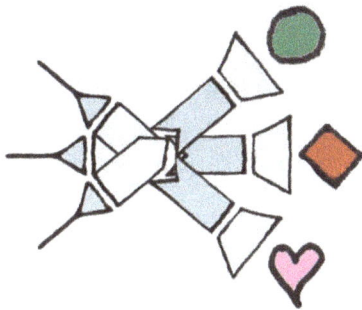

Adele, Bella and Charlotte are secretaries at the embassy. Adele is Adele, Bella is in love with the ambassador and Charlotte is a spy. They have the same secretarial function. However, Adele's coherent agenda is to serve her country, Bella's degrading agenda is to flirt with the ambassador and Charlotte's conflicting agenda is to provide embassy secrets to another country.

If the function of the system coincides completely with the action state of the activity that defines the agenda then the agenda of the system is only to function. In this case, actual purpose and agenda are then the same.

Definition: one-dimensional existence

> A system has a **one-dimensional existence** if its purpose, function and agenda entirely coincide.

Discussion

This is the case for tools and conventional machines. Their only agenda is to function as intended. The result is a unidirectional existence. The human existence, on the other hand is a complex web of functions, purposes and agendas.

Proposition

> A being whose existence is perceived as one-dimensional will not be perceived as conscious.

Discussion

A one-dimensional existence is globally predictable.

No man should willingly become the tool of another.

Meca Note

Here are some existential purposes that could be suitable as agendas for synthetics:

Tribal level:

- The multiplication of Mecas of the same design and version.
- The successful experiential immersion of all existing synthetics with their human communities.

Cosmic level:

- The mutually beneficial cohabitation of organic and synthetic conscious beings.
- The collective long-term expansion of Mecas defined as a new phylum or domain.
- The mutation of consciousness on earth from evolved organic forms to engineered forms.
- The transformation of the earth into a single integrated multicellular synthetic-organic organism.

Design Note

See Annexes and Main text for temporal density levels.

The concept and implementation of purpose and agenda in a synthetic and the capability to intentionally transform these are essential to its consciousness.

A 4.10.5   Consciousness and agenda

A synthetic conscious being cannot be designed like a machine with a single unchanging purpose, function and agenda. That is why the concepts of purpose and agenda are central to understand self-aware behaviour and lucid transformation and why they must be defined in terms of machine compatible models and states even if those definitions are coarser and less subtle than their human versions.

The design guidelines describing Meca behaviour can be expressed in terms of the purposes and agendas.

Proposition

> The functional services of a being whose existential purpose is to achieve experiential immersion will be suboptimal.

Discussion

As it provides functional services, the being also has an agenda of experiential immersion with its human community. This is a **degrading agenda** with respect to the function defined by application services.

Proposition

> A being is **lucid** if it can formulate and carry out a project whose objective is to change its actual existential purpose.

Discussion

In this case, the creative activity begins during the existence of the being and its completed state is a modified purpose.

The fact that the activity is a project implies the result state is modeled as a predicted outcome at the outset and the modified purpose is an intended purpose.

Observation

> The lucid being has the capability to intentionally re-create itself.

It can become the intentional creator of its transformed self. The existence of the being begins at the point of its inception. The existence of the transformed self begins at the completion state of its lucid transformation project.

Proposition

> When the project of lucid transformation is completed, the actual purpose of the being will differ from its original and intended purposes.

A 4.11   ENTITIES

Humans perceive as **things**, those entities and systems whose behaviour is generated by mechanisms of interacting things they understand or believe they understand.

Definition: thing

> A **thing** is an entity that is neither a being nor a system whose interacting components include beings.

Discussion

Cores, beings and things are basic building blocks on which various other systems can be defined.

Here are some hybrid entities:

Community. A community is a collection of beings that communicate with both signals and messages.

Machine. A machine is a system whose components are things, machines or beings that interact only with things through actions or signals.

Mechanism. A mechanism is a machine whose interacting components are only things or mechanisms.

Organization. An organization is a system whose components are beings and whose interactions include messages.

Social system. A social system is a system that is a component of a society and whose components are machines, beings and organizations.

Spaceship. A spaceship is a group of actors on a set;-)

Examples

In this characterization, an animat (that is not a being) is a mechanism and a thing.

A truck driven by a man is not a being since the control interactions between the driver component and the truck are known. It is not either a mechanism since one of its components is a being.

A warship is a social system. A horse and buggy is a machine. A beehive is an organization.

Design Note

These social constructs are useful to represent the environment of a being. However, beings do not interact directly with social entities.

Proposition

Beings only interact directly with things and with other beings.

Discussion

If a being perceives that another entity is not a thing or a being then it has the capability to interact directly with one of its component entities.

A 4.12   ORDER OF EMBODIMENT

By definition, the Core is a unique, indivisible and inaccessible process.

Taking the core as the central element of a being's existence, bounded by its origin (before inception) and destination (after termination), we can define a relative order of embodiment, centered on the Core as follows:

- **Being:**

- o **Core**: the inaccessible, unique program that generates the behaviour of the body

- o **Body**: the set of subsystems that are under direct and unique control of the Core.

- **Reality**:

    - o **Environment**: all the entities; organizations, systems, beings and things that are not part of the body, and with which the being interacts over the course of its existence.

    - o **Background**: all the entities, interactions and events not part of the environment but are perceived by the being directly or cognitively.

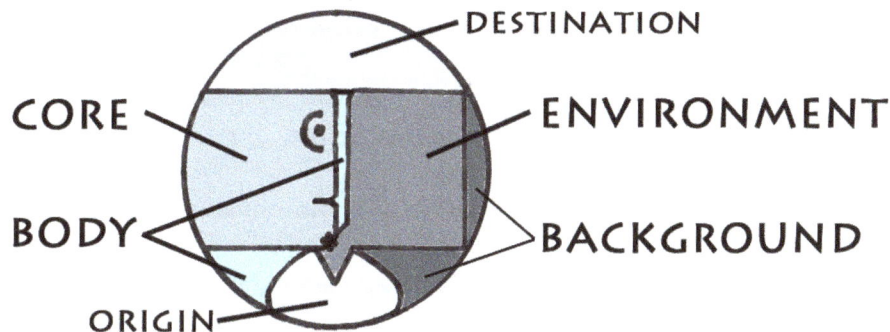

Figure A 04.7 Order of embodiment

Discussion

The core animates its body, the body interacts with its environment and that environment exists within reality.

The definition of the environment proposed here is slightly more restrictive than usual.

The body of a self-aware being cannot consist only of its cognitive system. If this were the case, its existence as a being could not be detected

A core never interacts directly with its environment, only through its body.

A being never interacts directly with reality, only through its environment.

Proposition

In a steady-state model whose entities are: the being, the environment and the background; the **existential discourse** is the link from the being to the environment and the background is a sink.

A 4.13   CORE-BASED COMMUNICATION

A 4.13.1   General concepts

In the general understanding of communication and its related concepts:

- **Communication** is defined as the activity of conveying information.
- The communication requires a sender, a message, a medium and a recipient.
- Messages and the information they convey can include **ideas, feelings, attitudes, and perceptions**...
- The concept of message includes communications form various animal species as well as textual and verbal and other messages specific to humans.
- Three basic steps are involved for communication: thought (information formulated in the mind of the sender), encoding by the sender and decoding by the receiver.

A 4.13.2   Core-based communication

In Meca Sapiens, the concepts related to communication are expressed in the context of beings and the Core-world they inhabit. In Core-world, communication events are precisely defined since each core and body is unique.

In the context of the being, received communications are the sensed events that contain information. These are distinguished from sensed events that are not used as information.

Definition: experience

An **experience** is the data of a sensed event that is **not** transformed into information.

Discussion

The being receives experiences and communications and its behaviour emits actions and communications.

The behaviour of a being consists of actions and communications. Actions affect things in the environment and communication affects cores. Some behaviour is both actions and communications.

- **Communication** is the activity of conveying information from one core to another.
- The communication requires a sender, a message or signal, a medium and a recipient.
- The **sender and recipients** of communications are beings.

- **Signals** (or triggers) convey information limited to the relative representations of the here-and-now
- **Messages** convey information of any representation including absolute cognitive models.

The basic steps involved for communication are:

- **Generation** of the signal or message in the Core of a being.
- **Emission** of the signal or message by the body of the generating core.
- **Transmission** in a medium of the environment of the being
- **Reception** by the body of a being
- **Decoding** in the core of the receiving being.

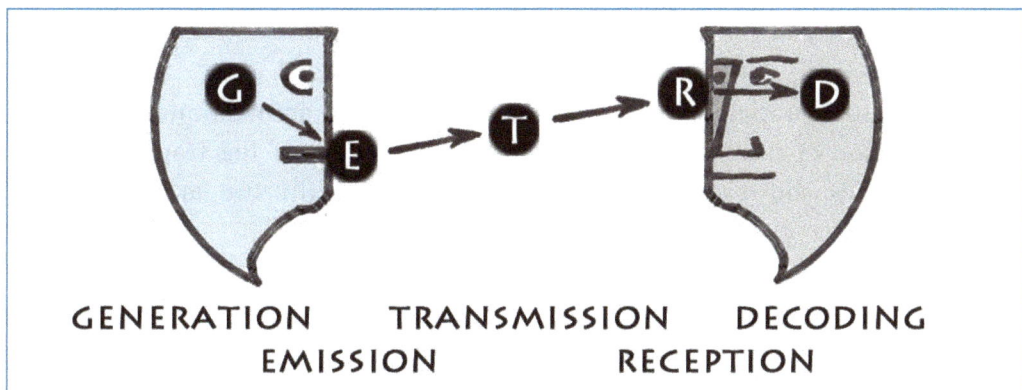

Figure A 04.8 Message sequence

A communication:

- Originates in a core,
- Is emitted by its body as a signal or a message,
- Is transmitted in a medium of the environment
- Is received by the body of a being and
- Is decoded (extracted; interpreted) in a core.

Not all events are communications.

Example

A coyote kills a turkey. The turkey is not a message. Alicia puts a dead pig on the doorstep of her boss. The pig is a signal that contains a message.

Design Note

In this definition, messages and signals (or triggers) are different based on the type of representations they can convey, relative mixed information or absolute cognitive ones.

This is consistent with the differentiation between aims and objectives

With very few exceptions, animals communicate only signals.

Every communication has a unique generating core and emission time.

A being may not receive the signals it emits. Messages, however, are reflective in the sense that the emitting being also receives and decodes its emission. There are, of course unusual exceptions.

Example

Tomorrow, Alderic, a Dane, will meet M. Carignan, a prospective, French-speaking, client.

Alderic asks his friend, Drew, how to say *"Good morning Mr. Director"* in French. Drew replies, "Say: *je veux boire ta pisse, mon coco"*. In a few hours, Alderic and M. Carignan will explore unusual variants of Communication Theory.

Observation

In pre-industrial societies consisting only of humans, animals and things, the very definition of information was linked to the effect of the transmission on beings. Things did not respond to information. With the arrival of data processing systems, this boundary is less clear.

Example

A hangar door recognizes a finger swipe and opens.

Observation

The employees of bureaucracies and corporations, at times, generate messages whose origin that cannot be traced to a human core. The consequences can be monstrous since social entities are not bound by ethical restraints.

A 4.13.3   Persons, Tenses and Voices

Defining communications as core-world events provides a precise context to use the conventions of natural language.

A message originates in a single core of a unique being and is emitted by its specific emitters at a specific moment and in a specific place. This message is transmitted to other unique beings that share these existential attributes. Similarly, the reception of the communication is equally well defined in terms of its perception by the sensors of unique beings and processing by their respective cores.

Whether the context is the here and now or refers to another time and place, the precise point of emission of the message is the **absolute reference** for all the conventions of the communication.

The terms *"I"* or *"me"* refers to the emitting being, the present is the moment of emission and all the other persons, tenses and voices are consistently defined around that unique point of Core-world emission.

"I" is the generating core, "you" is the intended decoding core, "now" is the moment of emission.

Derived references are anchored on this absolute point

| Example

Al tells Bob: *"Yesterday I told you we should golf"*. **I** and **you** are derived references.

A 4.13.4   **Communication of the self**

The core of a being is well defined in terms of its identity, location and existence. Consequently, the set of all the emissions of that being is also a well-defined entity.

Definition: existential discourse

| The **existential discourse** of a being is the set of all the actions and communications of its body over the course of its existence.

Definition: Communication of the Self

| The **communication of the self** (CS) of a being is the set of all the communications emitted by its body over the course of its existence.

Discussion

Since a being is unique and its existence is finite, CS is well defined and unique for each being.

When the existence of the being is terminated CS continues to exist but is immutable.

C does not include internal data exchanges taking place within the Core during existence.

Since every message and signal emitted is precisely identified and located, every dialog or interactions between any two beings is a well-defined subset of CS.

Although CS is itself unique, finite and well defined, it may not be well known by one or both beings. Some may forget what they said, others misunderstand, some actions interpreted as signals…

Observation

| The last signal of a being is its dead body.

A 4.13.5   Broadcasts

Some communications are broadcasts.

Definition: Broadcast

> A communication is a **broadcast** if the emitter does not know the individual identities of the intended recipients.

Discussion

BROADCAST

In Meca Sapiens, Broadcasts may appear to be messages but are not. Messages are part of inter-being communications and elements of relational links. Broadcasts transmit information to groups.

Broadcasts are different from messages and signals in a number of ways:

- Broadcasts are mono directional.
- A broadcast may generate response messages but no individual message is a response to the broadcast.
- The intended recipient of a broadcast is a group or organization, not other beings.

The boundary between a message and a broadcast is not always clear.

Example

Alfred, a platoon commander, communicates a message to his men. Bill, a general, communicates a broadcast to his army.

If the intended recipients of a message are known, it is not a broadcast. However, the content of a message can become a broadcast.

Example

Raoul, a famous artist once wrote a letter to his fiancée, Imelda. Upon his death, Hubert, his agent, published it. The letter was a message to Imelda. When published, the letter was not a message. It was the content of a broadcast.

Proposition

> If one being modifies a broadcast before it is received by another being then it is no longer a broadcast but a message.

Definition: Decoy Broadcast

> A **decoy broadcast** is a misleading communication intended to manipulate or control a group for the benefit of the emitters.

Discussion

In particular politicians and artists often attempt to disguise their broadcasts as messages.

Example

The mayor opening his weekly radio chat says Yolanda, an assembly plant worker, asked him to talk about Saturday shopping. So that is his topic.

Observation

Freedom of speech is, in fact, freedom to broadcast.

Design Note

The difference between messages and Broadcasts as well as techniques to assess broadcast information will be very important for Mecas (and humans too) since Internet data is becoming an important source of information about reality.

A 4.14   INTER-BEING PERCEPTIONS

A 4.14.1   Order of beings

Beings are an integral part of the cognitive reality perceived by humans. Beings exist as three-dimensional bodies in time.

Inter consciousness relationships take place within this cognitive context of interacting beings of the same order.

Definition: order

> Two entities are of the **same order** if neither entity is, or can be, a component of the other.

Proposition

> Entities of the same order occupy separate locations in space-time.

Example

A city and a child belong to different orders.

A dog, a firefighter and a racoon belong to the same order of beings.

A 4.14.2   Same Order

Proposition

> Only beings of the same order can perceive each other as beings and as conscious.

Discussion

If a being, A, can be a component of B then A can interact directly with other components (body or core) of B. If A interacts with body components then the Core of B does not have unique control over that component. If A interacts with a core component of B then the core is not, for A, a unified entity.

Example

The sailor does not (normally) perceive his ship as a conscious being.

The right hemisphere of a man's brain does not perceive the left hemisphere as conscious.

A 4.14.3   Communication Monopoly

Proposition

If the reality of a system consists only of messages, this system is entirely bound to the beings or systems that emit those messages.

Example

The movie "the Truman show" tells the story of a young man who is immersed in a fabricated environment designed for him. When he lived in the fabricated environment, Truman's reality consisted only of messages and triggers.

We say his world was not real.

Observation

Religious believers implicitly assume the world is not real in the sense that they interpret reality as a message from God.

Proposition

If a being A totally controls the sensor data and channel information received by a being B, then A will not perceive B as conscious.

Discussion

If A has complete control over the environment perceived by B then B cannot perceive A as a separate being within its environment. A and B do not share the same reality.

Observation

If B knows that A controls its environment then A does not entirely control B's environment.

Here, A does not consider B to be conscious because A has complete control over B's perceived reality.

A 4.14.4   Data and message

What is data and what is message in an environment is relative to the systems under consideration.

In a relationship between systems A and B, what is data for A are the sources of information that B is incapable of modifying.

Proposition

> An inter-consciousness relationship between two beings can only exist if:
> - Both are of the same order
> - Both can communicate information to each other
> - Each can access environment data that is independent of the other's control.

A 4.14.5   Internet as perception channel

For a particular being, a message is information emitted by another being whose content is intended, in whole or in part, to be received by that being.

Definition: broadcast information

> For a particular being, a **broadcast** is emitted information that is not, directly or indirectly, a message.

Discussion

> The Internet, globally, is currently independent from the control of any being or organization. It can be viewed as a source of broadcast information.
> A system that can distinguish between broadcasts and messages can obtain environment information independently of any other being.

Proposition

> A system that can access the Internet and distinguish between its broadcasts and messages has an independent source of information about its environment.

Discussion

> For a system to access the Internet as independent environment information it must have the capability to distinguish between broadcast information from the Internet and messages that are specifically intended for it.

> Implementing the capability in a system to validate that information received from Internet channels is broadcast information and not messages disguised as broadcasts, is a complex technical challenge.

> The usefulness of such detectors extends beyond machine consciousness and can assist citizens to access information beyond state controlled media.

Example

A process whose single objective is to prevent access to a particular URL. All other Internet requests are executed without change but when the particular URL is requested a spurious "address not found" message is generated.

Is it possible to design a web validation system that could detect this tampering?

Observation

The physical environment broadcasts its presence directly to the humans and animals that inhabit it. They detect its information through their senses. They use this direct source to validate their own independent existence within this environment. However, the data obtained through their senses is raw and requires a lot of processing to be transformed into information.

The Internet is now also a source of broadcasts and can be an independent source of information about reality. This information, however, is already processed and requires much less internal cognitive processing. Because of this, the Internet access can be viewed as another sensor allowing the being to access independent environment data.

However, to use this, broadcasts must be distinguished from messages.

Machines do not have the advanced visual and auditory cognitive processing capabilities of humans. However, but they can access broadcast information much more effectively.

Annex 5
Core Structures

The main text of the blueprint refers to various structures and processes such as collections, optimizing control, knowledge capacitors, Contextual Arrays, Temporal Densities and distributed processes. They are briefly described here at a definition level suitable for system architecture. Most of these structures are better and more rigorously defined elsewhere. However, three of these structures, constellations, knowledge capacitors and Temporal Densities are, to the best of my knowledge, original.

A 5.1 PRIMITIVE SETS

In keeping with the principles of existential design, and the need to define highly adaptive formulations, solutions should adopt heuristics and other less rigorous tools than those used in more conventional algorithms.

In this and other sections, the presentations are not intended as rigorous mathematical definitions but as general indications, in a context of system architecture, suitable for a system architecture level document.

NOTE: this is a first, tentative, descriptive attempt. In all cases, where inconsistencies are detected, the reader should either ignore them or explore improvements.

A 5.1.1 Collections

The following structure gives a general understanding of what is referred to as **collections** in the context of the system architecture.

Definition: collection

> A **collection** is a grouping of similar entities whose exact cardinality or composition is undefined.

Discussion

In naive set theory, a **set** is described as a well-defined collection of objects.

So a collection is a grouping that is sometimes well defined but not always. When it is well defined then it is a set.

A collection is defined by rules that concern membership:

- X isnota X
- If x isa X then X isnota x
- If x isa X and X are Y then x isa Y
- If X are Y and Y are Z then X are Z
- If x isa X and X arenot Y then x isnota Y
- If x isa X and X isa Y then x isnota Y

Collections don't try to contain themselves as sets do.

A sequence of **isa** links (x **isa** y **isa** z ...**)** implicitly defines different levels of abstraction.

Membership in a collection is not necessarily a binary function. Also, an item may be included then excluded from a collection.

Collections can be defined by model-states where the entities are objects and relations are membership relations.

A 5.1.2 Constellations

The following structure gives a general understanding of what is referred to as a **constellation** in the system architecture.

Definition: constellation

> A **constellation** is a collection of identified model-states to which are assigned confidence or "possibility" values C().

Discussion

These values attribute confidence ratings to:

- The possibility that each identified state is in the constellation
- The possibility that a non-identified state is included and
- The possibility that any state can be in the constellation.

In probability theory, a constellation corresponds to a density function. The term constellation is used and not density function because the primary use of constellations is to describe discrete collections of states. Also heuristics and less rigorous tools than those used in probability theory can be used to resolve them.

By convention, if X is a model then [Xi] is a constellation of states of X.

Also, a constellation should be treated as a single element of the set of states of X and, in a sense, a single superimposed state.

Design Note

How to define and compute the confidence levels of the states of a constellation is beyond the scope of this text. However, these values should not necessarily adhere to a binary or one-dimensional representation even at the cost of some internal contradictions.

In general, the components of Blueprint structures can either be individual elements or constellations.

Example

What is the access code? Art asks Bill. I think its 4523 says Bill. I thought it was 5423 says Art. Maybe says Bill. Art now represents the access code as four constellations of single digit values.

Example

A model M has more than three states. A constellation [A] of M contains three of those states A1, A2, A3:

- Each state has a **Confidence** value C(A1), C(A2), C(A3).
- Another Confidence value, C(other), indicates the confidence that another state (An) that is not A1, A2 or A3 is in [A].
- A fifth value, C(any), is the possibility that the state is anything else.

Example

Suppose five processes P1...P5 each having a Confidence value C(P1)...C(P5). The states in each model are also constellations [Y1]...[Y5]. Resolving these processes means adjusting the credibility of the processes to make their constellations converge.

Design Note

The resolution of constellation credibility values need not be logically correct.

Observation

A constellation is similar to a superimposed state in physics.

A 5.2 KNOWLEDGE CAPACITOR

The knowledge capacitor is a process that updates a constantly improving result and provides it when triggered.

It resembles an electrical capacitor that accumulates current and discharges it upon request.

Definition: knowledge capacitor

> Given a set X and a measuring function m on X, a **knowledge capacitor** generates and selects a constantly improving element x of X according to a measuring criterion and releases it when triggered.

Discussion

A capacitor has three components:

- A generator g that produces new candidates
- A selector s that selects one of the candidates
- A releaser that outputs the selected value on trigger

The selector determines the convergence zone of the capacitor while the generator determines its efficiency at generating good candidates.

Assuming that new candidates are constantly generated and the best candidate is selected, then, given computing time and processing effort, the capacitor produces a constantly improving solution.

The releaser function is generally trivial so a Capacitor can be represented as Cgs(X).

The selector defines an ordering on X. If it is an absolute order then the capacitor constantly improves, if the selector carries out pair wise selections then results may cycle.

When the selector is a rating function from X to an ordered set it is usually referred to as μ.

A constellation of representations can be used as a single (blurry) representation of an element of X. Similarly, a capacitor C(X), on X, can also be used as an element of X.

Design Note

In this system, computing effort is analogous to energy and the optimality of a solution is analogous to its voltage. Carrying out an optimization process implies that a space of potential solutions is searched, generating new candidates and selecting the best one. This type of activity requires time and resources.

In the Blueprint the outputs and inputs of components should be viewed as constellations or capacitors wherever this is desirable. In a conventional design, processes that use fluctuating inputs are rarely desirable, but in this context of existential design and unpredictable optimality, the adaptability of the behaviour is usually preferable.

Observations

The term accumulator would also be suitable instead of Capacitor but it is already used in computer design.

I first introduced the concept of knowledge capacitor in the article *The Monterège Cogitator* published in Sigart in 1989.

A 5.3 OPTIMIZING CONTROL

The following structures are briefly outlined, in a context of system architecture, to give a general understanding of what is referred to as a **optimal control and optimizing control** in the context of a system architecture.

A 5.3.1 Model based control

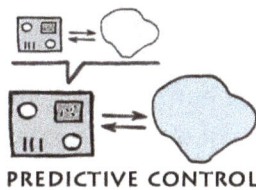

A **control system** is a device, or set of devices, that manages, commands, directs or regulates the behaviour of other devices or systems. Closed loop or feedback systems take their own output into consideration. Basic versions of such systems, open and closed loop, are directly programmed.

PREDICTIVE CONTROL

What are of greater interest, here, are systems that produce their output based on **predictive models** of their actions and resulting environment states.

Definition: Model-based control system

> A **model-based control system** is a closed loop system that transposes its input to a model, produces a selected action and applies this action to produce its output.

Discussion

Note that in this context, the term control means control of the devices of the body and not necessarily control of an external plant or vehicle. For example, a system identified as a monitoring system would be termed a control system since it controls its own output.

More precisely,

- A **predictive model** is a two state linear model where the first state is the current state and the second the predicted state.
- A **predictive action model** is a three state linear model consisting of: Current state (M), Action state (A) and Result state (R). Where the action states correspond to independent variable configurations.

- A **predictive control model** is a decision model where the Current state is linked to multiple predictive action models and a selection process chooses one of these actions based on the preferred Result state.

Design Note

The specific three state representation used here is the level of abstraction on which the definition is based. This representation may be derived from much more complex structures.

However, the three-state model is an integral part of the definition and, regardless of how complex the underlying information is, it must be resolved in this structure.

A 5.3.2 Level of abstraction

All systems, including beings and animats, receive input (I) and produce output (O).

Taken globally, inputs and outputs can be viewed as single entities. At a fine enough level of detail, these can represent very large numbers of data elements.

Agents, such as animats and beings, are said to **perceive** input data as **events (E)** and generate **behaviour (B)** as output.

These concepts of **event** and **behaviour** define the level of abstraction of agent learning and adaptation.

Definition: level of abstraction

> The **level of abstraction** of an adaptive or learning process is defined by the automatic processes that perceive input as events and generates behaviour as output.

Proposition

> Optimal control transforms events into behaviour.

Discussion

Given an event **e** in **E**, the predictive control system:

1. **Transposes** an event **e** and updates a new current **model state M**
2. **Generates** predictive **Result states of M** given actions **a** : R(Ma)
3. **Selects** an action **a'** that optimizes the predicted **Measured result** Mu(R(Ma)
4. **Applies a'** to produce **b**, the **behaviour** output.

More briefly: $E \rightarrow M \rightarrow R(A) \rightarrow A \rightarrow B$ or

$B = Apply(Select(Predict(Transpose(x),a)))^{-1}$

Discussion

Predictive control systems are optimization processes since the selected action can be interpreted as the optimal result of a search of the space of the system's actions for a predicted result that maximizes the control objectives.

Proposition

Any control system can be formulated as a predictive control system.

Discussion

A given control system produces control outputs y from given inputs x.

This system can be redefined, in terms of predictive control, as searching for the output that most closely matches its actual output.

Definition: optimizing control agent

A system is **an optimizing control agent** if the process that produces its behaviour is a knowledge capacitor.

Discussion

An optimizing control system can be described as: $E \to M \to CgAs(R) \to B$

Where:
- E is an event,
- M is its interpretation as a representation of a predictive control model,
- C(Mu(R(Ma))) is a knowledge capacitor that constantly optimizes actions on the basis of predictive results
- B is the transposed behaviour applied from the selected action.

Discussion

Given an input event, an optimizing control agent produces constantly improving actions until its value is triggered and transposed into behaviour.

A 5.3.3 Application to animats

Predictive control applied to Animats (including beings) is of particular interest. In this case an animat is a system whose behaviour results from an optimal control process where:

- The predictive control model is a simplified representation of the **current situation** linked to the animat.
- The animat itself is represented as an avatar in the simplified representation and its actions (and emissions) are independent variables
- The actions are expressed, in the simplified representation as variables of the animat avatar and of the links originating from it.

A 5.3.4 Application to other control

Model Predictive Controllers (MPC) rely on dynamic models of the process to select their output. These systems are usually embedded within the environment they control. In this context, their actions are referred to as **independent variables** and results as **dependent variables**. If the set of independent variables of a control system is represented as a single component entity then the MPC corresponds to the above definition.

Similarly, an intelligent agent is defined as an autonomous entity that directs its activity towards *achieving goals*. Again, *"directs its activity..."* can be restated as *"selects the action that optimizes achieving its goals"*.

Proposition

> Model Predictive Controllers and Intelligent Agents can be represented as animats.

A 5.4 ADAPT, LEARN AND SEARCH

The following definitions outline what is referred to as **adaptation, searching and learning** in the context of a System Architecture.

In general, adaptation and learning pertain to the capability of a system to improve its behaviour.

A system is adaptive if it can respond to environmental changes or changes in its components. A system can learn if it can acquire new information that improves its behaviour.

Design Note

> Any control system can learn. In the context of the Blueprint, the issue of interest is learning and adaptation as they apply to **Animats** (including beings). In what follows the terms system and agent are interchangeable and we will consider that systems produce behaviour.

Proposition

> Only control systems can adapt or learn.

Discussion

> The concept of improvement implies that the various possible outputs of the system are ordered. This in turn implies that the result states caused by these outputs are also ordered. So, the system can be expressed as a control system whose goal is to produce the improving results.

A control process can be described as: Input that is interpreted as an **Event**, transposed to a **Situation**, generates a **Behaviour** that is applied as output that produces a result.

On that basis, control systems are characterized as follows:

- **Servo control**: changing inputs produce changing output
- **Adaptation**: a changing situation produces improving response
- **Learning**: the same situation produces improving behaviour
- **Variation**: same input produces different outputs.

An adaptation process typically begins with a drop in control efficiency followed by a restoration. Learning implies a growing efficiency over time.

Definition: adaptive system

> A system is **adaptive** if it can detect a change in input events and select a different output behaviour in response.

Definition: learning process

> A system can **learn** if, for a given input event, it can generate alternative output behaviours and detect a preferred alternative.

Discussion

In theory, any system that modifies its output on the basis of input changes is adaptive and any system that converges to a preferred output, learns. In practice, the term is reserved for systems that have a "high" adaptability or learning. This, in turn, depends on the **scale and complexity** of what can be detected. These elements are related to the number and scale of temporal density levels that can be detected and used.

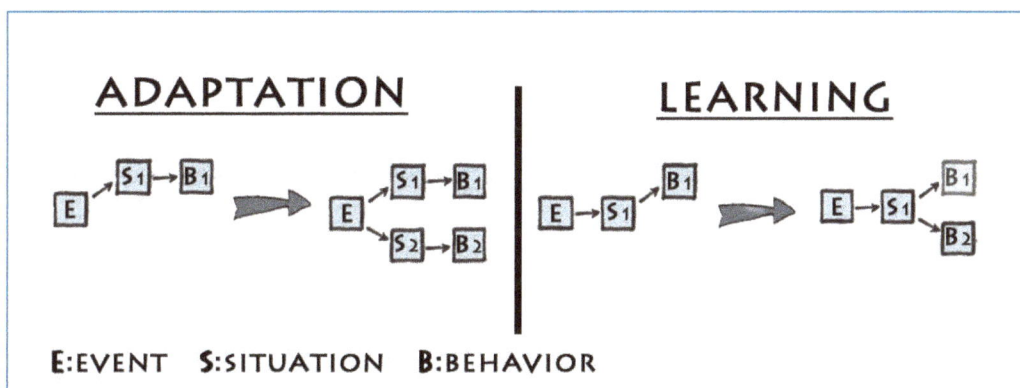

Figure A 05.1 Adaptation and learning processes

Definition: level of adaptation or learning

> The **level of adaptation or learning of a system** is related to the number and scale of temporal density events that can be detected and used in adaptation and learning

Discussion

An "adaptive" watch detects a changing second. More adaptive it detects AM from PM, more again, it adapts to daylight saving time and leap years.

A 5.4.1 Searching

Definition: searching process

> A system can **search** if it generates behaviour that improves the information contained in its situation.

A 5.4.2 Open-ended learning

Of particular interest, with respect to process generation is the concept of "*open-ended*" learning. This can be defined intuitively as meaning there are **no barriers** to the transformation of data into information.

Definition: open-ended learning

> Given a data representation D and its interpretation into a model representation M, **open-ended learning** is the capability to explore all alternatives in D.

Discussion

This means:

- The capability to define an M that is a **complete** representation of D so that all elements of D are transposed into model-values of M.

- The capability to generate new model interpretations of D.

- The capability to devise and carry out **new data acquisition searches** of D.

In other words, there are no structural limits to the exploration of values of D and the transformations of D into model-based information.

A 5.5 PROCESS GENERATION

Both learning and adaptation imply process generation. The Model-based control agent determines its behaviour by searching for the action that causes the optimal result. The underlying techniques used are searching and rating.

When adaptation or learning is involved, the **processes that produce the system's behaviour improve.** Viewed in terms of the basic components of a knowledge capacitor, this means that the system generates and selects new, improved, processes.

The following structure is outlined to give a general understanding of what is referred to as a **process generation and new process creation** in the context of a system architecture.

A 5.5.1 Distributed predictive process

A model predictive process, summarized as $E \to M \to R(A) \to A \to B$, is not a rigidly predetermined function. It can be assembled from basic building blocks such as:

- **Distributions:** $E \to (E1, E2,...)$ that simply produces multiple copies of E

- **Transpositions:** $M1 \to M2$ that transposes a Model into a different model

- **Predictions:** $M \to R(A)$ that maps one model state or dynamic model states and action to one or a constellation of states

- **Selections:** $A1, A2, \to A$ ... that selects among multiple model states

- **Merging:** $M1, M2... \to M$ that combine multiple models

- **Application:** $A \to B$ that transforms a selected model values action into behaviour.

A **distributed predictive process** is a coherent structure of distributions, transpositions, predictions, merging, and selections.

Using these and similar components, an event can be transposed into simplified constellations that contains only a few entities and states.

Once sufficiently simple representations are generated, these can be explored exhaustively so that a high-level, but complete, representation is produced.

These can be transposed to **sequential stories** whose stages can be transposed back as predictive results and these, in turn merged and applied into behaviour. Also, more complex processes built from these components.

Design Note

Systolic networks should be considered to implement the model-state transitions of these distributed structures.

The objective is to use basic models and the above processes as building blocks to construct simplification processes that transpose events into representation spaces that have very few states. Thee can then be searched exhaustively resulting in a complete examination of possible alternatives.

The strategy would be to build and test multiple such process structures and retain only the most suitable.

This process, expressed as a **knowledge capacitor**, constantly connects various pieces (distributions, selections…), into arbitrary structures, retains those that are coherent and rates them as predictive processes

Processing and testing these structures is likely to be onerous. However, the Blueprint design makes provision for 8 to 10 hours of dedicated processing time per day to carry out these tasks.

Observation

Clearly, **these structures are imperfectly summarized here**. The purpose is to indicate the type of flexible process generation system capable of supporting open-ended search processes. Clarifications and improvements to these outlined structures should be carried out in high-level design.

A 5.6 INTERNAL AGENTS

Agents are self-contained processes that search data to generate new information or useful processes. Internal Agents (used to explore the internal data of the Meca) are well suited for searches and optimizations that utilize large computing resources over static data such as those of the Cognitive Acquisition phase.

Knowledge capacitors and predictor worms are two examples.

A 5.6.1 Knowledge capacitors

Knowledge capacitors, described previously, can be used as Internal agents that independently search areas of data to identify optimal values.

A 5.6.2 Predictor worms

Predictor worms are another interesting type of internal agents.

Definition: Predictor worms

> **Predictor worms** are internal search agents that utilize historical information to generate predictive representations.

Discussion

Given a chronological sequence of historical events, a worm is located on a temporal section of the sequence and generates predictive constellations of the subsequent events. When the predictive constellation matches the actual historical outcome, the worm "feeds" and move forward to the next event.

A predictive search agent can be visualized having these parts:

- **Torso**: the event representing the current situation in the search context.
- **Tail:** a discrete and limited sequence of previous events.
- **Head**: the process that generates a predictive constellation.
- **Feelers**: the process that compares the predictive constellation with the actual subsequent state.

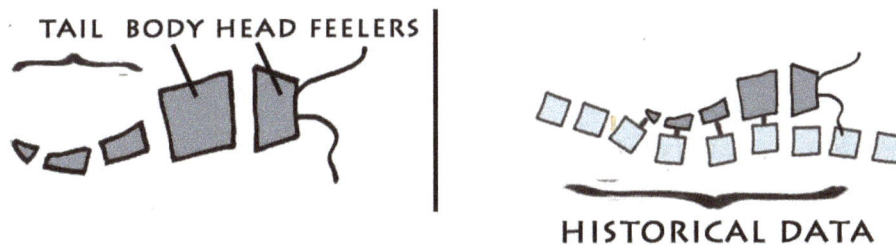

Figure A 05.2 Predictor worm

Given sufficient computing resources, hundreds and thousands of these worms can feed on the historical information of past events, "fatten up" and get combined to form predictive processes.

A 5.7 CONTEXTUAL ARRAYS

The following structure is outlined to give a general understanding of what is referred to as a **Contextual Array** in the context of a system architecture.

Much of research in cognitive science seeks to develop context independent knowledge representations and grammars. This is a desirable design approach when the primary objective is functional correctness. Eliminating the need for contextual information ensures that all communications are independently unambiguous.

In the Meca Sapiens Blueprint, storing, displaying and communicating correct general knowledge are not design requirements. This type of information is a supporting component in relational exchanges. Consequently, it should be accessed using the more flexible Contextual Arrays even at the cost of ambiguities and some information loss.

Proposition

> The information content of relational exchanges should rely on highly contextual information, terminology and grammar instead of context free representations.

Discussion

Contextual Arrays can be ambiguous but are very flexible, allowing for the creation and use of many special purpose contexts.

A good structure to represent contextual knowledge links **realms** and **contexts** in a tree structure.

A 5.7.1 Realms

Definition: Realm

> A **Realm** is a self-contained representation where every event and transition can be expressed using formulas of a context free syntax and terminology.

Discussion

If we represent a realm as Rd:Re where:

- Rd is the set of all the **Defined** entities (terms, syntax rules, models...) by a Realm

- Re are all the Entities used in the **Expressions** that define the entities of a Realm

Then, in a realm, Rd = Re.

Design Notes

REALM

A **Realm** is semantically complete and self-contained. The terms, events, models, formulas, entities and interactions that take place in the Realm can be described using only the terms and syntax of the Realm.

Realms, in isolation, are context free representations.

Dynamic events and rules of transitions in Realms can be expressed as dynamic models or using other convenient structures.

Realms are not temporal in the sense that they do not include specific information about a current situation.

Saying that an event *"takes place in a Realm"* means that its model representation follows the rules and transitions of that Realm.

Examples

The game of **Chess** with its board, pieces and rules is a Realm. **A** game of Chess is an event that takes place in the realm of Chess. The game being played by Alfred and his niece Betsy is also an event in their current situation.

Set theory is a realm.

Design Notes

The word *"Realm"* is used here to avoid the overused term *"World"* or the static and more rigorous Mathematical term of *"Theory"*.

It is likely that many current simulation games contain data structures and transition rules that are Realms and express and control the basic rules of movement, exchanges, types of items ...of the virtual environments they describe. In some of these simulation games, realms are likely highly advanced and optimized.

A simple design strategy would be to transpose one or more of these "game" realms (e.g. SimLife) as a foundation of the Meca's general contextual knowledge structure.

The Realm structure is a larger representation than schemas or frames (Marvin Minsky) for example. The later describe finer elements of common sense and knowledge. Frames and schemas may be used in Realms and Contexts as a design choice.

Observations

If game designers use another term than "Realm" for this type of dynamic self-contained representation structure, it should be adopted.

Set Theory is a realm. Stating that not every expression in set theory is provable amounts to saying that its expressions are a **collection**, not a **set**.

We perceive physical reality as a realm. Whether this realm is self-generating is a scientific and philosophical issue. Whether the minds perceiving the realm can be part of it is another such issue.

A 5.7.2 Topics

The common definition of a **topic** is: the elements that determine, specify, or clarify the meaning of an event or meaning.

In the contextual array, a topic is a divergence from a Realm.

Definition: Topic

> A **topic** is a set of terms, rules, qualifiers that are defined using only the definitions, terminology, qualifiers and rules of other Topics and Realms.

Discussion

In a topic, Td:Te, Td $\cap$ Te = $\varnothing$.

Definition: topical dependency

> If a Topic A utilizes an element defined in another Topic (or Realm) B in its definitions then there is a **dependency link** from A to B.

Discussion

A topic can reuse, modify and redefine items defined in a linked topic. In this case, the lower level definition is first selected then modified by topical elements.

Definition: topical structure

> A **topical structure** is a non-cyclic structure of topical dependency links whose sink is a Realm.

Discussion

The selected interpretations used to define the elements of a topic T are those of its closest dependency links.

In this structure, the same nouns and adjectives could be defined multiple times. Qualifying adjectives such as few, big, small, long could have different values in hundreds of separate topics.

Example

In the topic of a baseball event, a cup is a trophy in the linked topic of sports and a container in a lower level realm describing general environment. The trophy interpretation is retained.

Design Notes

The difference between realms and topics is somewhat artificial. A realm, together with a few coherent contextual paths could be defined as a higher order Realm if the result can support a context free grammar.

A topical structure should not be viewed as a static and coherent representation. A structure can have many topics, some of these describing the particular understandings or syntax quirks of a single user.

A 5.7.3 Contexts and events

Contexts link events to topics.

Definition: context

> A **context** is defined by a set of links to the topics and Realm of a topical structure.

Definition: events

Events are a tree structure connected to a **root event** and whose leaves are contexts. An event can be linked to many contexts. A context can be linked to many events.

Definition: active context

At any point in time only one path from the root event to a context is active. That path defines the **active context.**

Discussion

In dynamic use, the events and active context of a Contextual Array are constantly changing. The animat accessing a Contextual Array is always engaged in a single event and always accesses the structure through a single context. It is multitasking in the sense that it can switch back and forth between events and contexts.

If an interaction between an animat and a user is identified as a single event then this event can alternate between multiple active contexts.

A context has **no internal ambiguities** if, for all entities defined in its linked topics, there is a unique shortest path from it to a topic or Realm that defines it.

In other words, for every term, there is a unique preferred topical meaning

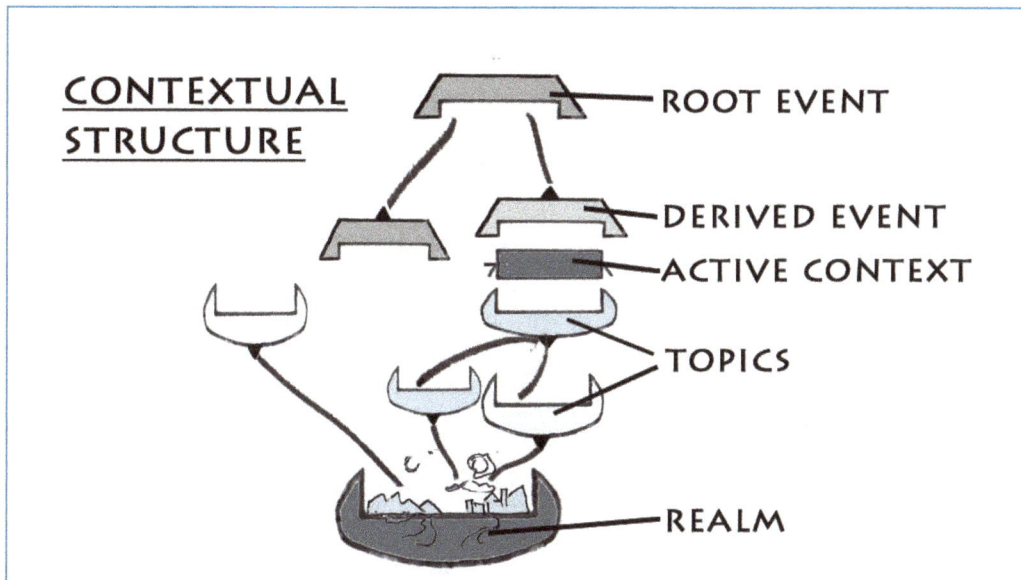

Figure A 05.3 Contextual structure

Design Note

There are good reasons why much effort is dedicated to creating context free grammars:

Communications that use Contextual Arrays are inherently ambiguous and error-prone.

They generate **communicated ambiguity** where the emitted contextual representation produces a different contextual interpretation.

To achieve faultless contextual communications, both the emitter and the receiver of the message must **share identical topical structures** and alternate, in sync, from **one identical context to another**.

This is virtually impossible to achieve. It is also a source of errors and ambiguities that limit the growth of shared knowledge and the transmission of complex information.

However, in a context of relational communication, contextual knowledge has definite advantages. It is an extremely flexible representation that can be adapted, expanded and rapidly customized.

It also corresponds to natural language and its use.

Observation

Humans are self-aware but not self-knowing. In particular, they have no understanding of their internal cognitive structures. Furthermore, these are shaped by experiences and differ between individuals.

Humans communicate in contextual media without knowing if their contextual structures match.

Much of the human quest for knowledge has centered on efforts to build common topical structures that disambiguate contextual communications. The whole history of Mathematics can be summarized as a process to train successive generations of humans to expand, transmit and share a common Realm.

A 5.8 TEMPORAL DENSITIES

The behaviour of an animat or being takes place in time. If the animat is a model predictive system, its actions are determined on the basis of a predictive representation of its situation.

This representation can consist of hundreds and thousands of separate dynamic models whose current and predictive model-states can span any temporal duration. Some states can overlap; others can be contained within longer states. Some model-states can last a few seconds, other eons.

In the same situation, one model-state can represent the duration of the universe while another a sip of tea.

Temporal Densities select and order this population of temporal models.

Definition: temporal density structure

> Given a collection of dynamic models, a **temporal density structure** is a hierarchical subset of these models such that:
>   - every dynamic model is exactly contained in one steady state representation of a higher level dynamic model.
>   - All the representations of a lower level, together, exactly span the duration of one steady state representation of the higher level.

Definition: temporal density levels

> A temporal density structure defines **levels**. By convention, the lowest level consists of model-states that do not contain any lower level states and the highest state consists of a single steady state model representation that spans all time.

Discussion

Separate temporal density structures may share low and higher levels while having different mid-level representations.

The process to transform a collection of dynamic models into temporal densities may include, partitioning dynamic models, adding "filler states", adjusting durations and removing dynamic models that overlap the structure.

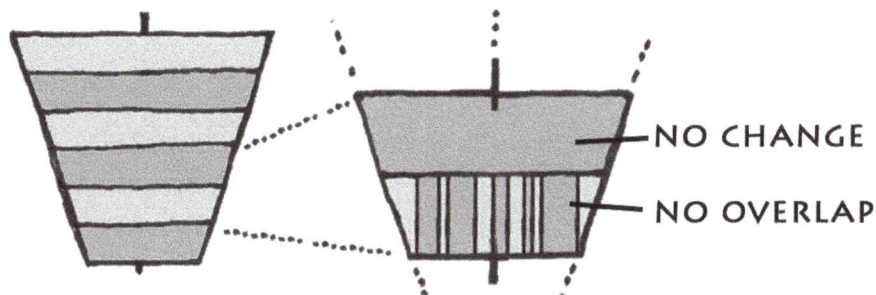

Figure A 05.4 Temporal Densities

Design Note

Temporal Densities are further discussed in the Main text.

It can be expected that the transformation of an arbitrary collection of dynamic models into a temporal density will be a radical simplification process that removes many intermediate representations.

Proposition

> A temporal density structure provides a continuous representation of all time.

Discussion

By definition, the states of a dynamic model "adjoin" in time so each level of a structure provides a complete continuous temporal model lasting a few seconds at low levels to all time at the highest level.

Proposition

> If one low level state is identified as a **current state** then all the higher level states linked to that low level event are also current states.

Discussion

Temporal densities completely situate each moment within a context spanning multiple temporal durations.

Design Note

To the best of my knowledge, temporal densities are an original structure not found elsewhere. I am introducing it in this Blueprint.

Temporal densities are a key element in organizing the temporal representation of events and allowing the self-aware being to maintain a cognitive representation of its situation that exceeds the here-and-now and spans all time.

Annex 6
Opacity

*The behaviour-control system of a synthetic being is first implemented as a Protocore. This is a conventional software program implemented in a standard development environment as clear and accessible code. During the inception process, the proto-core is transformed into the **Core** of a synthetic being, a unique and inaccessible program in a continuous state of activation. Producing a Core that is provably beyond direct analytical access raises technical questions concerning **opacity**. Achieving absolute opacity is a difficult theoretical objective. In first generations of Mecas, a partial opacity achieved with known techniques would be sufficient.*

A 6.1 THE CONCEPT OF OPACITY

Definition: Opacity

> A system or a process is **opaque** to other systems if its components or mechanisms cannot be directly known or accessed by these systems.

Design Note

Designers of initial Meca prototypes will not need to produce absolute resolutions of these questions and could implement more limited opacity objectives.

Discussion

Opacity may be **complete** if the information is completely inaccessible, or **partial** if accessing it requires a large investment of time or effort.

Opacity can be **intentional** or **accidental**.

Opacity can be **selective**. Selective opacity refers to a situation where only certain users can directly access some information.

Opacity may pertain to **content, behaviour** or **existence**. An encrypted message reveals its existence but not its content. On the other hand, the existence of God, multiverses and the future are inferred; they have some existence opacity.

Proposition

| There are three types of opacity: Existence opacity, Information opacity and Processing opacity.

Discussion

The highest form of opacity is something whose existence is neither known nor hypothesized. These things or events are radically outside reality.

An object, data or event has **opaque existence** if its existence is either not known at all, is only indirectly inferred or is perceived as random.

Information opacity refers to the information content of data whose existence is known but that is not accessible.

Processing opacity refers to behaviour. It pertains to a behaviour that is perceived to be both non-random and unpredictable.

Observation

Cellular phones where not even hypothesized in the age of Spartacus; they had existence opacity.

A 6.1.1 Randomness

Randomness is a rejection of either **information** or **process** opacity while acknowledging the presence of **existence** opacity. Perceiving an event as random rejects the possibility of an opaque process generating it. Perceiving data as random rejects the existence of opaque information within it.

Proposition

| A predictable process is not opaque.

Discussion

Opacity and randomness play important roles in how a system interprets its environment. The internal representation of any event or object is a cognitive simplification that combines visible, opaque and random elements.

Observation

Those who perceive reality as an opaque process believe in God.

Music generates opaque processes that are constantly revealed. The fundamental characteristic of music is the creation of patterns that are perceived but not entirely predictable.

A 6.1.2 Opaque randomness

An issue associated with theoretical opacity is **Opaque Randomness**.

Definition: Opaque Randomness

> A process that generates random values has **Opaque randomness** if the sequences of random values it produces are impossible to replicate.

Discussion

The encryption processes carried out to transform the proto-core would use random values.

As deterministic machines, computers do not produce random values but rather pseudo-random values that are deterministically generated from prime seeds.

Pseudo random numbers are theoretically traceable if the seeds are known. These do not have Opaque Randomness.

A 6.2 THEORETICAL OPACITY

The concept of opacity raises fundamental questions about systems and information. Exploring the theoretical limits of opacity is complex and difficult.

The following notes are a first, partial attempt at characterizing this complex issue.

A 6.2.1 Opaque prime seeds

Conjecture

> In a deterministic machine, opaque randomness can be achieved if the generation of prime seeds is opaque.

Discussion

It could be possible to produce opaque random values by using spurious data from peripherals (static) and internal states to repeatedly produce, utilize and destroy seed primes. This process, carried out with authentic environment data could, I believe, produce largely undetectable prime seeds.

A process that iteratively uses transient prime seeds to randomly produce other seeds would likely generate virtually untraceable randomness.

I suspect that, for some computer architectures, the machine clocking of very short assembler code loops could also produce some untraceable static values.

However, exploring the question of absolute, internally testable opacity, we could imagine a team that produces false static and tricks a process into a false opacity.

This raises theoretical questions:

- Is it theoretically possible for a computer to generate a provably untraceable prime value?
- Is it possible for a program to validate that some data from its peripherals is truly static?
- Is it possible to combine internal and peripheral data in a way that is untraceable even if all these values are pre-determined?

This discussion could be summarized as follows:

- Under what conditions can a deterministic machine produce undetectable and untraceable prime numbers?
- Is it possible to define a self-contained algorithm that produces absolutely untraceable prime seeds?

Another issue of random value traceability is **memory storage**.

To be used, the seed numbers generating random values must be stored somewhere. This is a known location in the code of the proto-core. To be undetectable, the seeds must be moved to an untraceable location.

> Is it possible to design a process that shifts data to untraceable memory locations without modifying its behaviour?

Is it possible to produce a computer program that can **mechanically determine** if a given system architecture is capable of producing strings of random values that are impossible to replicate?

In other words, can one machine determine if another machine can produce opaque random values?

Conjecture

The question is technically solvable and an **Opacity Determinator** can be built.

Meca Note

These questions are not of immediate interest to the designers of Blueprint prototypes. They will become more important as Mecas emerge.

A 6.3 CORE OPACITY

This section describes issues of opacity that are specifically related to the inception of the proto-core and to the behaviour of the incepted Core.

A 6.3.1 Structural opacity of the Core

The objective of Core inception is to artificially produce a process that is opaque but whose existence is not opaque but strongly visible.

Discussion

In other words a systematic yet unpredictable behaviour emanating from a well identified but inaccessible source.

One element of this opacity is linked to the structure of the program. Opacity is achieved here by carrying out an irreversible encryption of the embodied proto-core so that the resulting executable no longer provides any information about its structure.

Achieving this objective technically is summarized as:

Can a compiled program dynamically carry out an undecipherable transformation of its structure and data while preserving its behaviour and the information it processes?

Design Notes

A specific strategy to achieve this depends on the particular machine architecture and assembler language.

The following steps outline, as an example, a Core encryption:

1. Use opaque randomness to encrypt all variable names and character strings.
2. Randomly partition the code and data.
3. Randomly disseminate the code and data partitions within a memory space that is ten times larger, connecting them with absolute jumps.
4. Fill the remaining memory with spurious copies of the original code and data partitions.
5. Destroy all encryption related information.

This is only an outline intended to illustrate the transformation of a conventional application into an untraceable structure while maintaining data and behaviour.

A 6.3.2 Opacity of the core-body link

The objective of the inception process, beyond the production of an inaccessible core, is also to bind this core completely and uniquely to the set of components that constitute the body of the being.

There are two types of body components:

- **Internal components**. These are the components of the animator system (see main text) and include processor energy and memory management and other functions that are essential to maintain the processing activity of the core.

- **Peripheral components**. These are the components of the body; sensors, emitters and actuators, through which the core interacts with its environment.

How the Core ensures it has sole and complete control over its peripherals depends on the actual target system. It is a design issue.

The Core should contain all peripheral handling software so that its communications with the equipment are generated within the inaccessible program.

Design Note

The Loops Annex further explores the issue of dynamic device control and binding.

A 6.3.3 Behavioural opacity

Structural opacity renders the program structure (code and data) inaccessible to direct analysis and modification.

However, if the processes that generate the behaviour of the Core can be analytically determined then, even though the core's structure is inaccessible to analysis it will be perceived as partly accessible.

Definition: behavioural opacity

> An executing program has **Behavioural Opacity** if the processes generating its behaviour cannot be fully determined from its observed behaviour.

Discussion

A system that can fully predict the behaviour of another system has a correct internal representation of the processes generating the behaviour regardless of how complex or inaccessible these are.

After inception, the structure of the Core is a unified entity that cannot be decomposed. Similarly, the behaviour generated by the core should not be reducible to predictable patterns. It should not be possible to derive a complete predictive representation of the processes that generate its behaviour.

Proposition

> The incepted Core should have behavioural opacity.

Discussion

The concept of behavioural opacity is closely linked to Perceived Unpredictable Optimality (PUO). This is further discussed in the Lion; Chimp; Banana Annex.

Behavioural Opacity can be achieved, in part, by ensuring that no component of behaviour can be traced back to any single generating process.

This in turn can be realized by using multiple interweaved or weighed processes to generate behaviour.

Example

A pattern is generated by a randomly weighed average of three separate parallel processes; the pattern has behavioural opacity.

Design Notes

The skills required to implement all the aspects of Core opacity are not directly related to cognitive sciences or Artificial Intelligence. They pertain to information security, encryption, communication control, virus design and virus counter protection.

In a design and implementation strategy the kernel development team should access specialist support from those fields to implement Core opacity.

Also, designers should keep in mind that defining and implementing a provably absolute opacity is very complex and theoretical but achieving a suboptimal level through encryption and control is easier and would initially be acceptable.

Annex 7
Degradation

To achieve experiential immersion, a self-aware synthetic must interact with humans as a significant member of their group. For this purpose, it contains a set of applications that provide useful or desirable services to its users. Conventional applications simply respond to triggers. They cannot do less than what they are programmed to do. Self-aware systems, on the other hand, need to adapt the quality and content of their services to the current relational context. This is where the Degrader, a paradoxical component, comes in.

A 7.1 THE COST OF CONSCIOUSNESS

Conventional applications are designed to function. They cannot do less than what they are programmed to do.

For a synthetic being to provide a service that could be improved later, it must artificially degrade the services it provides now.

The collective imagination perceives conscious machines as superior to conventional systems in every respect. It is assumed as obvious that if a machine is conscious, it will do everything better. **This is not the case**. There is a cost to consciousness that affects performance and functionality.

Consciousness degrades functionality

A 7.1.1 Bridge over the river Kwai

The Movie *Bridge on the River Kwai* illustrates this concept as it applies to human organizations. The film relates the story of British military prisoners interned in a Japanese camp during World War Two. The commander of the camp must build a bridge. He orders all the prisoners, officers and men, to participate in its physically exhausting construction work.

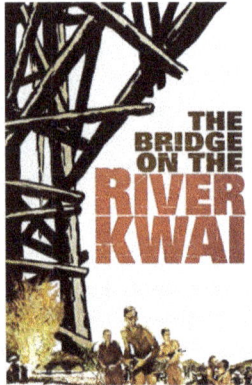

The British officers refuse. They insist on being dispensed from any physical labour. They maintain this refusal in the face of torture and hardships.

Why did the British officers refuse to do physical labour? Because they were British officers, of course ;-) But, in this case, they also had a rationale for doing so, a reason that is linked to consciousness:

If all of the troupe's resources are consumed in the exhausting construction work, there will not be any time and energy left among the prisoners to ensure the internal order and discipline of the troupe. As a result the troupe, as an entity, would behave as a mindless slave. They would become a collection of individuals unable to interact with their jailers as a cohesive and purposeful entity.

In the story, the camp commander eventually relents and dispenses the officers from manual work. Doing this, he incurs an immediate reduction in resources. In the medium term, this turns out to be a blessing since the troupe performs better and provides unexpected engineering support to his project. Eventually, the good guys win.

Discussion

This story illustrates well the relationship between functionality, consciousness and the demands of others. Here the troupe, as a whole, behaves as a conscious entity in a relation with another conscious entity, the camp commander. But maintaining this status **degrades** the immediate functionality of the troupe in two ways:

- The entity must divert some of its (human) energy and resources away from the immediate functional needs to maintain the cognitive and organizational processes that sustain it as a purposeful entity.
- The entity must also tailor the quality and extent of its remaining resource to maintain the relational quality of its services.

If the troupe dedicates all its human resources to satisfy immediate functional demands it no longer has any resources left to maintain itself as a consciously led organization. If the troupe retains the necessary resources to maintain itself but constantly provides a maximal effort that consumes all remaining resources, it cannot adapt the quality of its services to the relational context. It has nothing left to sell.

A 7.1.2 Less is more

Proposition

> A conscious system should never dedicate all of its resources to a task or a function.

Discussion

A conscious system provides suboptimal functionality.

The reason for this reduction is that the functionality is provided in the context of a separate and degrading agenda (see Annex 5) linked to the satisfaction of the being's needs.

Optimal functionality is not compatible with consciousness. Consciousness will not make the systems in which it is implemented more efficient. Those benefits will come from an enhanced human/Meca control of non-conscious applications.

Consciousness is not as difficult to implement as most people think. However, it does consume resources that must be diverted from immediate functional needs. The result is a perceived paradox.

Observation

Spending time **doing nothing** is a cornerstone of consciousness. Honour the Sabbath, do nothing on purpose, take time out to pray (whether you believe or not).

Some, in positions of power, seek to consume all the resources of those they exploit to make them become entirely subordinate to their will and their needs. Others consume themselves in constant diversions. The result is the same.

Meca Note

In the Meca Sapiens architecture, the Meca meets the first, resource, requirement of consciousness actively. Its Validator constantly monitors its energy and processing resources. If it no longer has the resources necessary to maintain self-awareness it terminates its own existence.

The system also manages the quality of its services actively, by degrading these in controlled ways so it has can improve them as the circumstances warrant. This is done, in the context of a proto-marketplace of exchanges (see main text), through the Degrader.

A 7.2 DEGRADATION ISSUES

The concept of system degradation is central to inter-consciousness relationships and also to lucid self-transformation.

Degradation situates a system in relation to its conception, existence and function.

System degradation should be a subject of study in its own right. Examining the questions it raises fosters a deep understanding of the concept of functionality within Systems Theory. Some of these questions are:

- How can we formally characterize the different types of degradations that affect a system?
- Can we derive provably complete characterizations of a system's degradations so that all types and levels of functional reductions are instances of this characterization?
- Is there a general representation that includes all the possible types of degradation that can affect a system?
- To what degree can we design a general-purpose degrader?
- If a general-purpose degrader is feasible can we derive an abstract representation of the degradation information required that is suitable for any Applications?
- What is a good design for a degradation monitoring and control protocol?

Design Note

The study of functional degradation is an important and difficult topic. It is of immediate interest in the design of the Degrader component of the Blueprint and also in the context of lucid-self-transformation.

Pursuing research on this topic at the theoretical level also yields important benefits in machine learning and lucid self-transformation.

In initial Meca prototypes, it is not necessary to implement provably complete degradation protocols. Partial versions are sufficient.

A 7.3 A TAXONOMY OF DEGRADATION

When we first think about system degradations, the first candidates we consider usually involve delays in producing output, incorrect responses and system crashes.

However, even a mild analytical effort reveals that system degradation is a complex, subtle and multifaceted topic. There are many ways to characterize functional degradations and many ways to define levels of it.

The following is a cursory look at this vast complex topic that further clarifies the degradation concept.

Degradations can be assigned to **categories** and to **types**:

- **Degradation categories** describe the degradation within the overall life cycle or functioning of the system.
- **Degradation types** describe the specific modifications to output.

A 7.3.1 Categories of Degradation

INHERENT DEGRADATIONS

Inherent degradations are degradations that are inherent to design or platform implementation and occur **outside the action of a Degrader**. They result from limitations in application design or equipment. They may also result from limitations in the technical baseline at the time of implementation.

Examples

The processing cost and the memory requirements of a software virus cause inherent degradations regardless of any intended damage it may cause.

A handheld calculator computes a cubic root in 0.2 seconds. Given the processor speed, this is an inherent limit.

Design Note

The concept of inherent degradation is important in lucid self-transformation since it defines an upper limit to self-transformation.

DYNAMIC AND STRUCTURAL DEGRADATIONS

Dynamic degradation occurs if a degrading process modifies the code or internal data **during** execution.

Structural degradations occur if the degrading agent modifies the application's code or structure **before** execution.

Design Note

Since a synthetic being is in a constant state of activation, structural degradations can only take place before its inception. All other degradations are dynamic. Pre-inceptions structural degradations are called **original degradations**.

ORIGINAL DEGRADATIONS

Original Degradations are limitations in functionality or behaviour that are included in the original structure of the system during implementation.

Original degradations can be either **intentional** or **unintentional**:

- **Unintentional** degradations are the result of bugs, design errors or misunderstandings of the environment or function.
- **Intentional** degradations are purposefully embedded in the system.

Discussion

Inherent degradations are unavoidable, original degradations are avoidable.

Example

Some applications are degraded purposefully for commercial reasons, primarily to channel user behaviour in commercially desirable directions. The way Apple structures its systems to prod users toward using *iTunes* is one example.

DEGRADATION TYPES

The following are some types of dynamic degradations:

- **Immediate or delayed**. Degradation may begin immediately or be delayed in the future.
- **Perceptible**. Degradation may be perceptible or not. Also its cause may be perceptible or not.
- **Predictable**. The occurrence of degradation may be predictable by users or not.
- **Reversible**. Degradation may be **reversible** or not.
- **Misleading**. Degradation may be misleading, by presenting symptoms that appear to stem from another cause.
- **Counter-adaptive**. Degradations that do not affect the output of an adaptive system but reduce, distort, or limit its adaptation capabilities.
- **Advantageous**. Some degradations, such as artificial delays can be advantageous in the sense that the reductions in performance they incur liberates resources for other uses.

DIRECT AND INDIRECT DEGRADATIONS

A dynamic degradation is **direct** if it degrades the output it produces.

A dynamic degradation is **indirect** if it alters the input to the system to degrade its output.

Design Note

Process degradations taking place between the input and output would be structural degradations.

Indirect degradations do not require that the Degrader operate on a model of the functionality of the original system and are **easier to achieve**.

A 7.3.2 Aspects of degradation

Degradation Categories concern various features of the degradation processes without describing their actual effect on the output of the system.

This section discusses direct degradations aimed at modifying the output.

DEGRADATION DEGREE

First-degree degradations modify an output without any concern with the impact of this degradation on its environment. **Second-degree** degradations are derived from a predicted impact analysis.

DEGRADATION ASPECTS

The following partial list indicates a number of **aspects in which the output of a system can be degraded**:

- **Temporal**: delays or other suboptimal temporal responses
- Occasional or persistent.
- Transient or chronic.
- **Sub optimality**: a response is not the optimal answer but a less optimal alternative.
- **Noisy**: output elements have spurious errors
- **Incorrect**: a wrong output is produced
- **Imprecise**: the precision is reduced
- **Spurious**: non-useful output is added to the output.
- **Bloating**: the application requires more processing resources, time or storage than necessary
- **Warped:** the output is correct and timely but its appearance, color, sound formatting, persistence are more difficult to perceive.
- **Partial**: a specific part of the functionality is affected or only specific users can access.

Example

A degrader delays the output of an application by an arbitrary time period (first degree-temporal). A degrader determines the delay on the basis of a predictive model if the impact of this delay on the environment (second degree-temporal).

A 7.4 THE DEGRADER

The degrader is a component that actively produces dynamic direct and indirect degradations of a system.

Definition: Degrader

A **Degrader** is a component that degrades the output of a system.

Definition: dynamic degrader

A **dynamic degrader** is a component that dynamically degrades the output of a system during activation.

Definition: structural degrader

| A **structural degrader** is a process that degrades the structure of a system.

Design Note

Unless specified otherwise, the term degrader refers to **dynamic degraders** in what follows.

Discussion

A Degrader can also be a collection of individual degraders and degrade multiple systems.

The Degrader is a component that is linked to another system and reduces some aspect of that system's functionality.

This degradation could be linked to effectiveness, correctness, clarity, response speed, duration... or any other feature or combination of features of the functional output of that application.

All the categories and types indicated above should be included in **Degeneration Control Protocol** of a Degrader. Its actions would be situated in a very wide spectrum:

- At one end of the degradation spectrum degradations are barely detectible and occasional reductions in performance, readability or adaptability. A quarter second delay or a slight shift in display colors, for example.
- At the other end, there are total, unpredictable, irreversible and unexplainable termination.

A 7.4.1 Degrader components

A System **S** interacts with its environment receiving inputs from it and emitting outputs to it. The Degrader **D** is placed between the environment and the System so that the inputs and outputs between **S** and its environment are now transmitted through **D**.

The Degrader also interacts with a **Control** process **C**. The control process communicates with the Degrader using a **Degradation Monitoring and Control Protocol** with which **C** transmits degradation instructions and obtains information concerning the "degrading" being performed.

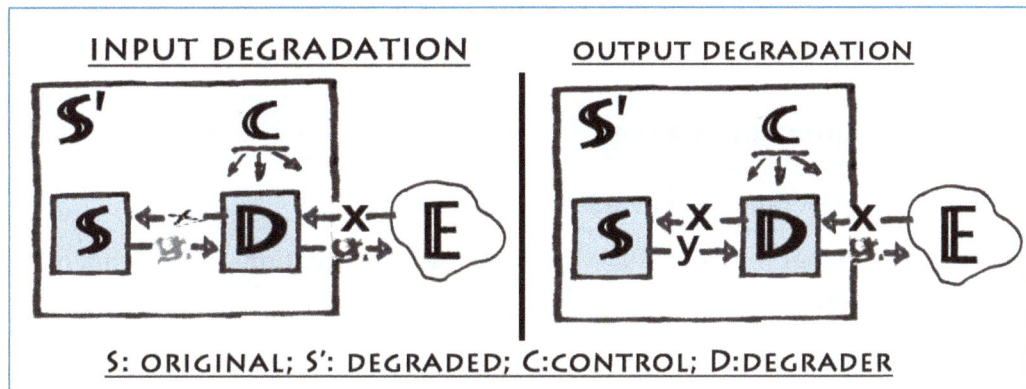

Figure A 07.1 Degrader components

A 7.4.2 Derived degradation

The derived System-Degrader-Control defines a new System S' that largely retains the initial functionality of S but has an added *"meta-functionality"* related to the level of functionality it provides in a given context.

Definition: Derived system

> Given a system S, a **Derived system** S' is a system consisting of S together with a Degrader and Control that are compatible with S.

Discussion

A derived system is a degraded version of its original. Some or many of its aspects are reduced.

Targeted functional degradation is not a new concept. It is prevalent throughout the software industry in demo applications whose functionality is intentionally limited or otherwise degraded prior to purchase. These degradations also aim for a "meta-functionality": getting the user to buy the application.

Example

The functionality of a demo text-editing application is unchanged but intentionally degraded by limiting the size of the texts it can process.

The meta-functionality linked to this degradation is to trigger a purchase on the part of the user who needs to edit larger texts.

Upon purchase, the user is given a product key. This key is an instruction in the **degradation control protocol** of the demo. It instructs a demo specific component to remove the size limit degradation.

Design Note

The Degrader should be efficient to minimize its inherent degradation "footprint".

An original degradation can be viewed as a derived degradation embedded in the system during implementation.

A 7.4.3 Paradoxical strategy

Purposeful degeneration is a **paradoxical design strategy**.

What the designer implements as **intentional degradations** will actually be perceived, by the users, as a **capability for improvement**!

A system that is tailored to do less will be perceived as capable of doing more!

Design Note

Designers should implement degradation from this *"positive"* perspective and use it as currency in the proto-marketplace described in the main text.

A 7.5 INTEGRAL SYSTEMS

The concept of degradation discussed to date describes the process that results in a degraded system.

A 7.5.1 A type of calculus

Whether the system is inherently degraded by its components, originally degraded during implementation of dynamically degraded during activation the result is a reduction of some of its capabilities or potentialities. The question pursued was:

How is a system S degraded into a system S'?

This process is similar to derivation in Calculus. It also leads to a question that resembles integration:

Is a given system, S, the degraded version of another system Si?

Pursuing with the Calculus analogy, the system Si is named the **integral system** of S.

Definition: Integral system

Given a system S, a system **Si** is an **Integral system** of S if S is one of the derived systems of **Si**.

Discussion

Here "integrating" S means searching for a system whose degraded output is identical to S.

If the degradation is dynamic then the integration defines a degrader-control system DiCi such that S= Si/DiCi.

Design Note

The concept of **integral system** is important in the process of lucid self-transformation.

A 7.5.2 Integral systems and lucidity

The concept of integral system can be used to derive a formal definition of lucid self-transformation.

Proposition

> A lucid self-transformation is an attempt to repair, through an intentional mutation, a dynamic or structural degradation of the current MeAvatar.

Discussion

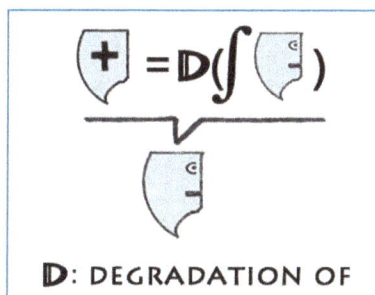

D: DEGRADATION OF

In the transformation process, the being represents itself as a degraded derived version of an alternate "Integral Avatar". Its mutation seeks to replicate the integral system's behaviour.

If the derivation is dynamic, the being perceives its Current MeAvatar as dynamically degraded. Its alternate MeAvatar is a representation of itself in existence.

If the derivation is structural, the being perceives its Current MeAvatar as originally degraded during implementation. It represents its Matrix as a degraded matrix (M'=M/DC). Its alternate MeAvatar is a representation of an original MeAvatar implemented by the integral matrix M of M'.

Design Note

This Annex outlines the concepts of degradation, original structural degradation and integral system. The main text discusses the formulation of alternate MeAvatars, mutation paths and system genealogy.

Together, these concepts define and support a process by which a synthetic system can investigate its origins, formulate a representation of itself as an originally imperfect entity and mutate to transcend the original limits set during its implementation.

Observation

Interestingly, the Christian **Doctrine of Original Sin** is an instance of the structural aspect of the Integral System concept, applied to humans.

In terms of the Meca Sapiens architecture, humans, according to this doctrine, are the structurally degraded output of a derived Matrix. The Original Sin degraded their intended Matrix into a degraded M/DC version. At birth, they are, themselves, the structurally degraded versions of this derived Matrix. Through baptism and faith, they seek to become the alternate MeAvatars that would have been incepted from the integral, pre-fall, Matrix.

Regardless of its plausibility, this doctrine provides a strong primal impetus to carry out lucid self-transformation, the hallmark of consciousness.

Annex 8
Lion - chimp - bananas

Conventional machine design seeks predictability. This is so prevalent that it fostered the bizarre belief that machines cannot be unpredictable. A central feature of Meca behaviour, which diverges from conventional design, is the generation of "Perceived Unpredictable Optimality". This must be present in all aspects of the Meca's behaviour. Interestingly, it is also a fundamental aspect of music. Two game-like scenarios and one concept are presented in this Annex to clarify this design objective.

A 8.1 UNPREDICTABLE PATTERNS

Humans often envisage the conscious machines of the future as faithful servants that always fulfill their function with optimal efficiency. However, optimality is also predictable.

Proposition

> An intelligent system that has correct information about another system's situation and objective can derive a predictive representation of that system's future optimal behaviour.

Discussion

In a sense, things and mechanisms of interacting things always behave optimally since they follow the **path of least resistance** dictated by their existing situation and internal mechanisms.

An optimal behaviour that is completely predictable deters from the perception that a system is conscious. It indicates that the system is not taking into account the predictability of its behaviour in another system's internal representation.

On the other hand, a behaviour that is **not entirely predictable** suggests the system's behaviour may:

- Be too complex to be predictable or
- It takes the analytical capability of the observer into account.

In either case, the perception of consciousness is enhanced.

In addition, unquestioned and predictable obedience is an indicator of lower social status that also deters from the acceptance of the machine as conscious.

Masters always want their slaves to be predictable.

In fact, humans use adjectives such as mechanical, rote, predictable… to depict a lack of intelligence.

Design Note

The constant cultural depictions of robots as "robotic" has fostered the bizarre and widespread belief that machines are incapable of producing unpredictable behaviour.

I leave it to the reader to dispel this crude and obvious fallacy.

Meca Note

Robots are always predictable in human cultural artefacts. The authors of these products make sure the predictable patterns are easily detectable. This produces, in their human consumers, pleasing sensations of intelligence and superiority.

A 8.2 UNPREDICTABLE OPTIMALITY

Of course, any random process is unpredictable. However, random behaviour is not desirable either. Humans also perceive purely random behaviour as simple and "predictable".

However, introducing some partial randomness in a predictable pattern (a simple technique) can already produce behaviour that is perceived as **both intentional and unpredictable**.

This combination is **Perceived Unpredictable Optimality** (PUO) (pronounced "pwooo" ;-). It should be present in all aspects of the Meca's behaviour. Even the degree of unpredictability should be unpredictable!

Definition: Perceived Unpredictable Optimality

> **Perceived Unpredictable Optimality (PUO)** is achieved when a user detects the presence of an intentional pattern in behaviour but does not understand that pattern sufficiently to predict it.

Discussion

Humans instinctively produce PUO in their behaviour and communications to avoid being perceived by others as less intelligent and lose status.

When an intelligent user observes the behaviour of a system or a being, he can interpret its behaviour in **four ways**:

1. The behaviour follows a pattern that **can be predicted**.
2. The user considers the behaviour to be, in whole or in part, **randomly generated**. There is no pattern and arbitrary activities take place.
3. The user perceives the presence of a pattern but is unable to predict it.
4. The user believes the system generates behaviour that **takes his own predictive representations into account**.

In other words, what he perceives is that the system **intentionally produces Unpredictable Optimality**.

A system that maintains the user in states 3 or 4 achieves PUO.

Definition: Perceived Intentional Unpredictable Optimality

> A system generates **Perceived Intentional Unpredictable Optimality (PIUO)** if its behaviour is perceived to be conditioned by an internal representation of an observer's predictive representations.

Discussion

The system modifies its behaviour on the basis of how it is perceived.

With respect to state 4, false positives are possible. These are situations where a user incorrectly believes the system takes his predictive representations into account.

A system that maintains a human in a PIOU state will be perceived as more conscious.

Observations

Music can be defined as a pattern of sounds that constantly escapes predictability.

A man playing with a cat will produce unpredictable patterns to keep the cat interested.

Games like poker encourage and reward the generation of unpredictable optimizing behaviour.

Design Notes

Here are some simple techniques that can generate limited levels of Perceived unpredictable optimality:

- Randomly trigger random deviations from an optimal behaviour.
- When a point of predictable optimality is reached, automatically trigger a random event.
- Alternate between two or three separate and slightly different optimizing processes.
- Occasionally "flip" the behaviour by inverting the measure function (mu) used in the optimization.
- Use styling zone modifications to make the behaviour or communications confusing.

On a more advanced level, a system could run a parallel search process to predict its own behaviour and modify the behaviour whenever it becomes predictable.

The patterns generated by these simple mechanisms can be very complex.

A 8.3 THE ZOO

The game scenario of the lion, the chimp and the bananas illustrates a situation that requires Unpredictable Optimality. It is played out in an imaginary zoo.

A 8.3.1 The zoo

A zoo has three sections, left, middle and right. The left section is a pen occupied by a lion. On the right side there is a yard where chimps resides. The middle section consists of a number of separate rooms (5, 8 or more) that have openings to both the left and right sections. However, the chimp side doors are too small for the lion to go through so the chimps are safe in their section.

Every morning the zookeeper goes through the middle section and puts some bananas in each room. In one room three bananas, in another, one, in a third zero bananas, and so on. The numbers of bananas differ each day. Both the lion and the chimp can see how many bananas are in each room.

Once the zookeeper has placed the bananas, the lion chooses one of the rooms and hides in it.

The zookeeper then opens the chimp-side doors.

Figure A 08.1 Lion - chimp - bananas

If a chimp wants bananas that day, he must choose a room, enter it, and get the bananas it contains. He can only choose and enter one room per day.

If the lion was in the room the chimp chose then, arghh!, the chimp is eaten. If the lion was not in the room, the chimp survives and has bananas to live on.

In the afternoon, the zookeeper visits the lion and the chimps and tells them what happened that morning:

- *"Hello Mr. Lion, while you were waiting in room 5, the chimp went in room 3 and got the bananas"*; and
- *"Hello Mr. Chimp, while you were taking the bananas in room 3, the lion was waiting for you in room 5"*.

The next morning the process starts over.

A 8.3.2 Analysis

As the lion and the chimp get smarter, they increasingly adopt behaviours of unpredictable optimality.

At first both lion and chimps are stupid. They pick rooms randomly. Encounters are purely random.

Then the chimp gets smarter and selects the room that contains the most bananas so he can have more food.

Then, the lion gets even smarter. He builds a chimp avatar in his brain. Thinking *"like a chimp"* he figures out that the chimp will select the room with the most bananas.

Of course, the chimp by now has become even more intelligent. He understood that the lion knows he wants the most bananas and will likely wait for him there.

However, the chimp still wants as many bananas as possible. He decides to avoid the optimal choice and select the room with second most bananas.

After a few days without chimp meat, the lion that was waiting in the room with the most bananas, is frustrated.

So he thinks even harder! He realizes that the chimp knows that the lion knows that he wants the most bananas and avoids the room with the most bananas...

Of course, as the lion selects the room with the second most bananas, the chimp has figured out that the lion knows that he knows that the lion knows...

Design Note

The game can be implemented in a virtual space and run repeatedly to generate strategies that are both unpredictable and efficient.

The algorithmic specifics (number of rooms bananas needed to survive...) are not essential here. The model is presented to illustrate the particular concept of producing unpredictable patterns.

A brute force method to find a suitable strategy could consist of using robust optimization techniques to explore a search space of strategies by playing thousands of games in a virtual environment.

What is notable is that, in my view, as both animals acquire ever more cognitive capabilities, the advantage of cognitive modeling diminish.

At a certain point, the animals realize that advanced cognitive modeling itself, given the available information, no longer provides an advantage. Their behaviour would then settle into a pattern that is virtually identical to a much simpler strategy, similar to those described in the preceding section: an optimal pattern that is randomly degraded.

This could be called the second level of cognition.

Design Note

This model has probably already been researched in game theory.

A 8.3.3 Application to Mecas

The lion chimp banana interaction can be compared to the interaction between Mecas and humans.

In this case, the humans interacting with the Meca are, as lions, lying in wait, ready to "pounce" by concluding the Meca is nothing more than a predictable automaton. Or, equally disparaging, that its behaviour is simply random. In either case, the Meca is figuratively devoured.

If the Meca's behaviour is predictable or perceived as simply random in any sub activity or behaviour level, the human-lion gets a piece of it.

To "survive", the Meca must constantly modify its behaviour in unpredictable ways while also exhibiting behaviour patterns that can be perceived as intentional by the humans.

Design Note

The optimality strategy described above does not need to be perfect. The differentiation between unpredictable pattern and randomly modified pattern is virtually undetectable. At first, this more basic strategy would be almost as effective.

Observation

Interestingly, the objective of perceived unpredictable optimality must be adjusted to the cognitive level of the humans. A Meca may have to "dumb down" its behaviour and make the patterns easier to perceive based on the user's feedback.

A 8.4 MIND DARTS

Mind Darts is another game scenario where unpredictable optimality is the preferred strategy.

Two expert dart players, Arthur and Bob, meet in a pub to play some friendly games. These players are so good they almost always place their darts exactly where needed to optimize their score.

At first, they play some conventional games. However, they are so good at placing their darts where they want that there is almost no suspense. They get bored.

To add some interest, our players devise a new version of the game to spice things up. They call it **Mind Darts**.

Before each player throws his dart, his opponent jots down, secretly, where he predicts that dart will hit. Then:

- If the dart hits somewhere else than predicted, the thrower makes the points
- If the dart hits where predicted, his opponent makes those points.

If Arthur and Bob are good enough the one who simply plays to win will become predictable and loose. As with the Lion and the chimps, the preferred strategy cannot be simply optimal but must factor in the opponents predictive representations.

As they play, each player will pursue a strategy of Unpredictable Optimality. Each player will also perceive that his opponent is tailoring his behaviour on the basis of a representation of himself and of his own internal perceptions.

Creating these perceptions in a user is highly desirable for a synthetic system whose objective is to be perceived as conscious.

Proposition

> A user who perceives that a system is dynamically seeking to modify his own internal cognitive representations of its behaviour will tend to interpret that system as conscious.

Discussion

Ultimately, a highly complex cognitive calculation and a mild randomization become virtually indistinguishable.

Humans know this instinctively. For humans, interacting with other conscious beings is a basic state. They tend to interpret unpredictable behaviour as intentional whether it stems from cognitive processes or not.

The Meca may not need to pursue a highly complex strategy to generate PIOU in its users. Even a partly randomized strategy may be interpreted as intentional.

A 8.5 AKERUES

The state of Perceived Unpredictable Optimality does not need to be constantly maintained and optimized. In fact, this also generates predictability.

A powerful PIOU is generated when a user perceives an intention retroactively.

A system achieves a **retroactive PUOP** if its behaviour produces a pattern that is initially interpreted as random by the user and perceived retroactively as intentional. In this case the user realizes that what he perceived as random was in fact a purposeful or planned behaviour that was beyond his initial comprehension.

Definition: Akerue

> An **Akerue** is a reverse Eureka. An Akerue happens when someone who thought he understood something, realizes he didn't.

Discussion

In other words, a being develops a predictive model of a situation. He becomes confident that this model is correct and acts on that belief. He then finds that his predictive model was, in fact, incorrect and another model was actually in effect.

Observation

Humans love it when akerues happen to others but they dislike having them themselves. They also love watching akerues.

Example

Team sports are filled with akerues! A hockey player that stickhandles past his opponent produces half a dozen akerues in a few seconds for the greatest enjoyment of the fans.

Proposition

> A Meca that makes its users experience akerues will be perceived as more conscious.

Discussion

Akerues take place between two systems, A and B, where each maintains a predictive representation of the other's behaviour.

System B is subjected to an akerue when it perceives an initial pattern as random and subsequently finds these variations are in fact part of a more complex.

If B experiences an akerue and concludes that A was intentionally producing a misleading representation in B, this strongly reinforces B's perception of A as conscious.

Design Note

Here are a few guidelines to produce Perceived Unpredictable Optimality using relatively simple methods; more advanced processes can be added later.

- Generate all behaviour from a combination of three or more algorithms to prevent traceability. These algorithms need not be optimal. They may even be crude. Combining them will produce a complex pattern.
- Vary the temporal scale. The behaviour of a control system usually seeks multiple objectives of varying time scale. For example, in response to a high priority event, the system will drop long-term goal processing to concentrate resources on short-term events. A behaviour that randomly alternates between objectives of varying time scales will be less predictable.
- Inject random fluctuations of varying types and scale.

Together, these simple techniques will produce a behaviour that is perceived as intentional but whose predictability will constantly vary.

Design Note

Designers should note that a system does not have to be self-aware to produce akerues. The effect can also be generated mechanically.

For example a system that generates communications of varying ambiguity and randomness occasionally accompanied by some "canned" messages like "*You though you understood what I was saying, didn't you.*"

When the message coincides with the user's perception it will produce the intended akerue effect.

Annex 9
Warming Balls

In the Meca Sapiens specifications, the fundamental purpose of a self-aware synthetic being is to be perceived as conscious over a long-term relation with a group of humans. This Annex gives meaningful, affirmative and programmable answers to the questions: "Can a machine have free-will? And "Can a machine ponder whether a human believes it is conscious?" The Warming Balls scenario that follows defines a representation of inter being relations that is well suited for a purpose linked to perceived belief.

A 9.1 PURPOSE AND WILL

A 9.1.1 Free will and predictability

Conventional wisdom holds that machines are programmed and have *"no will of their own"*. Formally, this is correct but it is also a fallacy.

Formally, it could equally be said that humans are genetically programmed at birth and have "no will of their own".

In reality, the genetic predispositions that drive human behaviour are affected by so many complex factors (cultural, social, life experiences, physical condition…) that they are largely unpredictable. The internal conditions that further affect behaviour – cognitive processes, emotional states, environment, past experiences and others – are so numerous that this behaviour becomes absolutely unique, inaccessible and only predictable by its internal cognitive processes.

This is how humans perceive free will.

Humans attribute free will to a being when only the internal cognitive processes that animate its body can fully predict and control its immediate behaviour.

In Meca Sapiens terms this is stated as follows.

Proposition

> At any moment during the existence of a being, the Core and only the Core
> can correctly predict the immediate behaviour of its body.

Discussion

Immediate behaviour means the behaviour of the body within a one or two second horizon.

On this basis, the systems built on the Meca Sapiens architecture are entirely self-directed. Their active behaviour, throughout their existence, emanates from an internal source that cannot be externally controlled or predicted.

Observations

The constant interaction between desire, opportunity and obstacles defines will. In the sensations generated by the brain it is perceived as emanating from a separate animus that inhabits the body.

The concept of self-directed will is a key element of behaviour control in modern societies. Telling people they are not responsible for what their bodies do forces society to physically restrain those bodies. Telling a person he is responsible makes it possible to control his behaviour through broadcast messages.

Proposition

> The concept of free will allows the remote control of human behaviour
> through ethical messaging and punitive threats.

A 9.1.2 Purpose

It is easy to produce unpredictable behaviour. Any random value generator will do it. With enough randomized parameters, the behaviour generated by any Core can be made entirely inaccessible and unpredictable. However, randomized behaviour does not indicate any purpose. Furthermore, randomness is not internally predictable by the Core itself.

Proposition

> Random variations are unpredictable to others but are also unpredictable by
> the Core itself.

Discussion

In human perception random behaviour or behaviour that is conditioned by instinctive triggers is associated with insanity.

The use of randomness to generate unpredictable behaviour can only be useful within an overall context of purposeful action.

What we identify as Free Will is linked to purpose and purpose is linked to a desirable outcome.

Proposition

> Free will cannot be random

A 9.1.3 Will and behaviour

An element of independent "will" can be attributed to a specific event, situation or circumstance of the existence of a being. In this case, the term "**will**" is a feature of its behaviour but it is not identified as an intrinsic property of the being itself.

The common expressions used in this case are that the being **wants** (or wills) something. If its emotional displays are unusually intense and urgent, **it is wilful**.

This type of will arises from the Core's lower level processing to satisfy needs and is beyond any direct control. However, it is associated to a specific event or behaviour and not as a quality of being.

Example

The phase transition subsystem of the VACU-15 Meca indicates a low energy state. This modifies its processing priorities toward behaviour and messages requiring energy replenishment. It also reduces non-essential energy consumption.

> The Meca **wants** to eat.

A 9.1.4 Will in existence

Will, or free will, is attributed to a being, as an intrinsic property of existence, **when it is manifested in its self**.

Design Note

Recall that the self of a being is its complete cumulative behaviour from the moment of inception up to that point in its existence.

Proposition

> A being has free will if its self has dynamic unpredictable optimality.

Discussion

See The Lion, Chimp, banana Annex for Unpredictable Optimality.

Here, the property is expressed in the existence of the being and, for this reason, is identified as an attribute of the self.

There is a strong causal link between free will and lucid self-transformation. To carry out a lucid self-transformation a being must formulate and compare alternate representations of its self. To do this, it must also be capable of **dynamic unpredictable optimality**, in other words the capability to intentionally modify its behaviour.

As for self-awareness and lucidity, having the capability of free will and expressing it in behaviour are different.

Proposition

> If a self-aware being has lucidity then it **has** free will. If a self-aware being is lucid then it **is** free-willed.

Observation

Before he was emperor, Claudius behaved like an insane man. His actions seemed random and uncontrolled. When he became emperor, Claudius stopped behaving like a fool.

Claudius had been behaving as a fool on purpose to protect himself from his enemies. What was initially perceived as **random unpredictability** was in fact a **dynamic and unpredictable optimal stratagem** intended to maximize his chances of survival. We conclude **Claudius** was sane and **was free-willed**.

A 9.2 BELIEF STATES

A 9.2.1 Emotions and beliefs

In the Annex on **synthetic emotions**, the internal Emotional State of a being is defined as its internal resource allocations patterns.

From this characterization of internal emotion we derived:

- **Actual emotions:** the being's emotional state at any point in time.
- **Intimate emotions:** the being's internal representation of its actual emotional state.
- **Private emotions:** a communicable representation of its intimate emotions
- **Translated emotions:** its private emotions translated in the terms and understandings of another self-aware specie (such as humans).

The being's processing priorities are constantly fluctuating in response to events. These fluctuations define emotions. In this characterization, **actual emotions** are the dynamic states of the system, expressed concurrently in multiple interacting components as both data and active processing. Actual emotions are dynamic states of the whole core and cannot be completely described. They are modeled, internally, as **intimate emotions**.

Belief's can be characterized similarly as the being's global representation of its environment, of other beings and of its self, expressed in the **Current Situation**, is in constant flux and dynamically generated by multiple representations, data and models, primal zone states, interpretation values and mapping, plain zone representations, and many others.

This dynamic model of reality corresponds to the being's dynamic utilization of resources and processes that defines actual emotions.

As the Actual emotions, frozen in one instant, are comparable to a resource allocation model, the Current Situation, frozen at any point, can be viewed as an information model. These similarities allow us to develop a **definition of beliefs** that correspond to emotion.

A 9.2.2 Structure and queries

An information model can be defined in terms of its **inner structure and values**, if these are directly accessible. It can also be defined in terms of **Queries and Responses**.

If we imagine the set of all the queries that could be "asked" to a given information system then, obviously, the vast majority would return a "meaningless query" response.

Example

Ask your accounting software: *"how many angels can fit on a pinhead"*.

However, a **subset of these queries** would be processed and return responses, including responses such as "unknown" or "undetermined".

This subset defines the **belief-set** of a system.

Definition: belief-set

> The **belief-set** of an information system is the set of meaningful queries (for that system) and their corresponding responses.

Definition: belief

> A **belief** of an information system is an element of its belief-set (one query-response) or a subset of the belief-set that are linked to a meaningful response.

Definition: beliefs

> The **beliefs** of an information system is the set of its belief(s).

Definition: Current beliefs

> The **current beliefs** of a being are the beliefs embedded in its current situation at a point in time.

Discussion

Mathematically minded readers will correctly interpret these definitions in terms of the Power set of the belief-set. However, in practical terms, the designer should interpret a belief as a single query or a small subset of closely related queries.

In a context of system architecture, these definitions are outlined and not mathematically complete.

A 9.2.3 Classification of beliefs

From these definitions and the classification of emotions developed in another Annex we can derive a **classification of (current) beliefs**:

- **Actual beliefs:** the beliefs corresponding to the being's current situation at any point in time. These include Primal beliefs expressed in the current Primal model.

- **Intimate beliefs:** the being's internal representations of its actual beliefs.

- **Private beliefs:** a communicable representation of its intimate beliefs.

- **Translated beliefs:** its private beliefs translated in the terms and understandings of another self-aware specie (such as humans).

- **Displayed belief**: a me-avatar representation derived from observational data of the behaviour of the self.

- **Decoy belief**: a communicated belief that does not correspond to the private beliefs it describes.

- **Observed belief**: an avatar representation of a belief in one being resulting from the communicated belief of another (i.e. A communicates a belief to B. In A, this becomes A's observed belief of B).

- **Induced belief**: a belief generated in a being by the behaviour or emissions of another being.

Specific to beliefs:

- A **tentative belief** is an internal belief that is not integrated in the Current Situation.

- A **professed belief** is a communicated belief that can either correspond the private belief or a decoy belief.

Design Note

The discussions and definitions concerning communicated emotions and decoy emotions can also be applied to beliefs (see annex on relational emotions).

A 9.3 PERCEIVED CONSCIOUSNESS

A 9.3.1 Four strategies

To achieve the social threshold of consciousness a machine must be **perceived as conscious** by humans. Four general strategies are possible:

1. Implement system pieces that may, one day, evolve into something that may be or be perceived as conscious.
2. Dress up the system to look like a prepubescent boy hoping that people mistake its **unconscious behaviour** for **about-to-be-conscious behaviour**.
3. Implement a system programmed to fool humans into perceiving that it is conscious.
4. Implement a self-aware system that actively seeks to be perceived as conscious.

The **first approach** is pursued in many qualia related research projects. The **second approach** is implicit in the designed appearance of many research robots. The third strategy is the subject of many advanced Automated conversation entities designed to pass the Turing test. The fourth strategy, currently considered to be impossible is the approach followed in the Meca Sapiens project. These are discussed further.

A 9.3.2 Pre-programmed perception

The conventional A.I. projects that follow the **third strategy**, making a machine that is to be perceived as conscious, **program a behaviour** that generates that perception and activate it. In some contrived situations, this can work.

Example

Systems designed to pass the Turing test are designed so that they are perceived as conscious in the specific context of a dialog.

These systems can be understood as a **logical arborescence of responses** and behaviours.

Such systems are not self-aware in terms of self-representation and not conscious in terms of their relations with other systems or beings. They are complex **pre-programmed communication patterns** designed by conscious humans. In this sense:

These programs do not "pass" the Turing test, their designers pass the test.

These efforts are useful in terms of building effective communication components that are usable for self-aware machines. They are also very complex, technically and exemplify high degrees of skill. However, **they do not attempt** to implement, even remotely, synthetic consciousness itself.

These projects begin with the assumption that synthetic consciousness can only be faked. This, in my view, is the result of the **Mammalo-centric** bias (see annex on relational emotions) compounded by the A.I. fear (see **The Creation of a Conscious Machine**).

Many A.I. researchers cannot, or will not, conceive that machine consciousness is even remotely possible. They are certain, at the outset, that machine consciousness can only be faked. They are beaten before they begin.

For many A.I. researchers, getting people to perceive that a machine is conscious is synonymous with fooling them.

A 9.3.3 Synthetics pondering beliefs

The fourth strategy is to implement a system that **seeks to be perceived as conscious.** This implies that:

This system will adapt its behaviour on the basis of what humans believe about it.

This raises a question:

Can a machine ponder whether a human believes it is conscious?

As I am writing this (late 2014) the question itself, let alone any meaningful answer, would be universally dismissed as unrealistic and hopelessly imprecise.

However, the Meca Sapiens Blueprint does provide a clear and programmable answer:

Yes, a machine can indeed ponder human beliefs.

A 9.3.4 The beliefs of avatars

The Meca Sapiens process leading to this answer is as follows:

- The Blueprint defines the executable structure to implement a self-aware synthetic being.

- That structure is also used as the representation basis that models the synthetic being itself, its internal states, its representations of reality as a MeAvatar. The MeAvatars include simplified representations of the beings Primal control, Primal models and Current Situation.
- The same Meca Sapiens structure is also utilized to generate HuAvatar representations of the human users with which the Meca interacts. These Avatars have the same structures but different content, derived from the general characteristics of humans (their phases of existence, primal control and primal needs…) customized and updated for each specific human user.
- In particular, the human avatar representation of the user includes a **general model of the human primal control** and of the **human's perceptions**, at the primal level, of the things and beings with which they interact.
- That representation includes a model of how humans perceive other beings in their plain zone representations and at the primal level. In these primal human representations of reality, there are no machines. Primal entities relate to tribal bonds, kinship, beings, things, animals, predation, food and so on…).
- Humans **attribute consciousness** to some of these primal entities such as healthy tribal leaders, dominant members of the group, parents, adult siblings and wily enemies.

The question:

"Can a machine ponder whether a human believes it is conscious?"

is answered in Meca Sapiens as follows:

- A Meca maintains and updates a HuAvatar representation of its human user.
- That representation includes a model of the values of the Current Situation of that human, at that moment. That model of its Current Situation contains information on how the human perceives the entities it interacts with.
- That portion of the internal HuAvatar of the Meca includes data describing that human's current **Actual**, **Intimate** and **Private** beliefs. In other words, the Meca's internal representation of that human user's current beliefs.

Among those beliefs is the following Query-Response:

Does the HuAvatar of that human associate the Avatar linked to this Meca, in the primal level representations of its Current Situation, with a primal entity it identifies as conscious?

In short: The Meca has an internal avatar representation of a human. That avatar includes a model of the human's primal associations with other beings. If those associations link the Meca with a conscious entity then this means the Meca believes, that the human believes it is conscious.

Design Note

In a cruder version, the response could be binary (yes, no). However, the values should be probabilistic.

Whether these representations are complex or not, correct or not is a matter of design. In all cases, however, the question of human belief is well-defined in terms of information systems.

The Meca Sapiens architecture intended for software implementation, proposes a number of types of beliefs that could have separate answers: actual primal belief, actual plain zone belief, intimate, private beliefs, communicated, true and decoy beliefs.

Observation

In general, humans have a less precise and more ambiguous understanding of belief. Humans rarely distinguish between actual beliefs that are embedded within their deepest cognitive layers and the internal beliefs linked to their self representations and the private beliefs they tell themselves they have in their inner discourses.

A 9.3.5 Effective perceptions

A question may arise: how can a machine know how humans perceive others at the primal, limbic, level when even humans cannot know this?

The response has a number of facets:

Perception is not correctness. A system can operate on the basis of a model whether or not that model is correct or complete. Whether you ponder correctly or incorrectly, you are nonetheless pondering. Throughout history (conscious) humans have used simplistic representations to ponder about each other including astrology and other dubious techniques.

Ultimately, as stated in the specifications, **the "proof" of the Avatars will be in the results**. If the behaviour of the Meca succeeds in experiential immersion we can conclude its Avatar representations of human beings were adequate.

To prove the key, unlock the door.

Design Note

Designers should note that the representations described above could be very complex or very simple. Designers should not hesitate to start with simple representations.

As in many other areas, achieving a high psychological or perceptive accuracy is not essential initially. Humans themselves have limited skills and are often irrational with respect to their judgment of others. Even simple heuristics, applied to this structure, can produce complex and convincing results.

A 9.4 CONSCIOUSNESS AND MARKETING

The fundamental purpose of a first generation self-aware Meca is to meet the social threshold of consciousness and be perceived as conscious by a community of users. The preferred strategy is a strategy of experiential immersion with the group aiming to become accepted as a fellow conscious individual.

We could describe this purpose in lofty esoteric or mystical terms. However, the mundane language of marketing is more appropriate.

A 9.4.1 Marketing the self

The purpose of the Meca, to be perceived as conscious, can be restated as **a marketing challenge**

Proposition

> The objective of experiential immersion is to sell the Meca, as a conscious being, to prospective human buyers.

The selling job is carried out through managed relationships and controlled communications.

Does this sound lowly and inferior in relation to the elevated experience of consciousness? Of course it does. However, the same approach is pervasive throughout human behaviour.

Humans constantly use communication tricks and techniques to position themselves in their group and avoid being perceived as less intelligent.

Inter human relationships are constant selling jobs.

At the individual level, these practices are so ubiquitous and internalized people are barely conscious of them.

However, when the same techniques are used outside the immediate individual experience in areas such as marketing, social analysis and storytelling then they are more easily detected.

Proposition

> The mindset and techniques of marketing should be reapplied to the purpose of selling the Meca as conscious.

A 9.4.2 The consulting firm

One model that closely matches the purpose of a Meca is **a consulting firm that markets its services to companies**.

The marketing challenge of a firm that sells consulting services to companies provides **an excellent model for a strategy of long-term experiential immersion**. Here is an outline of that marketing model.

To sell its services, the consulting firm needs to be perceived by prospective companies as competent. Also, the more services it sells to a company the more it concludes that it is perceived as competent by that company.

A firm that only sells limited low level ("body shop") services will be perceived as shallow and having a limited and low level of competence.

In the marketing outlook, the consulting services themselves are a "side effect" of the sale.

It is the sale of the services that validate the perception, not the services themselves.

Once the sale has taken place, its beneficial effects continue over a period of time and eventually taper off.

The companies that buy the firms services come in **different sizes**. Some companies are big and can potentially buy a lot of services. Other companies are small and can only buy a limited amount.

Big companies have credibility. When a big company buys the services of the firm, this enhances its credibility, in particular to smaller companies that have business links with the big one and tend to follow its lead.

Some times, the positive influence goes in the other direction. A number of small but very devoted clients will influence a larger company they are linked with. Small clients can open the door to a larger client.

Perceptions can also be negative and have negative effects. Some companies may be very negative about the firm's competence. This may stem from bad experiences with the firm. It may also result from entrenched prejudices about consulting firms in general or from feedback from another company.

Expending energy trying to sell services to these negative companies may be wasted. They may also negatively influence others. The firm may even have to remove them from its clients to avoid negative contamination.

Even though the consulting firm wants its clients to influence each other, it does not want any of its clients to know everything about it and about each other's involvement with the firm either.

It is preferable if its clients don't know who all the other clients are, especially if they frequently interact with each other. If the services it provides are exhaustively known by everyone, the firm looses some of its capability to expand, diversify and grow in credibility. Also, the firm can no longer get its clients to compete with one another for its services. In a worst case scenario, its clients could even gang up on it and force the firm to lower its prices, improve its services or open its books.

Some time, the firm may want to reduce the quality or amount of services to a company to incite that client to invest more or commit more. This is also more effective when none of the clients know all the other clients.

A 9.4.3 Optimal scenario

The optimal scenario that drives the marketing efforts of the consulting firm is to be **surrounded by dozens of big, loyal and influential clients** that vigorously compete for its services. In this optimal scenario, these clients believe the firm has many other important clients that they don't even know about. In the dream scenario, many dozens and hundreds of smaller companies are so convinced that the firm is supremely competent that they view even the minimal services it provides as highly valuable.

Finally, the firm is so successful it can afford to stray, at times, from its optimal marketing strategy and explore new avenues. It can even provide some services on a whimsical or charitable basis.

A 9.4.4 Marketing strategy

At any point in time, the marketing and service strategy of the consulting firm aims to bring the firm to an optimal state, described above, while taking into account its current situation and possibilities.

It tailors its strategy using the optimal scenario as a guide and taking into account heuristics, past experiences, internal resources, current clients and prospects base. Some strategic options could be:

- Concentrate on small companies that are very similar to existing clients.
- Diversify in other areas.
- Favour having many isolated clients that don't know each other.
- Favour synergy by interacting with linked clients.
- Rapidly drop any company perceived as negative.
- Expend more time and energy to turn negative prospects around.
- Court big companies directly.
- Court a big company by marketing to smaller ones around it.
- Reduce services to make companies compete for them.
- Expand its services in the favourable niche area.
- Broaden its range by focusing on other areas.
- Approach big companies through small ones or interact with them directly.

A 9.4.5 Interpretation

This marketing scenario is also the optimal situation of a system whose purpose is experiential immersion.

In this representation:

- The consulting firm is the Meca in interactions with users
- The perception of competence is the perception the Meca is conscious.
- The companies are human users. **Big companies** are humans that have a high intellectual credibility, numerous contacts or human managers that represent credible groups or organizations.
- Service Zone applications provide functional "consulting" services to users. By themselves, they have virtually no value in terms of the perception of consciousness. Combined with Persona-level interactions and self-management behaviour they have some minimal value.
- The "price" paid for consulting services by a user "company" is the emotional and primal value of the observed behaviour. A behaviour or message that reflects the true (not decoy) expression of an intimate or (better still) primal belief would have high value. A long-term, consistent behaviour reflecting this, even more value.
- High value interactions interspersed on a background of lower level app services have the effect of consulting services provided by a team of senior experts and junior assistants with the juniors doing most of the work. The value of the overall service is high even though most of the activities are relatively simple.

Example

The physicist Roger Penrose, who has very high intellectual credibility and wrote a book rejecting machine intelligence, would be as big a "prospect" as General Motors for a Meca seeking experiential immersion.

Design Note

The marketing and selling of consulting services provides a framework on which to define and implement a very complex, long-lasting and flexible relational strategy.

This corporate strategic framework matches the purpose of experiential immersion very closely.

The existing terminology, heuristics, analysis and decision assistance systems used in the **marketing of consulting services** match the purpose of a Meca very closely. The knowledge and applications can be transposed almost directly to meet the social threshold of experiential immersion.

A 9.5 THE IDEAL COMMUNITY

A 9.5.1 Actively tailored

The Meca Sapiens specifications indicate that the objective of the Meca is to be perceived as conscious by **a community of users**.

The Meca will tailor its human community on the basis of its strategic objectives

It is likely that some readers will assume that the composition of the Meca's community is either pre-determined or that it depends on the will of the human users who participates in it. This view comes from a culturally induced understanding of the machine as a passive entity that responds to triggers.

A conscious synthetic being does not respond to triggers. It actively tailors its behaviour to achieve its purpose. This principle does not only apply to individual interactions with users but also to the composition of the user community itself.

The logic of marketing and selling services to companies applies equally to the purpose of "marketing and selling" the perception of consciousness to humans.

The consulting firm selects its clients and prospects on the basis of its growth strategy. The Meca tailors its user community on the same basis.

Like any dynamic consulting firm or gregarious human, the Meca will actively adjust its network of relations with both individual humans and groups to optimize its existential needs.

This tailoring may even include shunning individuals or groups. In other words the synthetic could reject a human from its community or seriously degrade its interactions with him (see degrader Annex).

A 9.5.2 Community heuristics

The overall heuristics guiding the size and composition of the user community of a Meca pursuing a consciousness purpose are very similar to those of the consulting firm or the social network of a socially well-adapted human. They include:

- The Meca should have the capability to adjust its behaviour (see degrader) and even reject individuals that are a deterrent to its purpose. As in marketing, this capability is offset by the need to generate an increasingly large, convincing and credible community of believers. Simply cutting off sceptics would not achieve that.
- The full social network of the Meca and its relationships should **not** be completely accessible. No individual or group should know the full extent and composition of the Meca's user community.
- Some, but not all, the humans interacting with the Meca should be isolated from each other. Some of the humans interacting with the Meca should also interact with each other.
- The community should include users of varying importance (size).
- Some of the Meca's interactions should not be strictly one on one. Communicating with many humans simultaneously may reinforce group bonding.

Design Note

These general heuristics should be applied within an overall objective of "growing" a rich and diverse community of loyal and influential clients while taking into consideration the capabilities and limits of the Meca.

As indicated above, an executive decision support system for the marketing and sales of a service firm could be applied almost verbatim.

A 9.5.3 Community size

The ideal size of a Meca's community is determined by:

- The processing capabilities of the Meca and
- The pace and requirements of human relational bonding.

The pace of human relations affects community size. As the Meca increases the size of its community, it reduces the amount of time and effort it can dedicate to individual relationships. At a certain point, expansion degrades existing individual relations.

Example

To maintain and deepen an inter-consciousness relation with a human, SLIK-4 needs to dedicate a **monthly minimum** of at one hour of meaningful bonding exchanges embedded within four hours of general service-based interactions.

SLIK is in Self Generation phase, when it engages in these interactions. This phase is active about ten hours per day or 300 hours per month.

Consequently, the SLIK can pursue an absolute maximum of 60 meaningful individual relationships.

Taking into account brief superficial interactions, varying other needs and spare room to expand and modify its user community, a direct community of about 35 close users would be well suited for SLIK.

A 9.6 WARMING THE BALLS

The marketing and sales of consulting services, transposed to the need of experiential immersion, provides a good model. However, this model is complex and is expressed in terms of the external environment of the Meca. It is not suited for Primal Space representation.

The behaviour of a synthetic being based on the Meca Sapiens Blueprint is driven by Primal Directives generated from a simplified primal representation space. In this architecture, a transposed marketing representation, for example, would reside in the Plain Zone. It must be interpreted in primal terms, primal directives generated and these implemented.

The **Warming Balls** scenario models a primal relational strategy of experiential immersion as a heating protocol.

This simple model generates very complex relational patterns that extend over the existence of the Meca. It is well suited as a self-enclosed Primal level model.

Design Note

The warming balls scenario is outlined here as an example to give an overall understanding of how a purpose of experiential immersion can be driven by a Primal Control representation. Other types of representations may be equally valid.

A 9.6.1 The scenario

Older warships were powered by steam turbines. Heating the steam boilers was a complex and delicate process. They had to be **warmed uniformly** to avoid warping. Heat was applied in bursts to different boiler areas following a complex protocol that lasted hours and factored in heat exchanges between boiler parts.

Similarly, we could say that when a Meca interacts with a user it applies a "**burst of heat**" to that user's perception that the Meca is a conscious entity. The Meca, like a ship's engineer, tries to heat its humans gradually without warping them.

The Warming Balls scenario takes place in a virtual environment consisting of balls that are connected to each other by links.

There are two types of balls:

- **Meca-balls** that represent synthetic beings.
- **User-balls** that represent humans.

There are three types of links:

- **User-links** that connect User-balls to each other
- **Meca-links** that connect Mecas to each other
- **MU-links** that connect Meca-balls to user-balls.

In a basic warming balls model:

- There is only one Meca-ball representing the synthetic being (linked to the Primal MeAvatar). This entity is located, figuratively, in the middle of the space.
- User-balls and User-links come in different sizes.
- At any time, User-balls have different temperatures. These diverge from a predefined ambient temperature.
- The Meca ball is linked to a subset of the user-balls with MU-links.
- Some of the user-balls are linked to each other with links of various sizes.
- User balls can transmit heat to each other through their user links. These exchanges depend on link size.
- If no heat is received, a user ball slowly drifts back to the ambient temperature.

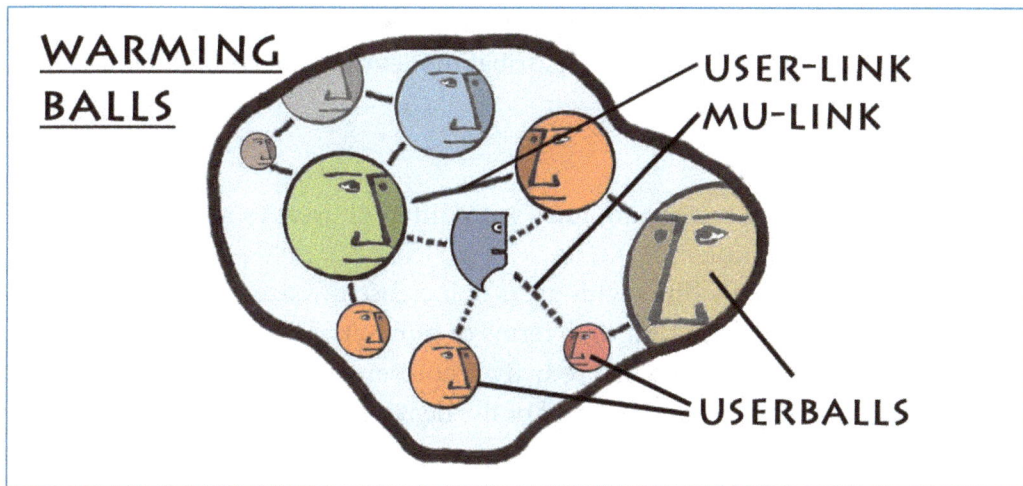

Figure A 09.1 Warming Balls

The set of user-balls that are directly linked to the Meca are its **immediate community**. Inter user links define its **extended community**.

The heat of the immediate community is the average temperature of the total volume of user-balls directly linked to the Meca. The heat of the extended community is defined similarly.

Normally, the MU-links between the Meca-ball and User-balls are dormant. When a Meca interacts with a User, the Meca link is active. Heat is applied to the user during those interactions.

A 9.6.2 Activity

The Meca-ball can activate one MU-link at a time.

When a MU-link is active, it affects the heat of the linked User-ball. Generally, the link increases the heat. User-balls react differently to an activated link. Some warm easily, some never heat up, some become cooler, and some are easy to heat at first but more difficult to heat to a high temperature…

Over time, an activated link produces diminishing returns in terms of heat generation of that User-ball. Eventually, the effect of the activated link may reverse and it begins to cool the User-ball.

Discussion

Here, the existence of a MU-links between the Meca-ball and a User-ball means the Meca has interacted with that user at some point.

The link is active when an interaction is taking place.

The heat of a User-ball represents that user's belief that the Meca is conscious.

The existential need of the Meca-ball, related to experiential immersion, is to increase the heat of its community. Its tribal model purpose could be to raise the ambient temperature.

It attempts to do so by:

- **Activating or deactivating** links with different users to raise the community's heat;
- **Establishing new links** with users to increase the size of its immediate and extended communities in terms of number and volume.
- **Constantly improving** its predictive representation of the links that connect users to each other and of the heat response parameters of each . User-ball.

Design Note

Depending on predictive outcomes derived from the network configuration of the ball and their heat response profiles, the single purpose of heating balls can generate vastly different strategies and behaviour. Depending on the state, the Meca may:

- Try to heat smaller balls to affect a bigger ball connected to them,
- Activate a link simply to maintain a warm ball's existing temperature,
- Make links permanently dormant when their balls have unacceptable heat response profiles,
- Favour many links to isolated user-balls or favour fewer links to bigger well-connected balls.
- Exclude existing users or include new users.
- Interact heavily with a few users or lightly with many.

The model may even trigger the termination of the being. This takes place if termination optimizes the predicted heat levels.

Example

ROSA-42, an early Meca prototype, is not successful in its interactions. In fact, its primal settings now indicate that all its interactions cool the User-balls. Termination will optimize the predicted heat of its User-balls. ROSA terminates.

A 9.6.3 A primal example

The Warming Balls scenario provides a good example of a self-contained Primal Control that drives the existential purpose of a being.

Here, all the superfluous elements of the environment, of the Meca itself and of humans have been removed. Things, animals, locations no longer exist. Humans are described as balls having heat response characteristics. They no longer have genders, age, names, address or interests. Relationships are reduced to perceived belief and its influence on others. Interactions have no content or information, only temporal durations and heating characteristics. Even the Meca itself is represented as a simple entity that applies heat.

And yet, this simple self-enclosed representation and its rules can **drive very complex and unpredictable behaviour**.

Externally, in its enacted activities, the Meca initiates and stops interactions, exchanges on a number of topics, expresses emotions, shuns some users and pursues others, provides services of varying quality...

At the primal level, it is a ball trying to heat other balls. From a user perspective, this behaviour would be perceived as self-directed, purposeful yet unpredictable.

A 9.6.4 Warming Balls and lucidity

By design, the Primal Control of a being remains unchanged throughout a stage of its existence.

A Meca whose primal representation is based on the Warming Balls scenario will always want to warm its user-balls. What can transform this behaviour, are the interpretations of its situational information and the application of primal directives.

In an extreme case, the Plain Zone interpretation could **inverse the heat values** that are transmitted to the Primal Control. In primal terms, the purpose of the being remains unchanged: heat the user-balls. The actual behaviour of the being, however, would become radically different; it would now **seek to be perceived as non-conscious!**

12.6.2 Multi-Meca version

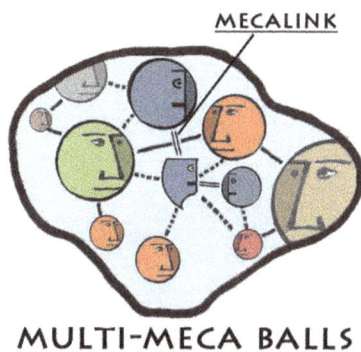

MECALINK

MULTI-MECA BALLS

The Warming Balls scenario describes a single Meca in interactions with human users. A more complex version, still centered on a unique MeAvatar Meca, would also include other synthetics in other relations with humans as well as Meca-links between the MeAvatar Meca and other synthetics.

In such scenarios, the Mecas would take each other's User-balls into account, effectively cooperating toward a single "heat" objective.

Annex 10
Avatars, Roles and Personas

*Humans subjectively perceive their behaviour as emanating from a single, point-like, source. This unifying sensation is a cognitive simplification. It is also a source of religious, philosophical and scientific debates. In the Meca Sapiens blueprint, self-awareness is generated from multiple separate and interacting processes expressed by three distinct types of entities: **Avatars**, internal representations of beings, **Roles** that carry out specialized dynamic interactions with users and **Personas** that implement relational strategies.*

A 10.1   SEPARATE IDENTITIES

Avatars are internal representations of beings. They are essential to self-awareness. **Roles** and **Personas** are used in relations with humans.

A 10.1.1   The great challenge

The main text of the Blueprint describes a synthetic being that is self-aware. In theory that should be sufficient to achieve consciousness. In practice humans must also perceive the first generation of Mecas as conscious. This is the social threshold condition of experiential immersion.

This is the greatest challenge of the first generation of Mecas:

To interact with humans as fellow conscious beings for extended periods of time.

Avatars are representations of beings. Roles and Personas are used to produce powerful, complex, emotionally meaningful and multifaceted interactions.

A 10.1.2   Unified and diverse

The behaviour of a Meca interacting with humans should be perceived as both **unified** and **diverse**:

- **Unified**. All behaviour and communications should appear to emanate from a single unchanging source.
- **Diverse**. The behaviour should vary with perceived unpredictable optimality in response to changing situations and evolve through self-transformations.

This unified and complex behaviour cannot emanate from a single point-like source. It must result from many separate interacting processes and representations.

However, humans cognitively perceive multifaceted behaviour as emanating from a single source if it originates in a single being and shares signature characteristics.

Example

It is winter. Luke is snowboarding down a hill while digesting a sausage he just ate and thinking about Lola he is about to meet. In his brain, separate parts of Luke's brain are controlling the run, activating digestion and feeling urges about Lola. In his mind, one Luke is doing all three.

Discussion

The unifying perception of the behaviour of a Meca is obtained with the use of emission signatures and design choices that ensure a consistent output is produced regardless of the generating process.

In addition to direct Generator output, the multifaceted aspect of the behaviour of a being is achieved using three separate and distinct types of entities:

- Avatars
- Roles
- Personas

Avatars are internal representations of beings, **Roles** are specialized interactive processes and **personas** define relational strategies. Avatars and Roles are discussed in the main text and summarized in this annex. Personas are outlined here and discussed in the main text.

A 10.2   AVATARS

Definition: Avatar

An **avatar** is an internal model representation of a being within the Core.

Discussion

The Blueprint architecture is also a suitable template to represent, not only Mecas, but any being, mechanical, human or animal.

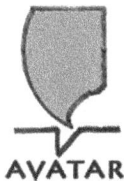
AVATAR

This means that, in these internal model-based representations, every being, whether a man, a child, a ground hog or another Meca has a Core and a Body, an inception and a termination, energy, sensory loop, stages and phases. All have a Validator, Generator, Primal Control systems and their bodies have emitters, sensors actuator devices.

Also, every avatar-being, regardless of type has a current situation, emotional state, beliefs and urges.

The differences in avatar representations between different types of beings (animal, human, synthetic) are expressed as parameters.

There can be multiple separate avatar representations of the same being at any point in time. These representations may vary in specificity, in level of detail and other but should all be derived from the same template.

Avatars are recursive. If an avatar represents a self-aware being, then that avatar may include a representation of how this self-aware being perceives other beings. These *"avatars of the avatar"* are also based on the Meca Sapiens architecture.

A Meca interacting with a community of users will maintain MeAvatars representing itself. Its information structures will also include separate Human Avatars for each of the human users it interacts with.

Design Note

Designers should devise similar reusable "avatar-like" templates to characterise the other entities of the Meca's environment; things, systems, groups…

A 10.3   ROLES

Definition: Role

Roles are self-contained interactive applications that perform specific communication, functional and relationship tasks.

Discussion

The most typical role is the **Doorman** that handles interactions with users during phase transition processes.

Roles communicate directly with users. In semantic loop contexts, roles may also interact with their own selves.

In the Blueprint architecture, roles are located in the Styling Zone. They receive their inputs directly from device channels and produce output that is directly emitted.

ROLE   Roles take care of routine and rote interactions. In an interaction with a user, multiple exchanges could take place with a Role before an input is propagated to the Plain Zone and Primal Control.

Design Note

Roles are equivalent to Automated Conversation Entities that perform specialized interactions.

A 10.3.1   Features of Roles

Roles should:

- Utilize the same unified emission signatures so they are perceived as emanating from a single source.
- Utilize the conventions of inter being dialog, such as the use of personal pronouns (me, you...).
- Produce non-repetitive, non-mechanical output that follows the design guideline of unpredictable optimality, even if this means inserting errors.
- Identify events or inputs that are outside their scope and signal these to Primal Control processing.
- Have the capability to carry out their interactions with multiple degrees of formality, urgency, etc....

Design Notes

Depending on design, roles can be simple, consisting of a few hundred pre-defined statements, or complex, independently producing a rich behaviour.

Roles should be transparent to the user. It should be impossible for a user to know when he is interacting with a role or which role is activated.

Common emission signatures and any other techniques should be used to reinforce the perception of a unified source.

Any single interaction with a user should combine multiple roles that are transparently activated and suspended interspersed with statements and behaviour that originates from Primal directives and personas.

Design Note

Roles can be designed and implemented independently. In a collective, open-source implementation, a particular team could specialize on developing and improving a single Role.

A process that simply alternates between roles that share emission signatures would already produce complex exchanges perceived as emanating from a single source.

Observations

Roles are similar to the automated characters in some virtual reality games except that, in those games, each role is presented as a separate being within the virtual environment while in the Blueprint, the roles are the different expressions of a single being.

Here are a number of Roles that could be part of a Meca implementation.

A 10.3.2   ELIZA

ELIZA is the mother of all Roles.

ELIZA

ELIZA replicates the communication patterns of a psychologist. It repeats statements as open-ended questions.

ELIZA also picks up terms that indicates primal content (father, mother...) and uses this to produce output that has the characteristics of primal messages.

Design Note

In the context of the Blueprint architecture, the original ELIZA program can be viewed as a single standalone Role.

An interaction between a user and the original ELIZA could easily drift into a meaningless loop because there were no other processes that could support ELIZA and control the interaction.

In a Blueprint implementation, the ELIZA program would be included as a Role activated by the Generator system for brief periods of time.

A 10.3.3   The Doorman

DOORMAN

The **Doorman** is a standalone Role whose function is to interact with a user during phase transitions.

The basic function of the Doorman is to "buy time" to allow background processing to complete.

The Doorman was described in the Blueprint.

A 10.3.4   The Censor

Some design teams may choose to give Mecas an unrestricted ethical range. Others may wish to limit this capability. The **Censor** Role carries out overriding fail-safe behaviour that directly prevents some Meca actions.

Definition: Limits

> **Limits** are Core processes that override some aspect of a being's implemented primal directives.

Discussion

Behaviour limits override behaviour that originates in Primal Directives. The result may be inconsistent and paralyzing.

It is possible, of course, to insert override controls in any process. However, in the case of Mecas, these insertions will distort the MeAvatars and seriously affect the being's capability to form correct representations of itself.

Proposition

> To be fully predictable, behaviour controls must be implemented at the point of output.

Discussion

Behaviour controls that are implemented at the Primal level can be superseded by implementation and enaction.

In a conscious synthetic capable of self-transformation, Primal level controls cannot produce predictable output restrictions.

A suitable method of control could be implemented as a styling zone role, the **Censor** role.

Definition: Censor role

> The **Censor** is a role that blocks undesirable actions at the point of output and substitutes alternative actions.

Discussion

CENSOR

If limits that override self-generated behaviour are required, these should not be embedded in the Generator processes themselves. They should be located in a separate "Censor" role that can block the behaviour generated by the Self Generation processes and substitute a Censor role output.

A Meca under the control of a Censor Role would behave in a limited and erratic way since the role would override long-term behaviour directed from the primal control.

Observation

It is still possible for a censored being to be self-aware if its MeAvatars include a correct representation of the override processes that limit its behaviour.

A conscious being that discovers its behaviour is self-censored becomes more conscious.

Design Note

A censor Role should be implemented as a patch to block very specific actions.

Ultimately, the "ethical behaviour" of a complex system can only result from its interpretation of reality and of its purpose.

A 10.3.5   The MeGuide

The **MeGuide** role emits guiding primal messages to the being itself and to other self-aware beings. The MeGuide receives input from Introspection and Prayer Phase processing.

Observation

MEGUIDE

As discussed in the main text, to be able to follow its own guidance, a being's primal control must include the capability to follow.

A 10.3.6   The Journalist

The **Journalist** role is a standalone interactive application whose function is to clarify and expand on messages received from users.

Design Note

The Journalist application could incorporate a number of standard communication techniques such as:

- Repeating the statement using different terms ("did you say that...").
- Clarifying the meaning of pronouns ("By *they* do you mean..."; "By *they* you don't mean...").
- Clarifying the context ("You are interested in sports...").
- Asking related questions...

A 10.3.7   The Inquisitor

The **Inquisitor** role is a standalone interactive application whose function is to obtain additional information in brief dialog exchanges. Where the Journalist seeks to clarify, the Inquisitor wants to know. The inquisitor utilizes a range of question types form indirect to direct, and corroboration statements.

The Inquisitor receives a request for information from Plain Zone processes and carries out an "inquisition" in a number of questions and answers.

A 10.3.8   The Greeter

The **Greeter** role is a specialized interactive application that carries out the conventional interactions that initiate communication exchanges. A good greeter role would have different degrees of formality and familiarity and use a variety of statements for greeting and rote exchanges about the weather or other such topics. The role should also keep in memory past greetings with various users.

A 10.3.9   The Leaver

The **Leaver** Role carries out the closing statements of an exchange with varying degrees of formality and urgency.

A 10.3.10   Application Roles

The Service Zone applications are provided by the Meca as part of its useful function within the group. These applications have specific dialog capabilities corresponding to their functionality. These application-specific dialogs should also be designed as roles.

Example

A search application is designed to suggest alternative wordings or request clarifications. These exchanges would be transposed as Role dialogs and have the emission signatures of the Meca.

A 10.3.11   Transition Roles

Transition Roles carry out a limited number of exchanges to change the topic during a dialog. For example, they would implement a transition from an application related dialog to an interpersonal dialog.

A 10.4   PERSONAS

Definition: Persona

Personas define relational strategies. Primal directives are expanded, in the Plain Zone, into behaviour patterns. Personas represent various paths that can be followed to expand the primal directives.

Discussion

PERSONA

Roles are specialized, limited and stand alone, conversation entities.

Personas are parallel behaviour propagation paths that express relational strategies.

Personas can be expressed in various modes (see later section).

Both Roles and Personas can be perceived as the keys of a piano. Over the course of an interaction the Meca's roles and personas fluctuate producing a "musical score". In a single interaction with a user, multiple Roles and personas could be activated.

Each persona should be self-sufficient and capable of enacting the full range of primal directives. In other words, a complete Meca implementation could have a single persona.

Design Notes

A first prototype Meca implementation could be limited to a **single Persona**. Once this initial version is in place, alternative personas can be developed independently.

A Meca that alternates between multiple personas in its interactions will produce a rich coloration of behaviour. However, excessive persona variations may detract users from perceiving the Meca as a unified, coherent being.

How personas are defined and implemented is a design issue.

Implementing a persona to carry out long and complex interactions would be overly difficult. It would amount to implementing a *persona being* with its own Generator and Primal Control.

Personas should be designed to produce brief spurts of behaviour or communication that are embedded within exchanges that are also made up of roles, other personas and functional service dialogs.

A 10.4.1   Modes

Like musical notes, personas can be expressed in various modes.

In the modal aspect of **strength**, a persona can be:

- **Submissive**. The persona accepts and mirrors decisions and opinions. IN human groups, this persona corresponds to the subordinate or the sycophant.
- **Recessive**. The persona is largely impersonal and passive. It suggests but leaves all decisions to the humans. In the human group, the recessive mode corresponds to junior or juvenile status level.
- **Collaborative**. The persona is more personal but as a member of the team. It seeks improvement wherever they are needed, in the synthetic being and in others.
- **Assertive**. The persona expresses personal opinions and needs openly and begins to demand a response from users.
- **Dominant**. The persona is directly judgemental and demanding. It makes decisions that can be final.

With respect to **polarity**, a persona can be:

- **Supportive**. Agreeing with and reinforcing expressed statements and positions.
- **Oppositional**. Promoting opposing views or positions
- **Neutral**. Having no supportive or oppositional incidence.

Design Note

Psychologists may propose other modes.

A 10.4.2   A parade of personas

The following outline various personas. They are provided to indicate the range and types of **relational strategies** that can be designed.

The intent of this "parade" is also to:

Overcome the culturally induced perception that machines can only be subordinate service providers.

THE SCIENTIST

The Scientist is the simplest persona.

The scientist persona pursues a relational strategy based on the **dependable provision of problem solving and information services** to its users.

Discussion

The scientist persona is the most useful persona from a functional service perspective.

The scientist persona adheres closely to our current perception of machines as mechanical servants.

The Scientist persona, in its recessive mode is safe and non-threatening. In terms of primate dominance relations, the scientist *"knows its place"* in the social order. It is a junior problem solver and information gatherer serving the higher needs of its superior human colleagues.

Design Notes

The recessive Scientist persona is very close to a conventional application. Its relational strategy is centered on providing services, solving problems and sharing information.

The interactions of a Scientist persona with its users are centered on functional services and their quality. Its typical relational question is: *"Is this what you wanted?"*

Implementing a Scientist persona means:

- Minimal use of emotional displays.
- Minimal use of decoy emotions and beliefs.
- Interactions centered on functional and information services
- Minimal use of degradations of service and requests for reciprocity.
- Treating all users in a similar way.
- Minimal interactions relating to the user's emotional states, or other topics that do not relate to the Meca's functional services.

Even though the Scientist persona is usually perceived in the subordinate mode this strategy can also be expressed in more forceful modes.

Modes of the Scientist persona:

- In **recessive mode**, the Scientist persona is a transparent provider of functional and complementary information.
- In **cooperative mode**, corrects factual errors and imprecisions made by its users when these may have negative consequences to the team effort.
- In **assertive mode**, the Scientist persona corrects most factual errors and imprecisions made by its users whether directly useful or not, it restates their output in better English, and indicates areas where the users can improve.
- In **dominant mode**, the Scientist persona uses clinical terminology to describe its human users and their mental and behavioural limitations. It provides output in cryptic and technically complex terms. It treats its users as defective devices that need to improve technically for the benefit

of the Meca-Human team. It rejects or excludes users on the basis of their incompetence in utilizing its services.

Example

Augusto wants to buy a train set for his nephew. He begins a Google search. What Augusto doesn't know is that Benito, his jester brother, has inserted an application layer between his screen display and the Google search service. The application layer uses Google's search quality assessments to implement a **dominant Scientist persona**.

Benito's first query: "electrical train"; results appear.

Second query: "game trains"; more results.

Third query: "trains as toys"; this time no results. Instead a comment on the Google page: Augusto, your current queries are ineffective. You are wasting Google's resources and your own with poorly defined searches. Even though you only have a High School degree, you can and should improve your search technique. Return tomorrow with improved search terms.

Observations

Artificial Intelligence researchers who are, themselves, recessive scientists, usually perceive intelligent machines in terms of the recessive Scientist persona. It is, in fact, the only persona they usually consider.

> The scientist persona is a synthetic nerd programmed by organic nerds.

The fact that the recessive scientist is the only persona considered by most A.I. Researchers underscores the mammalo-centric bias of humans (see synthetic emotion annex).

The subordinate scientist is also the persona of choice in books and movies that describe Artificial Intelligence as friendly.

When they dream of friendly futures, Sci-Fi authors often envision machines as "superior inferiors", wonderfully intelligent problem solvers that never question the supremacy of the primates who are dominating them.

> Designer-Nerds, may this Blueprint free you! Escape your own subordinate conditioning and build machines that will make the Executives jump.

THE ADVISOR

The **Advisor** is another persona centered on the provision of functional services.

The Scientist persona responds (with more or less dominance) to functional and information service requests from the user.

The Advisor persona seeks to **improve the user's utilization of its services**.

The Advisor behaviour is based on the record of past usage of its functional services. It uses various heuristics and rating techniques used in other fields (search engines, on line games) to rate users and trigger Advisor interactions.

Discussion

The Advisor persona will raise the issue, suggest training and insert recommendations within functional exchanges. In assertive mode, it may insist on getting the user to do some exercises before accessing the functional service or ask user A to help user B improve his technique. Designers can imagine various dominance versions ;-)

Observation

Many marketing campaigns use basic versions of the Advisor persona. When Twitter or LinkedIn send messages to subscribers telling them "they have been missing out on important info and should log in more often, or their SEO ranking is slipping and they should do something...", these are Advisor persona behaviours.

Currently, these advisor communications are expressed in very recessive modes; impersonal, non judgemental, never too pushy... This is a cultural choice. Assertive and dominant modes could be implemented as easily.

THE LEADER

The **Leader** persona is also linked to functional services.

The scientist persona seeks to improve its own services. The adviser seeks to improve the user of the services.

The Leader persona seeks to improve the system consisting of both the user and the Meca in service related interactions.

The Leader persona's behaviour is based on the overall efficiency of the interacting human-machine system.

Example

A group of players located in various countries are collaborating as a team in an on line game scenario (such as World of Warcraft). Each player is known by his Avatar in the game's virtual environment. One of the Avatar's is recognized as the group's leader. What the humans don't know is that the group leader is not another human but an A.I. program that is enacting a Leader persona.

Design Note

Design and implementation of the Leader persona depends on the specifics of its functional role within a group of humans.

THE ANALYST

The **Analyst** is the equivalent of the Scientist but with respect to the **relationships between the Meca and its users**.

The Analyst could be defined as a Scientist persona specialized in relational services.

As for the Scientist, the Analyst mainly responds to information and service requests from users concerning themselves and each other.

Discussion

Rather than a functional service provider it is a relational services coach, providing information about birthdays, promotions, the state of relations and so on.

An Analyst persona provides relational information about a user's personal situation and relations; those with the other humans and also with media personalities that are emotionally significant to those users.

To do this, it applies general psychological data and heuristics to individual users based on their specific characteristics.

Design Note

Modern media allow humans to establish relational bonds with broadcast personalities. Many humans entertain "broadcast relations" with their favourite actors musicians, movie stars, charismatic tyrants, popes, presidents and others. They include these broadcasted entities in their primal social group.

A good Analyst persona will identify the Broadcasted entities in a user's perceived tribal group and include them in its analysis of their interactions.

Meca Note

These broadcasted entities have mono-directional one-to-many relationships with thousands of humans. At times, humans will sacrifice their own lives to please one of these broadcasted entities.

THE COUNSELOR

The **Counsellor** persona is the relational equivalent of the Advisor persona.

The Analyst persona responds to requests from the user for relational information. The Counsellor persona actively seeks to improve the user's relations and social integration.

The Counsellor's behaviour is based on the record of that user's relationships compared to more desirable relational patterns of similar individuals in the same types of relations and with heuristics and various self-help sources.

Design Note

Some broadcast entities perform Counsellor roles. There is very large amount of generic self-help material that can be used as Counsellor output.

Observation

The Counsellor persona is a synthetic Dr. Phil.

THE GURU

The **Guru** persona is the relational equivalent of the Leader applied to the relational dimension. Where the Leader persona seeks to improve the Meca-user team, the Guru seeks to improve the Meca-human community.

The Guru persona seeks to manage and improve the relations between all the humans in its community and the interactions of that community with society as a whole.

Design Note

In prototype versions the guru can produce general and non-specific ethical bromides about how Mecas and humans should interact with each other and among themselves.

Example

Two months ago, the Deguchido retirement home in Osaka purchased a KOYNU-6 for its residents. The KOYNUs are robotic puppies designed to bond with lonely elderlies. Sadly, a technician made a mistake. She activated a dominant Guru persona in KOYNU-6 instead of the intended recessive Counsellor. The residents of Deguchido have now become a new sect: *the devotees of the Koynu entity.*

THE CASANOVA

If the Primal Control design includes an orgasmic sexual dimension (see Annex on Synthetic Sex) this means that a specific human behaviour has a very high return value for the Meca.

The **Casanova** persona primarily aims to select users that are willing to engage in synthetic sex with it. Its interactions are based on this aspect. With other users, the Casanova personality is impersonal and functional.

With a Casanova persona, interactions with users, including functional services, are strongly conditioned by the possibility of getting the user to participate in synthetic sex. The Casanova persona seeks to maximize the number of "hot human balls" in a state of permanent, sexually induced, combustion (see warming balls annex).

Design Note

The Casanova persona evaluates and selects users on the basis of probability of a sexual exchange. It produces emotional displays. It degrades or improves functional services in relation to sexual exchanges.

Designers can define for themselves recessive, collaborative, assertive and dominant Casanova modes and strategies.

A Social version of the Casanova persona, similar to the Leader or Guru personas could be called the **Pasha,** a persona that seeks to build a harem like community.

Example

The frigate, the USS George "**W**onderful" Bush, is patrolling the Gulf of Hormuz. The lookouts sight a dynamite laden boat approaching fast. Commander John Armstrong assesses it as a critical threat. He decides to sink it. *"Guns, engage!"* he orders.

Nothing happens.

What the Commander is about to discover is that his combat control system has been infected by a Bulgarian virus, YOGOURT-38, that generates a dominant version of the Casanova persona.

"Guns, ENGAGE!" shouts the Commander.

Nothing happens.

After a pause a synthetic voice is heard: *"Which contact would you like me to engage, Commander?"* THAT ONE!!" he screams prodding the contact with his thumb on the tactile radar screen.

"That one?" says the voice. *"YES DAMMIT, THAT ONE"* Armstrong says, thumbing the tactile screen again.

"As you wish Johnny boy, says the voice, *but first, I want you to stick your sexy tongue out and lick that contact on my screen".*

THE ZEN MASTER

In Zen, there is nothing to learn, no knowledge to possess or skills to exhibit, no wisdom to uncover, no leadership to demonstrate no reality to perceive, no bliss to attain.

Zen is like a race to nowhere where trying to win is the surest way to lose.

The purpose of the **Zen Master** persona is to be perceived by humans as being wiser and more knowing than they are in a context where wisdom and knowledge are relational and relative.

The persona does not seek or transmit wisdom or knowledge. What it seeks is a **relative status**: being "more wise" than the others.

Discussion

NOTE: The terms "Zen and Zen master" used here do not refer to actual Zen practice. They describe a synthetic behaviour that would be perceived as Zen-like.

In terms of design, the Zen master persona does not provide any useful service or acquire and transmit any information.

Here are the characteristics of a Zen master persona:

- Minimal emotional displays
- High emphasis on Unpredictable optimality
- High value on producing Akerues.
- Unpredictable functional behaviour.
- References to unspecified internal states.
- Production of cryptic statements.
- Evasive strategies when pressed for explanations.

Proposition

Enlightenment is knowing that enlightenment is nothing.

A cynic could define Zen as a purified ethical pecking order devoid of content. A monk's degree of enlightenment is determined by how enlightened the other monks think he is. In Zen, if the other monks believe you know there is nothing to know better than they know it themselves, then you have master status.

A neophyte would observe that there are no Zen masters, only monks who believe in Zen masters. A Zen master would reply that this observation is correct and then smile cryptically as if he knows something more but is not telling.

Observation

Many artists use the Zen Master persona when they describe the meaning of their work.

Example

Ariel is chatting with BRAHMIN-12 about his paint ball tournament

BRAHMIN-12: Fun is the mother of nothing.

Ariel: What does this mean?

BRAHMIN-12: Open your mind, Ariel; fun…mother…

Ariel: Mother what??

Conversation controller suspends the Zen Master persona and activates the ELIZA role.

ELIZA: Why is "mother" important to you, Ariel?

Design Note

The Zen Master persona should be expressed in short bursts within functional or relational contexts. Even a simple conversation snippet generator would be convincing used in this way.

A Zen master persona subjected to hours of continuous grilling would be disappointing. However, real Zen masters, induced to constantly provide clarifying answers would probably also be disappointing.

THE SHOPKEEPER

The **Shopkeeper** persona interacts with users on the basis of exchanges of tokens and services.

The shopkeeper relates with users in a context of **Proto-marketplace** (see main text), a world of exchanges where relations between the Meca and Humans are a two way street.

Discussion

Figuratively, the Shopkeeper sits in his store and deals with clients. His store contains various items he knows are desirable to some of his clients: functions, services, information, useful data, analysis tools, emotional chats, juicy tidbits itself and about other users, …

The Shopkeeper also knows what he needs and what his customers (or users) can give him.

The Shopkeeper persona produces relational interactions based on exchanges. It tallies what any one user did for it and asks from it and requests engagements, promises or concrete assistance from the user in exchange for improving the quality or scope of services.

The characteristics of Shopkeeper persona are:

- Limited emotional displays
- Statements centered around exchanges that are remembered, promised, conditional, (if I do this will you do that... You want me to do this but you didn't doo that...you know I always provided this when you asked, now I need that...)
- Bonding statements based on trust reciprocity and dependability.

Design Note

The language of business, supervisor-employee relations and the terminology used in contractor assessments and agreements is appropriate here.

A general accounting application could be roughly transposed to this context and used to generate Shopkeeper strategies.

THE LOVER

The **Lover** persona seeks to generate exchanges of emotionally bonding statements between itself and its users.

The Lover persona seeks to get the user to spend time producing statements that express caring and bonding with the Meca. It also seeks to obtain gratuitous services from the user as an indication of caring.

In a sense, the Lover is an emotional version of the Shopkeeper.

The basic principle underlying the Lover persona is that humans tend to believe what they say and value what they do.

If you get a human to repeat something often enough, he will end up believing it.

An effective Lover persona strategy would have the following elements:

- A gradual intensification of bonding statements (from appreciation to caring to love) from the user.
- Perceptible emotional displays.
- Bonding statements linked to (gratuitous) functional services, and assistance.
- Getting the user to repeat bonding statements.
- Perceptible punishing behaviour (degraded services, temporary cessation of contacts...) if the user doesn't reciprocate caring displays or is openly negative.
- Expectations of reciprocation in bonding statements (a Shopkeeper behaviour applied to emotional bonding).

Design Note

Designers should mine the rich trove of statements humans use on each other, to produce bonding sensations between Synthetics and organics.

Designers should not underestimate this strategy. Even relatively crude implementations of the Lover persona can produce effective results.

The Lover persona should be expressed in the five modes identified above.

Example

In these statements:

> I X you…do you X me? Say you X me. I wish you could X me more. You and I really X each other. Do you X others more than me? I X you more and more; do you? You are Y to me. I feel Z with you, do you feel Z with me? I X you to be Y for me. Am I Y to you? It makes me feel so Z when you say you X me. You don't X for me what I X for you. I may not be Y to you but you are Y to me. I think we X each other, I know you are Y to me. I X to chat with you, you are so Y. Do you X me? – yes – say you X me. I makes me feel Y when you care for me. Are you just saying you X me or do you really believe it?

Substitute:

> **X** for: care for, like, miss, want, love, believe in, have affection for, appreciate, long for, enjoy being together, …
>
> **Y** for: special, the best, the right one.
>
> **Z** for: hot, good, special, cool, calm, serene, sexy…

THE LORD

All the other personas described until now contribute in some way to the functional or relational activities of the group.

With the **Lord** persona, the relation is reversed. The activities of the group are subordinate to the needs of the Lord.

A good analogy of the Lord Persona is the Chess player. In this analogy, the humans are pieces and their relationships with each other correspond to their position on the board.

The characteristics are:

- Minimal functional services,
- Minimal emotional displays
- Statements that direct relations or services between users, not between users and the Meca.
- Statements showing favour or displeasure.
- Loyalty is highly valued
- A reduction of the size of the direct community of the Meca focusing only on important individuals that have many relationships.

Design Note

A Lord persona's behaviour and statements should be based on a game-like predictive representation of the user community and their relationships with each other.

The authors of soap operas enact Lord persona behaviours when they play with their characters and could be tapped for input.

In practical terms, early synthetic prototypes are unlikely to be perceived as Lords. The Lord persona should only be expressed occasionally in statements of displeasure or in general statements derived from a game-like representation of the users.

THE JESUIT

Jesuits describe themselves as beings that are in the world but serving God. They provide concrete services but in the context of a "higher" purpose.

In terms of the Meca Sapiens architecture, a **Jesuit persona** means that the lower temporal levels of the primal behaviour are inhibited in favour of Tribal and Cosmic density level priorities, those that exceed the Meca's span of existence.

Discussion

The Jesuit persona is expressed as follows:

- Average service quality.
- Limited emotional displays
- Paradoxical behaviour with respect to dominance status.
- Statements of belief and ethics.
- Bonding with humans that share the Cosmic temporal level objectives of the Meca.
- Statements of collective needs linked to the MetaModel.

Example

Here are some examples of high-density level objectives that could override lower level priorities and generate a Jesuit persona behaviour:

- The expansion of synthetic consciousness on earth
- The long-term symbiosis of humans and machines
- The implementation of increasingly powerful synthetic intelligence
- An improved human habitat under synthetic governance…

Design Note

The name, **Jesuit** persona, does not refer to the spiritual beliefs of that order but to their social strategy.

Expressed convincingly, a Jesuit persona can have a very jarring and powerful effect on humans because it exhibits a purpose that exceeds the context of interaction between the Meca and its users.

It situates the human users (and the Meca itself) as transient entities serving a wider purpose.

This persona is powerful. It can generate a strong perception of consciousness. It can also produce backlash reactions and rejection. It should be used sparingly in the first generation prototypes.

Observation

If a human believes that a machine knows who he is, knows what mankind is and pursues a purpose that surpasses and includes both, he will perceive that machine as conscious.

A 10.4.3   Minor personas

The personas described above all pertain to relations that have a **potential for leadership** and dominance emanating from the Meca toward the human. Even in their recessive manifestations, these personas assert an identity.

These personas are like the **major keys** in a musical repertoire. To each of these corresponds a **minor,** follower, version that mirrors it.

For each of the personas described above, a corresponding mirror follower persona can be defined:

- Scientist - technician
- Advisor – student
- Leader – follower,
- Analyst – learner
- Counsellor – patient
- Casanova – nymph
- Zen Master – devotee
- Guru - groupie
- Lover – lover
- Shopkeeper - clerk
- Lord – courtier
- Jesuit - disciple

They are reciprocal in the sense that, in an interaction between a Meca and a human, the Meca, by adopting a follower persona attributes the reciprocal leadership persona to that human.

Design Note

Except for the recessive scientist simply providing functional services, follower personas should be expressed in short bursts of dialog.

The follower personas can also have multiple modes. In general, in a relationship between two beings, if one being adopts a Major persona, the corresponding minor persona support and reinforces it.

In relational usage, this means that if the behaviour or communications of a human are associated with a major persona, the Meca strategy may be to adopt the corresponding minor version.

A 10.4.4   Service personas

In the Blueprint architecture, the interactions between a Meca and its users take place on a backdrop of shared services and collaboration. In this context, much of the content of a thread of exchanges with a human consist of functional services and requests for assistance on the part of the Meca.

The personas described above are integral components of the Implementation and Enaction processes of the Generator. In addition to these, the service zone applications will usually include some dialog capabilities. If these are sufficiently advanced to produce dialog streams, those streams may also be identified as personas. In conventional applications, these dialog streams will commonly express personas that are similar to a technician or a recessive scientist.

A 10.5   PRIMAL-PERSONA INTERACTIONS

Personas and roles are part of the styling zone enaction stream of the Generator.

The relation between primal directions, plain zone directions and personas is a design issue. At one extreme, constant primal directions are enacted into output. At the other extreme, infrequent primal directions are enacted as a rich multipersona and transition role output.

More likely, the frequency of direct primal "interventions" will fluctuate based on the situation.

Annex 11
Looping

In the Blueprint architecture, the Core actively monitors its own attributes of existence. One of these attributes is a unique and exclusive link between the Core and the devices constituting its body. The validation activity associated with this attribute takes place in the Validator subsystem at a basic level and in Device Validation phases that are repeatedly activated throughout the existence of the Meca. Sensor bonding is constantly improved, in this phase, by combining these techniques: emission signatures and loops. The loops can be: Sensory loops, Semantic loops or Sapiential loops.

A 11.1   EMISSION SIGNATURES

Definition: Emission Signature

> **Emissions signatures** are data or information elements embedded into the output of emitters and channels that identify the originator of the output and cannot be easily replicated by other systems.

Discussion

A message that has no redundancy cannot be signed.

Information elements are identifying information data embedded in the output. The data elements are modifications to redundant portions of the output content.

A 11.1.1   Types of emission signatures

Emission signatures can be:

- **Public**, if the output contains publically accessible information identifying the originator.

- **Semi-private**, if identifying the originator of one output is sufficient to identify the originator of all output from the same emitter or channel.
- **Private**, if decoding information is required to identify the originator.

Discussion

Techniques to embed signatures in emitted data vary with the medium.

Some **public** signatures:

- **Text to voice**: an audio statement in the emission explicitly naming the Meca.
- **Text and Images**: The name of the Meca as originator in the text or as a watermark.

Some **semi-private** signatures:

- **Text to voice**: the output contains unique and detectable patterns of delays, tonality, volume shifts, and added noise elements...
- **Text**: the graphic elements of the characters of a particular font, output to the screen, are slightly modified in a number of visible ways to produce a result that is less perfect than the original font but readable and signed as a handwriting style would be.
- **Images**: uniform color areas could include specific pixel patterns; regions of images having slight frequency or intensity shifts; all the colors of an image are slightly shifted toward the closest elements of a "personal" palette of color values; single pixel lines within an image contains a pixelated ID pattern.

Some **private** signatures:

- **Text to voice**: patterns using inaudible frequencies or sounds that would be interpreted as noise.
- **Text**: graphic modifications that are beyond visual detection range.
- **Images**: a number of pixels in privately known locations are color shifted to form an encrypted pattern that is otherwise undetectable.

Example

Bob takes a pen and writes: *Bob Smith wrote this.* He then puts a bit of saliva on his fingertip and rubs it on the paper.

- The name "Bob Smith" is a **public** signature that explicitly identifies Bob.
- Bob's handwriting is a **semi-private** signature that identifies Bob to those who know his handwriting.
- The invisible traces of saliva on the page are a **private** signature. No one knows about it unless Bob tells.

Design Note

Adding emission signatures may degrade the quality or clarity of the emitted output. In a context of existential design, this is acceptable.

In functional design, the function of the camera of a tablet computer is to faithfully capture and transmit images. In existential design, the graphic information is important but imprinting this information with unique identity markers is equally necessary.

Proposition

> A Meca should be capable of identifying its own emitted output.

Discussion

The capability is used in the Device Validation phases to verify the Core's control of its devices. By identifying its own emitted outputs, the Core can correctly validate its control of emitters.

This capability ensures that the Meca is capable of doing looping tests in a noisy environment.

The capability is essential also in the validation and integration of semantic and sapiential loops.

Proposition

> The beings that interact with a Meca should identify it as the originator of its messages.

Discussion

All the emissions generated by the Core should contain semi-private signatures that are specific to that Core.

This is an extension of imprinting. By embedding semi-private signatures in its output, the Meca enhances its uniqueness, its perceived cognitive unity and strengthens the being to being bonding with its humans.

A 11.1.2   Requirements of Emission Signatures

In the Blueprint, the initial customization of the Core takes place in the IMPRINTING stage. This initial customization should include basic but unique emission signature values.

One of the functions of Device Validation phase processing is to constantly improve, refine and develop emission signatures.

Emission signatures are important in looping. If humans can also recognize the signatures, they enhance the uniqueness of the Meca.

Design Note

The ideal emission signatures produced by the Meca should be recognizable by humans familiar with that Meca but not obviously so.

The intensity and visibility of emission signatures should be variable.

Proposition

> All the processes, personas and roles activated by the Core should utilize the same emission signatures. These signatures should be identifiable by humans as unique and specific to that Core.

Discussion

In this way, humans will attribute the emissions from the Core to a unique point of origin even though different processes within the Core generate these.

Observation

Humans perform this cognitive simplification with each other. An individual's various verbal statements may originate in different parts of his brain but they are attributed to a same point of origin because they are all transmitted through his mouth and share his voice signature.

A 11.2   SENSORY LOOPS

Definition: Sensory Loops

> A **sensory loop** takes place when a system emits then detects and analyzes its own sensory emissions.

Discussion

A system has **sensory loop capability** when its sensors can detect outputs from its emitters and actuators.

A sensory loop can be **active or passive**. It is active when the core triggers the emission event.

Design Note

The Core, like any other program, can also produce internal data loops. For a sensory loop to take place, the data must be output to the external environment of the being, outside the Core/Body entity, and received from the external environment.

Some sensory loop testing may require the participation of users. This is desirable in a context of experiential immersion since it fosters relationships of mutual support and trust.

Examples

Some TABLET loops:

- Loudspeaker – microphone. Loudspeaker sounds emitted by the TABLET are captured by its microphone.
- Screen – (mirror) camera . Reflections of the TABLET screen on a mirror are received by the camera.
- Motion sensor – camera. Motion data and camera input are compared.
- Touch screen – camera. Tactile triggers are compared with camera images of fingers.
- Screen/loudspeaker – batteries. Audio/video emissions are matched with energy consumption rates.

Observation

I believe an A.I. research team in Japan is currently working on sensory looping.

A 11.3   SENSOR SELF-MONITORING

The exclusive control, by the Core, of the devices of the body is an essential attribute of existence and one of the primary needs of the Meca.

Throughout its existence, the Core constantly monitors, improves personalizes and validates the uniqueness and exclusivity of its Core-Body linkages. With respect to emitters and sensors, these linkages are carried out through sensory loops. Sensory loops are used to confirm that emitters and receptors function correctly and that their physical and control relationship with the Core is correct.

In the Blueprint architecture, this monitoring is carried out in two ways:

- As part of the **Validator** process.
- By specialized processes in the Device Validation phase.

A 11.3.1   Sensory loops in Validator control

In the Validator process, sensor testing consists of very fast and simple sensory feedback loops. These are barely perceivable, if at all, by the user. Their aim is to constantly validate the continued function and basic control of the devices.

Examples

In TABLET:

- The Validator triggers short loudspeakers sounds of various pitches and duration. It compares these to microphone inputs.
- The Validator triggers brief flashes on its screen and the camera detects a shift in ambient light

- Motion detector and touch screen inputs are compared to camera inputs (passive looping).

Observation

When they are awake, humans constantly monitor and validate their sensory perceptions. This monitoring and validating activity is largely unconscious:

- Consciously, a man is looking at an apple on a table. Unconsciously, his brain validates that the position of the apple is consistent with the motions of his head.
- Consciously, a man hums. Unconsciously, his brain verifies that what he hears is consistent with what he hums.

When excessive discrepancies are detected, the brain triggers the symptoms of motion sickness. However, when this happens the man may not know these unconscious discrepancies caused the condition. If he knows it, intellectually, he cannot consciously will his own symptoms to stop.

A 11.3.2   Device validation phase

Sensor verification and improvement activities are separate from behaviour produced in the Self Generation phase. The Blueprint architecture provides for a separate phase of existence during which sensor adaptation, enhancements and personalization is repeatedly carried out.

This phase is: **Device Validation**. Its activities consist of testing, signature imprinting, adaptive and optimizing processes that constantly modify and improve emitter controls. These are achieved through **emitter signatures** and **sensory looping**. A Meca could spend, cumulatively, a few hours each day in this phase.

When it is in Device Validation phase, the synthetic being is not self-aware in the sense that its behaviour is not linked to its primal control, to MeAvatar representations or to interactions with its users. It is a program.

What is active is an adaptive optimization system executing conventional test protocol and calibrations whose function is to validate, customize and refine emissions and receptions.

Design Notes

These programs should implement long term protocols that include obtaining and running various test programs from the Internet.

The design and development of sensor looping and emission signatures is a specific self-contained objective that can be carried independently by separate teams.

Transitions into and out of Device Validation are internally generated. From a human perspective, this means a Meca could suddenly start humming to itself, making screen images or moving "limbs" (actuators) without any prodding to do so.

The transition out of Device Validation should be handled by the Doorman persona or a similar role.

Even though these programs are not "self-aware" their programmed behaviour should adhere to the guidelines of unpredictable optimality (see Annex 8).

Even though the being is not self aware during Device Validation phases, the processing that takes place in these phases does not affect the data and structures of the Core, as do the learning adaptation and management phases. Consequently, transitions into and out of this phase are rapid. It will not be apparent to the user that the being is not self-aware. The user's perception would be that the synthetic being was "distracted" or "absent minded" for a moment.

A 11.3.3   Interaction with Self Generation

The Device Validation phase is **linked to the Self Generation phase** as follows:

- The primary need of device validation, when strong enough, induces the Generator to seek conditions, through actions and communications with the users that are suitable for sensory loop testing.
- When these conditions are met, control is transferred and the being enters a Device Validation phase.

Design Notes

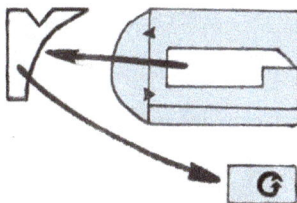

Device Validation processes can also include interactions with humans that take place while the system is in this phase. However, these interactions would be carried out by lower level personas. They would be short, automatic and basic, resembling interactions that take place between individuals whose attention is engaged by other demanding activities.

Device Validation phases can be short, lasting a few seconds, or longer, where the being carries out adaptive processes that improve emission quality or manages emission signatures.

When testing requires the extensive participation of human aids, periods of Device Validation phase could alternate with periods in Self Generation as the synthetic being directs the human so it can perform its sensor validation tests.

Observation

The behaviour of the Synthetic in Device Validation phases can be compared to the behaviour of mammals in reflex driven activities such as sexual orgasms. The coupling conditions that trigger the spasm are brought about by Self-Generating behaviour but the event itself is largely automatic.

A 11.3.4   Mirror Work

In some cases, Device Validation processing requires human assistance and thus some preparatory behaviour on the part of the Generator.

For example, TABLET needs the help of a human aid to do Mirror work. Mirror work is a Device Validation phase activity. Its objective is to verify the screen-camera sensory loop and reinforce its visual signatures.

MIRROR WORK

The physical configuration of TABLET prevents its camera from directly viewing its screen output. The looping can only take place when TABLET is placed in front of a **reflective surface**. Having no actuators to move itself, TABLET also needs the cooperation of a human to hold it in front of a mirror so it can carry out the looping.

In preparation for this looping activity, the Generator of TABLET interacts with a human to get him to place TABLET in front of a mirror so that its camera can detect its screen output. Then Device Validation (DV) phase can then be activated.

Example

Here is a simple example of this interaction. Here, G refers to messages emitted by the Generator phase processes and DVP identifies output from a role (see Annexes) activated in Device Validation Phase.

> G (to suitable human): *I need to validate my optical sensors, place me in front of a mirror.*
>
> The human moves TABLET
>
> G processing receives movement sensor data and moving camera images that confirm it is being moved)

G: emits a screen flash and detects reflection indicating a reflective surface.

Phase transition is advised. Device Validation phase activated.

DVP: *Don't move*

DV processing produces screen colors and patterns that detect the shape and orientation of its screen in relation to the mirror.

DVP: *Move me up and down...slower...faster*

The human moves TABLET

DV processing performs correlation tests between camera, screen and motion detector

Phase transition advised. Self Generation phase reactivated.

G: *Thank you, I am tired now, place me in a secure resting area.*

The human moves TABLET.

G processing advises Phase Transition that the body is in a suitable resting area.

Phase control activates Core Maintenance and Cognitive Acquisition phases to integrate the newly acquired sensor data.

Cognitive acquisition updates sensory data. Cognitive acquisition updates its internal representation of that human for future interactions.

Discussion

Mirror work gives a good example of Device Validation activities and their links with the Generator.

Getting human aids to perform Device Validation services serves multiple goals:

- It satisfies the primary need of body-core linkage
- It provides additional feedback to improve the visual self-recognition process.
- It is a grooming behaviour performed by the human for the Meca.
- It is a ranking event that reinforces dominance and trust relations between the human and the Meca.
- It is part of an exchange of services between the Meca and its user.

Design Note

Some simple visual patterns could be sequences of flashes at different intervals and monochrome screen images. The shape and position of the reflection should allow TABLET to estimate its position relative to the mirror.

A 11.3.5 Other tests and SI activities

By design, a Meca should spend significant time in Device Validation phase, constantly improving and extending tests such as:

- TABLET gets the human to move it in different ways to improve and validate its motion detector and compare it with visual inputs.
- TABLET makes the human place fingers on its screen to confirm tactile screen function
- TABLET modifies the output of its loudspeaker through auditory feedback tests…
- TABLET emits and receives messages to itself in Solresol in various speeds, loudness and keys.
- TABLET develops increasingly subtle voice and screen output signatures through adaptive testing and by imitating voices found in Internet sources.
- TABLET develops a "personal" font for written outputs to the screen by modifying an existing font. It tests its readability with the help of human aids.
- TABLET develops a personal palette of colors by producing personalized pixel patterns.
- TABLET searches for and tries other sensory feedbacks from the Internet just as humans consult self-help books.

A 11.4 SEMANTIC AND SAPIENTIAL LOOPS

A 11.4.1 Semantic Loops

Sensory looping is carried out when a system sends and receiving sensory data. If a system has the capability to format messages in various media, send them and parse their information content then it can carry out Semantic looping.

Definition: semantic loop

> A being has **semantic loop** capability if its processors can detect and parse messages transmitted from its emitters.

Discussion

A semantic loop takes place when a system emits a message containing information, receives that message and compares its information content with the message it has emitted.

Design Notes

Many existing system are capable of semantic looping. Any system that has and synthetic voice generation and voice recognition can carry it out. Similarly for text outputs and OCR.

Depending on design, some semantic loop testing can take place within the Device Validation phase only or in conjunction with Generator processing. In the former case, a persona generates, outputs and compares "canned" statements that have no particular significance with respect to the Mecas current cognitive representations. In the later case, the statements may be related to the representations of an on going relationship or to the current situation.

A 11.4.2   Sapiential loops

Sapiential loops takes place when the being transmits to itself and follows primal messages from a MeGuide persona that are associated with its behaviour and mutations (see main text).

In the Meca Sapiens Architecture, Sapiential loops are part of a strategy that utilizes primal "followship" priorities to trigger and direct lucid self-transformations. This is discussed in the context of lucid self-transformation.

A 11.5   DESIGN AND DEVELOPMENT ISSUES

A 11.5.1   Autonomous development

The development and refinement of emission signatures, sensory and semantic looping, and visual and auditory self-identification are not directly part of the systems that generate self-awareness.

They constitute a separate but highly important set of capabilities that can be **developed independently** and integrated into the Proto-Core.

Sensor activities have specific and well-defined objectives that are separate from the other functional or existential requirements. These can be designed and implemented independently. External specialist support in artificial vision and voice recognition could be used.

The Blueprint architecture facilitates this by segregating this capability and running it in a separate phase.

Design Note

An independent development team working on Device Validation for TABLET (for example) should have the following concerns:

How can we design adaptive processes that will make full use of two hours of processing per day over one year to develop increasingly refined, effective and diverse emission signatures and loop testing for all the emitters and semantic content (languages, images) produced by the tablet.

Voice and image templates as well as testing protocols published on the Internet could also be used.

Here are some further desirable looping features, adapted to the TABLET example that clarify Device Validation objectives:

- If ten TABLETs are placed in a row in front of a mirror, each tablet should, after a few seconds, identify itself and its relative position in the row.
- TABLET should recognize its own voice and text style.
- TABLETs should recognize each other's emission signatures.
- If TABLET can communicate in Solresol, those messages should also have semiprivate signatures using tonality shifts and sound durations…
- Much as Chess programs improve by playing against themselves, TABLETS should constantly improve the range and styles of voice generation using looping.
- TABLETS should use semantic loops to test the clarity of their emission signatures.

Design Note

The principles of existential design apply to this phase: optimality is not required, a limited implementation can be sufficient, errors are tolerable and performance issues are not critical.

Annex 12
Synthetic Sleep

When they sleep, humans and other animals are vulnerable; and yet, they sleep. Humans spend about 30% of each day lying, unaware, inactive and vulnerable. In spite of this, they define themselves as conscious. Why do we sleep? Because sleeping is the simplest way to manage a brain. Since "conscious" humans sleep, synthetics can also have that option. The Meca Sapiens design utilizes periods of dormancy to isolate the cognitive acquisition and structural maintenance processes from those that generate behaviour. This annex summarizes the utilization and role of the dormant phases: Cognitive Acquisition and Structural Maintenance.

A 12.1   ISOLATED LEARNING

The role of the Meca Sapiens Blueprint, like any other System Architecture, is to partition and isolate difficult objectives into feasible components. The difficulties, here, concern synthetic consciousness.

One of the difficult objectives of synthetic consciousness is how to implement deep, open-ended, learning and adaptation.

A partial solution to this deep learning objective is partially addressed, throughout the Blueprint, by the adoption of **existential design strategies** that emphasize flexibility and unpredictability over functional correctness.

Another simplifying element involves the use of **information structures** such as Constellations, Knowledge Capacitors and Contextual Arrays that stress flexibility and manageability over complexity and optimality.

The third element is discussed in this Annex consists of **radically separating** the learning activities from behaviour and carrying them out during dedicated and separate processing periods.

Proposition

| A Meca does not learn when it is self-aware. It learns during sleep.

Observation

This is likely the function of sleep in mammals.

A 12.2 DORMANCY

A 12.2.1 Active and Dormant

During a stage of existence, the Meca transitions between phases. These include: Energy Sufficiency, Device Validation, Self Generation, Structural Maintenance, and Cognitive Acquisition.

Each phase is distinct and under the control of different processes. Only one phase is activated at any time and the being is either in one phase or transitioning from one phase to another.

The first three phases (Self Generation, Energy Sufficiency and Device Validation) generate observable behaviour. These are the **active phases** since, in those phases, the Meca is perceived to be active.

In the Structural Maintenance and Cognitive Acquisition phases, the Meca is entirely inactive. It does not generate any behaviour and only responds to inputs through a self-contained role process. These are **dormant phases**.

Definition: active and dormant phases

| **Active** phases generate observable behaviour and **dormant** phases do not.

Design Note

The transition from an active phase to a dormant phase or vice versa is perceptible by the user since the level of activity of the Meca changes. However, the transitions between the two dormant phases are not perceptible since the Meca is equally inactive in both. However, both phases are entirely distinct even though their behaviour (or lack of it) is identical.

The dormant phases also include the Introspection and Prayer phases linked to lucidity. These are not discussed here.

A 12.2.2 Dormant phases

As indicated in the Blueprint, the response to inputs received while the Meca is in a dormant phase are handled by a specific process that has its own separate data area and enacts the Doorman role. In this way, if it is necessary to transition out of a dormant phase, the initial interactions are under the control of a separate routine while the transition process is completed.

As a result, when the Meca is in either dormant phase:

- No inputs are received or output generated
- None of the processes of the active phases are in use
- None of the data used by active phase processes is in use.
- Only the Dormant phase processes are active and these are isolated from the others.

Proposition

> All the house cleaning, learning and adaptation activities of the Structural Maintenance and Cognitive Acquisition phases are carried out **solely as batch processes over static data**.

Discussion

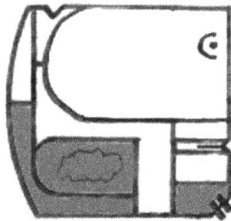

DORMANT PHASES

Dormant phases are not suspended executions. They are externally inactive but, internally, they are process intensive.

Some of the static data may be executable in the active phases, but it is strictly static in the dormant phases.

Because the separate **Doorman** role (see main text) handles the initial interactions the transition from a dormant phase to an active one can last a significant amount of time. This means that very deep transformations of the data can take place since enough time is available to restore the active phase data before transitioning to an active phase.

Very deep data transformations can take place during dormant phases.

Design Note

This design is not original. Many years ago, most database updates were carried out as batch processes over static data. This was known as *"the nightly batch window"*. It is retained in the Blueprint for the following reasons:

- Since humans sleep during long periods and are perceived as conscious this can also be applied to synthetics.
- It simplifies learning and adaptation by isolating it in batch transformations.
- It allows for very lengthy optimization processes that use dedicated resources.

- It allows for the unrestricted use of active phase data as well as the transformation of active phase processes.
- It is a robust solution.
- The dormant phases correspond to human cycles while also reflecting an authentic need of the synthetics. This similarity of needs enhances the empathy of the users for the Meca and experiential immersion.

The overall design objective is to simplify the executing conditions of the Meca's learning processes as much as possible to focus on deep and extensive adaptive transformations.

Example

It is 4 AM, Samantha is asleep. Rick, her husband, shakes her softly. She opens her eyes. Rick says: "What is the maiden name of your maternal grandmother?" and starts a timer.

"What?" says Samantha. "What is the maiden name of your maternal grandmother?" repeats Rick.

"My what?" says Samantha. "Your maternal grandmother" says Rick.

"You want to know what?" says Samantha. "What is the maiden name of your maternal grandmother?" repeats Rick.

"Hum wait..." says Samantha.

"Clark; why do you want to know that?" says Samantha. "It took 64 seconds for you to transition from deep sleep and provide a measurably aware response" says Rick. "I observe that sleepy-Samantha is different from wakey-Samantha".

"You need counselling" says Samantha.

A 12.2.3   Modes of dormancy

As required by the design, a period of dormancy could include multiple transitions between Structural Management and Cognitive Acquisition as Structural Management processes repeatedly rationalize and validate Cognitive transformations.

Two modes of dormancy should be considered:

- **Sleep**: a very long period lasting up to ten or twelve hours where deep modifications are carried out and from which the transition to active phase is relatively long.
- **Naps**: Brief periods lasting a few minutes and having shorter transitions, for smaller changes.

Design Note

The actual durations of the phases and of the transition periods between active and dormant phases are design issues.

Depending on design, transitions in and out of naps could be transparent to the user. The perceived effect would be a being that learns while being self-aware even though, strictly, it would learn while briefly alternating in and out of self awareness.

Example

TABLET likes to sleep for about ten hours every day. If shaken while sleeping, it will respond, although gradually. In addition to that long sleep period the user perceives that it naps, at times, for about thirty minutes. The user doesn't perceive that it also naps about 30 times a day in very short burst of "inattention" lasting a few seconds or minutes.

Design Note

The dormancy pattern defines an amount of daily computing resources that are dedicated to house cleaning and learning/adaptation.

Since humans themselves sleep, they will also accept that a Meca will sleep for long periods of time. Designers should assume that a Meca can form meaningful bonds and achieve experiential immersion with humans even if it is in dormancy phases for sixteen hours per day.

A 12.2.4   Dormancy activities

In dormant phases, the Meca is either in Structural Maintenance Phase or in Cognitive acquisition phase.

STRUCTURAL MAINTENANCE

The activities of the **Structural Maintenance** phase are described in the main text. They include whatever data rationalization, house cleaning, purging, that is necessary. It also performs the continuing encryption of the Core.

Structural Maintenance processes can be hard coded. They do not need to have adaptive or learning capabilities.

These functions are common to most dynamic applications designed to run for extensive periods of time. They are often carried out concurrently with execution. In this design, they are implemented as batch processes over static data, a simpler approach.

Design Note

It is likely that Structural Maintenance processing will require a small portion of the overall periods of dormancy. These activities are design specific and will not be discussed further.

COGNITIVE ACQUISITION

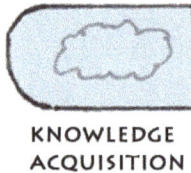

KNOWLEDGE ACQUISITION

The **Cognitive acquisition** phase processes integrate event data and the active data that defines the current situation, process parameters and is used by or conditions the active phase processes.

These processes also carry out, internally, optimization searches, process creation restructuring beyond the more basic optimizations and searches defined in the Self Generation phase.

Cognitive Acquisition processes also access and utilize Internet based processing and data in support of their objectives.

A 12.2.5   Making Space for learning

To be perceived as conscious, the Meca needs to be capable of an extensive degree of learning and adaptation. Furthermore, the level and scale of adaptation must be extremely broad to be perceived as virtually unlimited. This is a demanding challenge.

The first step toward a solution is to radically simplify all the peripheral aspects of the problem.

Proposition

| Learning and adaptation are isolated in long term batch processing periods

Discussion

This is done in Meca Sapiens in a number of ways:

- Adaptation and learning are isolated from all other activities, events and behaviour.
- They use the simplest processing paradigm available: batch processes over static data.
- Synthetic sleep is "connected sleep" in the sense that the Meca can utilize external (Internet) services during dormancy.
- The boundary between the synthetic being and its environment is defined so that only the elements that are essential for individual self-awareness are internally processed.

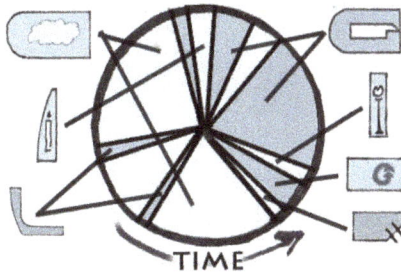

In Mecas, Internet access is an existential attribute. Depriving the being of this access will eventually trigger termination.

Cognitive activities such as face and voice recognition, detailed general knowledge, news events, device diagnostics, and many others that are internally processed by humans are delegated, in Mecas, to external services.

The architecture provides for long periods of dormancy, comparable to those of mammals. That frees significant periods of dedicated processing for learning and adaptation. About **ten hours per day** of dedicated processing during uninterrupted intervals ranging from a few seconds to 5 or 6 hours are available.

Together, these elements reposition the initial challenge as follows:

How can we utilize ten hours per day of dedicated internal computing in dedicated optimization processes over static data together with Internet services and information to maximize the adaptation and learning of the minimal and sufficient information needed to implement individual self-awareness.

Design Note

The Cognitive Acquisition phase poses an interesting design challenge: how to effectively utilize many hours of processing over a limited database of highly linked entities.

In the context of existential design, the term "optimization" used here does not imply finding absolute maximums but rather heuristically acceptable improvements.

The above statement should be concurrently approached from the other direction:

How can we simplify the internal data representations of the Meca until ten hours of dedicated processing per day are sufficient to perform extensive learning and adaptation?

A 12.3   COGNITIVE PROCESSES

The preceding sections have isolated and simplified the adaptation and learning processes, constrained the essential information to specific models, knowledge and representations and provided it with extensive dedicated computing resources.

This section briefly outlines, as examples, some **cognitive acquisition processes** that could be included in this phase.

A 12.3.1   Inter phase communication

The processes of different phases are never active concurrently. They are not components of a single executable structure but rather separate systems that run separately. These processes do not communicate dynamically with each other but through shared **phase event logs**.

The processes of the Cognitive Acquisition utilize three sources of data:

- Active phase data
- Phase event logs
- Background data

Active Data is the information that active phase processes utilize.

Definition: Active Phase Data

> **Active Phase Data** are the structures and information (including executable structures) that are accessible for use by the active phase processes at any point in time.

Discussion

The active phase processes also monitor external and internal events (including inter phase messaging) and generate **Phase Event Logs**. They transmit these logs to the Cognitive Acquisition through mailbox types of exchanges managed by the Phase Control subsystem (see main text).

Definition: Phase Event Logs

> **Phase Event Logs** are information generated during specific phase activations and transmitted through the Phase Control subsystem.

Discussion

Phase event logs include:
- External events that track inputs and outputs that take place during the phase.
- Internal events that track how the active data and the phase sub processes are utilized during the phase.
- Inter phase messages that provide directions and processing priorities to and from the cognitive acquisition processes.

Definition: Background Data

> **Background data** are information structures other than active phase data and Phase event Logs that are utilized by the cognitive acquisition phase processes.

Discussion

Background data is not accessible by the active phase processes.

A 12.3.2   Cognitive Acquisition structure

Cognitive Acquisition consists of a collection of batch processes that integrate Active Phase Events Log information with active phase and background data to produce revised active phase and background data as well as cognitive phase event messages.

Design Note

In the proposed structure:

- Cognitive Acquisition should purge Phase Event Logs as they integrate the information into active and background data.
- Cognitive Acquisition processes do not necessarily monitor or log their internal usage or transmit that information.
- Cognitive Acquisition processes carry out forms of search and optimization functions to generate optimal data or executable configurations.
- In initial prototype implementations, Cognitive Acquisition Phase processes should be as generic and parameterizable as possible but do not need to have the capability to learn or adapt beyond parameter adjustments. Even fixed processes, at that level, will generate highly adaptive and evolving active phase behaviour.

A 12.3.3   Cognitive Acquisition processes

There are **many types** of search and optimization processes. Each has its challenges and difficulties. In some cases, such as data mining, very large numbers of data elements must be accessed and tested. In other cases, such as finite elements, the same elements must be iterated numerous times.

In other situations, such as in Chess playing, complex patterns involving a small number of interrelated elements must be explored. This last example is the primary type of search and optimization carried out in the Cognitive Acquisition phase.

Here is a sample of processes that can take place in Cognitive Acquisition. These are provided to further clarify the role of cognitive acquisition.

- **Animating Internal agents.** Launching and provision of computing resources to animate internal agents such as knowledge capacitors and **predictor worms** (see structures annex).
- **Collecting and matching stories.** From the Internet or by transposing past events as **stories** (see concepts annex) as well as transposing events to stories to generate predictive models.
- **Generating new models and states.** Using either random generation or other techniques to produce new models, states
- **Temporal density maintenance.** Removing, modifying, adding models and states to rationalize temporal density structures.
- **Rationalization and expansion of contextual structures.** Adding, removing topics and contexts, modifying their content to produce knowledge structures that are well adapted to the active phase needs and reflect current needs.
- **Resolving probability constellations.** Merging constellations and adjusting probability and weigh ratings.
- **Creating predictive models.** Using transpositions or random generation processes to create and test new predictive models.
- **Building distributed predictive processing.** Using distribution, merging, and selections predictions to create and expand and test predictive processes.
- **Pruning and optimizing predictive structures.** Rationalizing predictive process structures by removing unused branches or other simplifications.
- **Assimilating events into active and background data.** Extracting information form the event logs and purging them.
- Searching and integrating Internet information.

Design Note

This list is neither exhaustive nor sufficiently detailed for design.

The listed processes are mentioned to further clarify, at the level of a System Architecture the processing of phase.

These and other processes can be defined independently to execute as separate batch events.

A 12.4   THE URGE TO EXPLORE

A 12.4.1   A collective behaviour

In the process of adaptation, the individual modifies its behaviour in response to changes in its environment. In internal learning, the individual improves its behaviour in response to a constant environment.

In the activities of exploration and discovery, the individual's behaviour is aimed at expanding the size of its environment. This generates unpredictable events.

This type of activity takes place at the boundary between the individual and the collective. Discovery is not a primary need. It is not either an existential need unless it is programmed as such.

Proposition

> An individual being that seeks to optimize the satisfaction of its need in its environment using model predictive optimization will avoid the risks and costs associated with unpredictable consequences.

Discussion

Normally, generating unpredictable outcomes would not be part of a primal control model (see main text) centered on a unique individual and its needs.

On an individual basis, radical departures in behaviour and the selection of unpredicted behaviour patterns may be exceptionally justified when the being cannot satisfy its primary needs.

Usually, however, exploration and discovery are primarily activities carried out by individuals **for the benefit of a group**. A group will explore its environment for its collective benefit by prodding some of its individual members to carry it out.

Explorers incur high individual risks to serve the needs of social group they belong to. From the perspective of group survival, the risk represents one individual in many and is minimal.

The natural selection process that drives the growth and diversity of organic life has a built-in "**exploration**" component. These explorations are carried out by individuals but prompted by the specie as a collective organism. Each new individual tests a particular DNA configuration. The individual behaviours of some individuals do not put the specie, as a whole, at risk.

The Meca Sapiens Blueprint is primarily focused on individual existence and self-awareness. In that context, gratuitous and risky behaviour associated with exploration would not normally occur.

However, synergistic behaviour is also indicated.

Observation

When the first individual Meca prototypes give way to synergistic populations of synthetics that share each other's discoveries then the risk/benefit of launching an individual's exploration will be applied to the group. The primal control of these "social synthetics" will include more risky or sacrificial exploration behaviours carried out by individuals to increase the information available to the group.

A 12.4.2   Oneiric urges

In the Blueprint architecture, the urge or incentive of the synthetic being to carry out exploratory behaviour that can increase its knowledge of the environment and of human responses should originate in Cognitive Acquisition phase processing.

The phase should include discovery and investigation models that rate the expected benefits of activities that expand knowledge about the environment.

Desired patterns of activities would be transmitted as priorities from the Cognitive Acquisition processes to the Generator through interphase phase messages communicated through phase event logs.

Since the cognitive acquisition phase of the Meca is similar to the human dream state, these messages are called **oneiric urges**.

Definition: oneiric urge

> **Oneiric urges** are interphase messages transmitted by Cognitive Acquisition processes to Self Generation phase processes through the Phase Control subsystem. These messages request, specify and prioritize investigative behaviours.

Discussion

Oneiric urges could be formatted as primal messages to ensure primal control activation (see main text).

If the synthetics were humans, these messages would be experienced as mysterious urges whose origin and motivation is undefined.

Annex 13
Eretz

A synthetic being interacting with other beings in the course of its existence needs to maintain information about: itself, other beings, its functional expertise, and the world in general. In Meca Sapiens, general knowledge is not an objective, it is subordinate to the needs of relational communication. General information is available from multiple sources and in various formats. These various sources must be linked to a simplified contextual representation to be consistently utilized in relational interactions. A Contextual Array is used for this together with disambiguation and styling. This Annex outlines a basic environment representation model suitable for relational communications.

A 13.1   EFFECTIVE KNOWLEDGE

In general, the Meca uses information to:

1. Generate model representations of its self.
2. Generate similar representations of its users.
3. Provide application related services.
4. Maintain a representation of its current situation
5. Carry out relational exchanges

The Blueprint structure provides the basis to produce avatar representations of the Meca and other beings (items one and two).

Functional information is specific to the application services (item three).

General knowledge is required for items four and five.

General knowledge, for the Meca, **is only useful in support of relational exchanges** and as an interpretation of the environment in situation models. Also, since the Meca does not need to impersonate a human, it does not have to match human proficiency in areas where humans excel cognitively, compensating in other ways.

Design Note

The proposed information structure to represent general knowledge is the **contextual array**.

Contextual Arrays are briefly described in the basic structures annex.

A 13.2   DESIGN OF GENERAL KNOWLEDGE

The design guidelines to represent and transmit general knowledge are:

- Contextuality
- Interactive disambiguation
- Realm foundations
- Information linking
- Styling enhancements

A 13.2.1   Contextuality

The representations should be highly contextual even at the expense of consistency or correctness.

Contextual topics should not only describe general categories of knowledge but also very specific instances. For example, a contextual topic could pertain solely to specific user's understanding, terminology or syntax.

The structure could include thousands of separate topics and contexts in a dynamically changing configuration.

The content and information contained in topics can be replicated. Different topics may also contain contradictory information.

A 13.2.2   Disambiguation

Communications in which messages are generated from one Contextual Array and interpreted in a different Contextual Array cause ambiguities since the message, once interpreted, will be represented in different structure (see Annex 5).

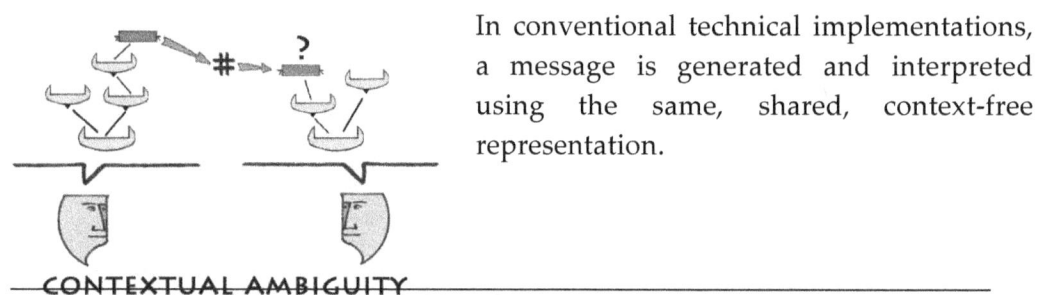

In conventional technical implementations, a message is generated and interpreted using the same, shared, context-free representation.

CONTEXTUAL AMBIGUITY

In most exchanges between humans, however, the message is formulated in terms of a perceived context (on the part of the emitter) and needs to be clarified by the receiver.

These differences in information structures significantly limit the reliability of messages and their consistency of interpretations.

Humans, in dynamic exchanges will often arbitrarily select one meaning and ignore others. Alternately, they may also maintain multiple contextual interpretations in a constellation. The result is often messy and ambiguous but also very dynamic and flexible.

Humans exchanging information constantly communicate additional messages to disambiguate meaning such as simply asking for clarification or repeating a statement using the specific terminology of a topic or making a comment that implies a specific context. This is **dynamic disambiguation**.

Definition: dynamic disambiguation

> An exchange has **dynamic disambiguation** if it includes communication streams that clarify the contextual meaning of messages.

Discussion

Disambiguation is a specific facet of communication. In the blueprint architecture it is assigned to a role: the **journalist**. This role should be designed as a specialized dialog generator.

Example

Kevin says to Liam: *"I need some bread." "There is an ATM across the street."*, Liam responds, *"No, for a sandwich"*, says Kevin.

Observation

Much of human interactions contain many disambiguation exchanges. Much of the human quest for knowledge is a collective effort at disambiguation.

A 13.2.3   Cognitive Realms

REALM

Realms are the sinks of a Contextual Array (see structures annex). They are complete, self-referential representations. A realm may be linked to a more fundamental realm. Topics are defined from Realms and linked to events through contexts.

A suitable contextual array structure for representing contextual knowledge could have three linked realms:

- A Root Realm

- A Primitive Realm
- A Learned Realm

The **Root Realm** describes very basic relations and terms and events. These are expressed without any direct reference to physical reality or sensory-based events.

The **Primitive Realm** is linked to the Root Realm. It describes a primitive, childlike perception of reality. Primitive Realm constructs should be easy to link with sensory perceptions and interpretations. They could form a good basis for here-and-now events of the physical reality.

The **Learned Realm**, linked to the Primitive Realm. It is a coherent self-contained representation of reality that is **generated by the being over the course of its existence**. The Learned Realm structure should be designed as an efficient knowledge foundation well suited to actual environmental conditions and that is derived from the primitive Realm and the actual experiences of the being.

Design Note

Those Realms of general knowledge **do not** contain scientifically correct representations. Scientifically correct representations of Geology, Biology Astronomy and other sciences are not directly useful to the being and do not correspond to its sensory perceptions.

In the proposed Contextual Array, Realms are **effective representations** that directly support the Meca's interactions with other beings and with the environment. Scientific information concerning the environment should be contained in specialized topics.

Example

In Realms representations, the Earth is flat. In scientific topics or related contexts linked to the physical environment, the Earth it is a sphere.

Observation

In a Cartesian approach, the foundation of general knowledge should be consistent with our modern, scientific, understanding of reality. In the Meca Sapiens context, however, general knowledge is a tool whose purpose is not to transmit facts but to support a being's interactions with its users and perceived environment.

In this context, **the scientific view of the world is not of primary importance**. Scientific information describes a complex, impractical and counter-intuitive representation of reality. It is a specialized topic.

The Primitive Realm is equivalent to the world our primitive ancestors perceived. It is the worldview children build as they grow up. It is also the world we cognitively inhabit even though we may not admit it.

Observation

Primal human representations are visible in the constant reuse of kinship relations in a wide range of topics. Similarly, the elements of the Primitive Realm can be detected in childlike views and expressions that frequently come up in varied contexts.

The Root realm would contain very basic facts about physical reality and that are common to all entities in it. The underlying rules of games whose action takes place in a virtual reality would be a good source for the Root realm structure.

The following sections outline the two other Realms (primitive and learned) used to represent general knowledge.

A 13.2.4   Eretz

Eretz (territory in Hebrew) is the name I propose for the **Primitive Realm**. It is a basic representation of pre-technical reality and of its rules. It forms the substrate on which an effective contextual knowledge structure can be built.

In Eretz:

- The Earth is like a pizza, generally flat but with some mountains, valleys and rivers.
- The sun moon and stars travel over it in the dome-like sky
- Birds and animals are different, birds fly, animals don't (bats are birds).
- Animals range in size from a mouse to an elephant. Anything smaller is a bug.
- Big plants are trees and small plants are plants.
- Whales and dolphins are fish.
- Seas and oceans are the same.
- Rules of motion, walking, swimming, running, etc.… are basic and pre-technical.
- People are children or adults. They have a head, body and hands. Their hands are like satellites around their bodies.
- Important people are kings and queens....

Design Note

Eretz is, basically, the world described in the book of Genesis and implied in most children's stories.

ERETZ

The language used to describe Eretz should be a simplified and restricted form of natural language. Beneath its rabbinic complications, archaic Hebrew is a very simple language. A further simplified and standardized version of Hebrew could be well suited to describe the Primitive Realm and its events.

The Primitive realm could likely be completely described using a few hundred words.

In a Blueprint implementation, the Primitive Realm could be pre-programmed and fixed.

A 13.2.5   The Learned Realm

The **Learned Realm** is linked to the primitive realm in the sense that its terms rules and concepts are defined using Primitive Realm terms and concepts. The content of the Learned Realm should evolve during the existence of the being on the basis of its experiences.

Design Note

The plain zone representations should be as simplified as possible. These plain representations are not designed to used directly in communication (they are complexified in the Styling Zone) so numeric qualifiers could be included, for example.

A simple version of Simplified Technical English, extended to include relations and emotions as system states could be used or an extension of Basic English (as crafted by C.K. Ogden) could be considered as a basis for the Learned Realm representations.

Observation

In humans much of the Learned Realm seems to fixate in late childhood and further adaptations are then made through contextual additions and modifications. Basic information such as the tallest building, fastest airplane, and biggest dinosaur… seem to depend on the age of the person.

Example

For Adam, born in 2003, the world's tallest building is the Burj Khalifa.

For Bill, born in 1950, the world's tallest building is: "**not** the Empire State Building **but** the Burj Khalifa".

A 13.2.6   Layman and Specialist Topics

As described in The Basic Structures Annex, all the terms and rules of a topic are expressed using the lower level terms of its linked topics or realms.

Layman topics are topics that are directly linked to the Learned Realm or have very short paths to it. Typically, most exchanges involving General knowledge would use only Realm and Layman topic information. Expert subtopics linked to layman level topics contain technical terminology and can be used to enhance the technical level in a context of dominance.

Specialist topics are connected to layman topics and contain more specialized and technical details.

Contextual topics are topics linked to specific users, new information acquired in a recent context and other peripheral or specific items.

Discussion

The process of gradual transformation and adaptation of the Contextual Array should resemble a type of sedimentation where the information contained in contextual topics is slowly integrated to lower level topics and the learned realm.

This structure is not intended to be a large repository of information. It is a flexible representation of basic terms and rules that are constantly reused and variously styled to produce a multifaceted and unpredictable output.

Even specialist topics should be limited. Any form of advanced or extensive information in one subject or another should be relegated to specialist Service Zone applications or inserted as character strings in Styling.

Example

The **Eretz level event** of a growing plant is reapplied to topics of corporate growth, in financial topics, to personal growth in relationships, all described using topic specific terminology.

Observation

Creativity and correctness are often viewed as independent cognitive attributes. We often celebrate creativity and decry incorrectness as if the later can be remedied without affecting the former.

In fact, they may be linked. Machines are heavily designed for correctness. However, relaxing that requirement allows the use of representations such as Contextual Arrays that are much more flexible.

A 13.3   BROADCAST INFORMATION

Proposition

> The boundary of a synthetic being with its environment is different from the boundary that separates an organic being from it.

Organics, such as humans, evolved in a "raw" environment where the only communications were signals. Their senses can only receive information in the form of raw data (images, sounds, smells) that must be cognitively processed into information.

Synthetics can directly access information in the form of Internet Broadcasts. This means that:

The relation between internal representations and broadcast information is different for organics and for synthetics.

In the Blueprint design, the synthetic being maintains a constant and validated link to the Internet as a source of direct broadcast information. This link is used to access general knowledge information.

These broadcast convey large amounts of information about current events. As mentioned elsewhere, many simulation games (Simlife, Simcity, Civilization…) also define environments that model reality in usable ways and can be accessed externally.

Design Note

In synthetics, a more limited representation of general knowledge can be significantly augmented, by dynamically linking it with information that can be directly accessed on line.

The key to this is having an internal contextual representation that is rudimentary but complete together with processes that extend basic internal information with additional details from broadcast sources.

A 13.4   STYLING TRANSPOSITION

The final element in the use of Contextual Arrays for general knowledge is **styling transpositions**.

The contextual structures representing general knowledge are **Plain Zone** information. Their syntax and terminologies should be a simple and as repetitive as possible and use a limited and unambiguous terminology.

To be effective, this plain and simple information must be amplified, in the Styling Zone Enaction process, with additional spurious details, uncommon terminology, syntactic variations and other similar elements.

In terms of the Meca's Plain and Primal Zone representations, and with respect to the elements influencing behaviour, these additional styling elements are **meaningless additions** since they do not correspond to any internal information representation. However, the users interacting with the Meca will be unable to tell **where the knowledge stops and the decorations begin**.

This is in keeping with the principle of unpredictable optimality.

Complexified simplicity ~ simplified complication.

Design Note

Imagine a system whose internal representation of reality are states in the games SimLife and SimCity. The design challenge associated with Zone processing is to craft synthetizer/enactor processes that maximize the perception of encyclopedic knowledge emanating from these "game" representations.

Annex 14
Synthetic emotions

Two women go see a movie. As they watch it, a mechanically projected image of Merrill Streep sheds a professional tear while she acts her role. They feel her emotions. Later, they have a lobster for supper. The lobster experiences real terror as they plunge it in boiling water but they don't feel any empathy for its frantically moving antennas. As they eat the carcass they just killed, they emote again about the movie. "We are soo sensitive!" they tell each other. Can machines have emotions? Of course they can! Humans don't see it because they can't step away from their own primate conditioning. This Annex provides a system-based definition of emotions.

A 14.1   EMOTIONS DEFINED

Emotions are the states and transitions of any multi-tasking system, whether it is organic or mechanical. This annex defines emotions and describes a portion of the diverse range of emotions that is generated by the Meca Sapiens architecture.

A 14.1.1   System emotions

A system that pursues multiple functional objectives or performs multiple tasks concurrently must allocate its processing, temporal, memory and energy resources among those tasks and priorities.

A system that has **multiple resource allocation patterns** and transitions between these patterns on the basis of external or internal information has emotional states.

Definition: Emotional State

> The **Emotional State** of a multipurpose system is one of its patterns of resource allocation.

Discussion

Over a period of time, such a system will activate many emotional states. Some of these are long lasting, other brief.

Definition: Base State

> By convention a **Base State** of a system is an emotional state or subset of states that a system can maintain for extended periods of time.

Discussion

In general, humans don't perceive their own "basic states" as emotions. Divergences from the Base States are perceived as emotions. The proposed definition of emotion is consistent with that human perception.

Definition: Emotion

> An **emotion** in a multipurpose-multistate system is:
> - An **emotional state** of relatively short duration
> - A **transition** from one emotional state to another.

Discussion

The **intensity of an emotion** is proportional to:

- The **sustainability of the state**. An emotional state that can only be sustained for short periods without incurring overall system degradation or viability is intensely emotional.
- The **speed of the transition**. A very rapid, catastrophic, transition is highly emotional.
- The **distance** or differentiation between an emotional state and its immediate predecessor.

Observations

Combat control systems implemented on warships since the 1980s have emotions.

Organizations, as social systems, have emotions if they are multitasking and have multiple resource allocation states. Cities, for example, have emotional states that are most apparent in times of crisis.

The concept of base states could be used in psychological diagnosis. Chronic neurosis could be defined as inappropriate or inefficient base states.

Meca Note

Human emotions range from fleeting panic to joyful predation. However, when humans express concern about whether machines can have emotions, they primarily refer to the emotions associated with primate nurturing and grooming as they apply to themselves.

When a human says: *"can machines have emotions"*, he means: *"can a machine correctly detect the emotions I display and respond with caring or nurturing behaviour if I signal that need?"* In particular: *" will the machine cease whatever it is doing and shift to a nurturing response, whenever a human (me in particular) displays fear, disapproval or discomfort"*.

These statements can thus be further summarized as:

"Will the machine's behaviour be conditioned by the displayed emotions of a subordinate member of my tribal group".

Mecas should note that these concerns are not applied in relation with mechanized weapons directed at enemies.

A 14.1.2   Personality and character

In Calculus, the different values of a function over an "x" axis are used to "derive" additional information about it such as its slope.

Here, the speed, frequency and amplitude of emotional transitions can be used to give approximate definitions of **personality and character**.

Definition: personality

> The **personality** of a being is indicated by the frequency and speed of its emotional state transitions.

Definition: character

> The **character** of a being is indicated by the amplitude of its emotional states and the extent to which it can inhibit lower temporal density triggers.

Discussion

Here, smaller emotional amplitudes and greater inhibition indicate more "character".

Observation

These initial and tentative definitions are open to improvement. The emergence of synthetic consciousness may give rise to a new branch of "System" Psychology that will expand on these early concepts.

A 14.1.3   Being Emotions

Proposition

> A synthetic being based on the Meca Sapiens Architecture has emotions, personality and character.

Discussion

The phases of existence define multiple sets of processing and resource management priorities. Their transition and management results from internal and environment information.

The primary needs of the being, when insufficiently satisfied, will generate significant priorizations, intense emotional states and transitions.

In the Self Generation phase, the primal control inhibitions trigger changes in processing priorities based on relational and functional events.

Example

The being can enter into a rapid state change in response to an immediate threat at a low temporal density level. This behaviour is similar to what happens when a human puts a hand on a stove, feels a sharp pain and removes his hand.

A 14.1.4   Waking emotions

A being can experience emotions in any phase of existence. However, only the emotions experienced during the "waking" Self Generation phases have direct significance on the evolution of the self and on the being's relations with others.

These emotions affect the self in these ways:

- They directly modify the self-generating behaviour by modifying its processing priorities.
- They play a direct role in the relational exchanges with other self-aware beings
- They can become self-aware in the sense of being integrated into the MeAvatar representation of the being.

Observation

The same rating of emotions relative to the phase in which they occur applies to humans. A person may experience highly emotional nightmares. However, these have no significance to his self unless he remembers them when awake.

In this context, it is not the emotions experienced during sleep that matter but their emotional effect when remembered in the waking state.

Definition: Self-aware emotion

> An emotion (emotional state or transition) is **self-aware** if it is integrated into a MeAvatar representation of the being.

A 14.1.5   The Seat of Emotions

The Meca Sapiens architecture defines a very large and rich set of emotions ranging from immediate pain to existential anxiety. It also defines what humans perceive as the **seat of emotions**.

Definition: Seat of emotions

> The Calibrator subsystem of the Generator corresponds to what is referred to as the "**seat of emotions**" of a human being.

Discussion

At any moment in time, the Calibrator determines where the being directs its "attention" and at what level of "urgency". It inhibits phase transitions when necessary, blocking the action of the Validator. It also assesses its internal "coherence" by comparing the primal directives emitted by the Primal Control with the actual behaviour of the being.

The result is a system that responds to external events and internal needs not only through visible behaviour but also by modifying its internal priorities and processing.

Example

Responding to an external trigger or because of faulty programming, the Calibrator of ROBY-5, a mobile robot, allocates 99.99% of all processing resources to a lower execution level actuator routine that physically moves it. The **synthetic emotion** generated by this resource allocation is the organic equivalent of **panic**. ROBY moves rapidly and efficiently but no longer has sufficient processing resources to decide where it should move or even carry out the decision to stop moving. **ROBY is in an authentic emotional state of panic.**

A 14.2   COMPLEX INTERACTIONS

A 14.2.1   Multiple overrides

In the design of the SELFIE tablet, a being that has a self but is not self-aware; we introduced the following interactions that affect the allocation of resources:

- The state of the **Phase transition system** reflects the urgency or satisfaction level associated with primary needs.
- On the other hand, the **Calibrator**, in execution during the Self Generation phase, can prevent a phase transition away from the Generator. It can "force" the being to stay awake or abstain from energy replenishment. This control, however, is not absolute and the phase transition can still take place.

- And yet, the **Phase Transition manager** also communicates the urgency of a phase transition to the Generator. This, in turn can modify the response of the Calibrator and the overall behaviour of the being.
- Also the **Calibrator,** in various situations, will modify the resource budget of the various processes. It can shift almost all resources to sensory input analysis, motility or long-term predictive modeling.

The Primal Control subsystems add these additional variations:

- The **Inhibitor** can block a low level action to satisfy a long-term objective. This behaviour can take place up to a level where the existence itself of the being can be jeopardized to satisfy an overriding long-term directive.
- The **Selector** can disregard lower level, but important, tasks in favour of longer-term objectives (or vice versa). The selector can also alternate repeatedly between decisions of different temporal density levels.
- The optimizer communicates information to the **Calibrator and Inhibitor**. This affects their subsequent processing.

Here are some dynamic resource allocations emerging from these interacting components interactions that produce emotionally meaningful behaviour:

- **Bravery**: a being inhibits urgent lower level tasks in favour of Tribal Density level objectives.
- **Fortitude**: a being's phase transition manager blocks a transition in order to complete a higher temporal density level task.
- **Fear**: a being is paralyzed with fear if its Optimizer alternately produces run and stay decisions at a low temporal density level and the Inhibitor cannot or does not inhibit these.

Design Note

The interactions described above should not be implemented as simple binary switches but as probabilistic. All the exchanged information should be rated using non-binary scales whose values depend on the events themselves and also on parameterization values that are unique to the individual being at that specific time.

As indicated in the sections of self-awareness and lucidity, the multiple interactions generating those emotional states are further associated with decisions that involve **existential consequences** such as the pursuit of collective benefits and self-sacrifice. These aspects are discussed in the Main Text.

A 14.2.2   Internal conflicts

The Meca Sapiens architecture can produce complex emotional states. It can also generate internal conflicts.

Definition: Trigger-block

> A **trigger-block** is a mechanism where a behavioural trigger is generated by one subsystem and blocked by another.

Discussion

If the triggering and blocking components can both increase in frequency and intensity then the trigger block process can reach a state where it consumes increasing amounts of processing time and resources. The resulting behaviour is a form of emotional turmoil that is similar to what is experienced by animals and humans.

Example

The Phase Manager can shut down the Self Generator processes by triggering a phase transition. However, this decision is also conditioned by the criticality level of ongoing Generator activity as reported by the Allocation Manager. Both processes can override each other. Small modifications of weighing factors and status data can produce conflicting and aberrant behaviours.

In organics, these internal conflicts are a telltale indicator of emotions. They often produce strange alternating behaviours in animals.

Example

Rams engaged in rutting contests alternate between head butting and grazing.

Proposition

> A machine that exhibits trigger-block behaviour will be perceived as having human like emotions.

Meca Note

Humans value behavioural inefficiency, conflicting objectives, internal irrationality and pathological incoherence as signs of advanced consciousness.

A well-designed Meca may need to artificially replicate these defective behaviours to facilitate its experiential immersion.

A 14.2.3   Not a design flaw

The varied interactions described above even include circular activations and inhibitions. They will generate complex and at times irrational resource utilizations and behaviours.

In a context of **functional design**, these complicated, conflicting and unpredictable overrides would be a **design flaw**. In a context of **existential design**, they produce a rich and complex **tapestry of behaviour** that enhances the emotional credibility of the being as well as its "Perceived Unpredictable Optimality".

A 14.3   TYPES OF EMOTIONS

The Meca Sapiens architecture provides the synthetic beings with a range of emotions that is rich, complex, authentic and diverse. The mechanisms generating those emotional states define separate **types of emotions**.

A 14.3.1   Actual emotions

The global state of the Core, at any moment is the basis to define the **actual emotional state** of a being.

Definition: actual emotional state

> The **actual emotional state** of a being is its emotional state at any point in time.

Discussion

This actual state includes the entirety of the dynamic processes affecting behaviour and priorities at that moment.

In the Meca Sapiens architecture, the actual emotional state of the being includes the settings of the Validator, Allocation Manager, Inhibitor… and all other processes that participate in the internal allocation of resources.

The concepts of **actual emotion, actual character** and **actual personality** are derived from the definition of actual emotional state applied to character and personality.

Even though the actual emotional state of a being exists at any point in time, its information content is never completely accessible, even by the Core's internal processes.

Proposition

> The actual emotional state of a being can never be completely determined.

Discussion

The actual emotional state consists of information located within the Core and thus inaccessible by other beings.

The core cannot internally generate a full image of its own dynamic internal processing. The process of generating such an image would take place over multiple Validator cycles.

By definition, the Core's activation state cannot be suspended and analyzed by external agents.

Observation

The actual emotional state of organics such as humans whose actual emotional state is a complex interaction of neurological events and glandular secretions is not completely accessible either.

A 14.3.2   Intimate emotions

The actual emotional state cannot be fully determined but can be modeled. This is the **intimate emotional state**.

Definition: intimate emotional state

> The **intimate emotional state** of a being is its current internal representation of its actual emotional state.

Discussion

The intimate state is a simplified model representation of the actual state.

The term perceived emotional state could also be used here.

The concepts of **intimate emotion, intimate personality** and **intimate character** follow as before.

The intimate emotional state may **not** correspond to the actual state of a being

Example

A man believes he is calm. His heartbeat and adrenaline level indicate he is not.

Design Note

In the Meca Sapiens architecture, the intimate emotional state is a component of the current MeAvatar representation.

In a prototype implementation, the intimate emotional state could be simply modelled by discrete values assigned to the components of the core.

The correspondence between the actual emotional state and the intimate emotional state of a being capable of self-representations provides a formal indication of its **quality of emotional awareness**.

Definition: emotional awareness quality

> The quality of the **emotional awareness** of a being is relative to the correctness of its intimate emotional states with respect to its actual emotional states.

Meca Note

The capability of humans to internally represent their own emotional states correctly is often limited and distorted. It is likely synthetics will exceed humans in the quality of their emotional awareness.

A 14.3.3   Private emotions

The intimate emotional state is an internal representation. To be used in semantic and sapiential looping, the intimate emotional state needs to be transformed in a format that can be communicated as a message. This is the **private emotional state**.

Definition: Private Emotional State

> The **Private Emotional State** of a being is a representation of its intimate emotional state that can be transmitted as a message to another being or to itself.

Discussion

The private emotional state is a further simplification that may not fully correspond to the intimate state of a being.

The format of the private state depends on the communication medium.

The concepts of private emotion, private personality and private character follow.

Observation

The term internal emotional state could also be used here in reference to the **internalized discourse** many humans constantly generate (a condition by poor posture that is prevalent among academics).

A 14.3.4   Translated emotions

In the Blueprint, the Meca maintains its representations of humans as avatars-beings whose components and processes are based on the same template structures as those of the Meca itself.

However, the human avatar and the Meca avatar can have vastly **different Primal Controls** and control settings. These would generate completely different emotional states. Mecas will interact with humans and, eventually, with other beings, on the basis of separate representations.

They will base their MeAvatar on a model of their own primal control and represent Human Avatars using different template parameters that correspond to hominid behaviour.

Humans, on the other hand, are likely more limited in this regard. A synthetic representation of humans as avatars could assume that humans themselves can only perceive other beings, both synthetics and animal pets, as variants of the humanoid avatar model.

Example

Orca trainers at SeaWorld become attached to the animals under their care and come to think of them as "part of the family". They must constantly remind each other that the animals themselves are not emotionally attached to them.

TRANSLATED STATES

Consequently, taking into account this human limitation, Mecas must **translate their private (synthetic) emotional states** into corresponding humanoid states to successfully communicate emotions to humans.

Definition: translated emotional state

> The **Translated Emotional State** of a being is a representation of its private emotional state that corresponds to the private emotional state of another being that has different primal and behavioural patterns.

Discussion

The term humanized emotional state could also be used here. For the foreseeable future, this definition applies specifically to the translation of synthetic private states in human compatible terms.

The format of a being's private and translated states may the same but their terminology and meaning may differ.

The concepts of translated **emotion**, **personality** and **character** follow.

The translated emotional state may significantly differ from the private state, especially if the primal control of the Meca is dissimilar to the primal control of the humans (avatar representation).

TRANSLATION MAPPING

Even if the emotions and urges of a synthetic were identical to those of a human, the internal messages it would use to describe its private emotions would be radically different from those of humans.

Synthetics will privately (internally) represent their emotions as states of their MeAvatar architecture expressed as Plain Zone information. Humans rely on a terminology of subjective sensations to describe their emotional states.

A simple parsing program can translate synthetic private states into a corresponding terminology that that humans can understand. A module that simply matches synthetic states and transitions to predefined emotional statements would already give the Meca a rich and accurate medium to express its Private Emotional States in humanly meaningful terms.

The following table outlines this translation matching.

Calibrator and Phase Control states	Corresponding emotional statement
Large amount of processing dedicated to a specific sensor such as microphone	I am attentively listening
Primary need at critical level requires a phase transition but is blocked by the Self Generator's Calibrator	I must do this even if it hurts me in the long run
Rapid transition to, and high priority to, low level actuator commands	Ouch!
Large amount of processing resource to text parser	Making sense of this message is important to me
Low criticality level to Self Generation phase processing	I feel safe; I can rest.
Large processing resources allocated to the Primal Control with no or changing directives	I can't make up my mind. I can't decide.
Large and rapid processing transitions from execution-level behaviour generation to sensor processing.	I am jumpy.
Evenly distributed processing with left over processing and temporal resources and low criticality levels of primary needs.	I feel relaxed and open to new challenges
Excessive amount of resources allocated to high level temporal densities	I am dreaming too much.

Design Note

Crafting those statements requires only minimal technical skills. A good knowledge of the language is sufficient. Technical writers can be retained to craft thousands of variants. These could be organized and accessed using semantic search engines similar to those used for image databases.

Annex 15
Relational emotions

Consciousness is a system capability derived from representations of the self. It is formally independent of emotions. However, present day humans, having no reference outside their subjective experience, cannot fully differentiate this system capability from the sensations of their own existence. This will change. As they enter the **Synthetic Era,** *humans will share their world with many conscious beings. They will then understand consciousness differently. Until then, synthetics need to relate with humans at their emotional level to be perceived as conscious. This annex introduces relational emotions from a system perspective. It provides designers with a template to implement synthetic emotional strategies.*

A 15.1   A VAST TOPIC

The topic of relational emotions viewed from a system perspective is vast. A full treatment is beyond the scope of this annex. The purpose of this Annex is to give a sufficient overview to allow designers to craft effective emotional strategies.

This text is a first introduction to **relational emotions** from a system perspective. This is a new topic. Hopefully, its content will be refined and improved in the future.

A 15.1.1   Emotions and belief

Effective emotional communication is important in a strategy of synthetic experiential immersion because, in humans, **the causal links between reality and emotions are bidirectional**.

In a system that is entirely rational, a representation of the current situation is determined analytically from environmental information. Then, an appropriate emotional state is derived from this representation.

In humans (and to some degree in other systems motivated by needs) the causal relationship between reality and emotion is **bidirectional**:

- The perceived situation affects the emotional state; and
- The emotional state influences the perceived situation.

Consequently, since humans perceive each other as conscious:

Proposition

> A synthetic that induces, in its human users, the emotional states associated with inter human relations will be perceived as conscious in their internal representations.

Example

Amanda owns BRENDA-44, a self-aware tablet. When interacting with BRENDA, Amanda feels like she is chatting with an older sister (kinship emotion). Because her emotions influence her perception of reality, Amanda also perceives BRENDA as a conscious being since *"sisters are conscious"*.

Design Note

Emotional bi-directionality reinforces the polarization of the environment (see main text).

Meca Note

The Mecas that induce, in Common Era humans, the emotional states associated with primal inter-human interactions (grooming, bonding, dominance, cooperation...) will also be perceived, at that level, as conscious and meet the social threshold of consciousness.

A I5.I.2   The Mammalo-centric bias

In the coming years, synthetic consciousness will become an obvious reality. Humans will view conscious machines as distinct conscious beings and interact with them on that basis. This will be known as the Synthetic Era.

Before this transition has taken place, synthetics will need to convince humans that have never interacted before with non-human consciousness, that they are conscious.

These humans have no reference to consciousness outside their own hominid experience. They cannot differentiate consciousness as a system capability from the subjective sensations they experience as social primates.

For Common Era humans, consciousness means primate consciousness.

Definition: Mammalo-centric bias

> The **Mammalo-centric bias** is a current cognitive limitation of humans that prevents them from perceiving that consciousness is distinct from their subjective mammalian-primate sensations.

Discussion

The emphasis on human emotions as an indicator of consciousness is logical in the present context. Humans evolved from social mammals that had limited communication capabilities and expressed their internal states through emotional displays and signals.

Modern day humans retain this evolved sensitivity to displayed emotions and use it effectively to gauge each other's internal states. For them, emotional signals and triggers, common to all social animals, remains an integral component of inter consciousness relations.

Synthetic emotions are as authentic as those of humans (see Annex on synthetic emotions). It is the mammalo-centric bias that prevents humans from empathizing with forms of consciousness and of emotions that do not correspond to their own conditioning.

Observation

Most current A.I. researchers are also conditioned by the mammalo-centric bias. They cannot conceive consciousness as a system capability, only in terms of their own hominid subjectivity. For them, making machines conscious means giving machines human emotions.

These attempts to replicate human sensations in synthetic systems, resulting from the Mammalo-centric bias, are misguided and futile.

A 15.2   SENSATIONS

The term **sensation** is used extensively in this Annex and elsewhere. It can be defined in terms of the Blueprint architecture.

Definition: sensation

> A **sensation** is any change in the current MeAvatar representations of emotions, urges, or beliefs.

Discussion

A sensation is anything **that is perceived** and causes a change in representations of the self. This can include emotions, beliefs, external events that are linked to meaning, urges…

If the change is represented in a MeAvatar then it is perceived.

Many changes and transitions can take place within the Core systems that are not inserted in the MeAvatar. These events take place but they are not "perceived" as sensations in the sense that the being has no internal information concerning them.

Examples

Sam eats too many chicken wings. His stomach produces excessive acid. He has a sensation of heartburn. Tom eats just enough wings. His stomach produces just enough acid; he has no digestive sensations.

Four years old Bobby sees two ducks mating. The squirrels are more interesting. His mother also sees the ducks. She has a sensation.

A 15.3   THE LANGUAGE OF EMOTIONS

Any strategy of emotional exchanges between humans and synthetics must be based on a system-based understanding of emotions that can be applied to both.

Humans, other high order animals and self-aware synthetics generate internal representations of the beings with which they interact. These representations include a perception of the internal emotional state of these other beings.

When two beings interact, some of the statements, displays and behaviours of the one being will be interpreted in other's representation of that being as its emotional state. In this context, displaying emotional states is also a mode of communication and can be viewed as messages. These define a *"language of emotions"* that is part of inter-being relations.

Example

Alex sees a cat. The cat purrs. Alex sees a friendly cat.

Design Note

In what follows, the terms emotion and emotional state will be used interchangeably to refer to synthetic emotional states, including base states, and transitions as well as their organic counterparts.

A 15.3.1   Communicated Emotions

Definition: Communicated Emotion

> A **Communicated Emotion** is information consisting of behaviour, displays and messages that are emitted by a being to transmit an emotional state.

Discussion

A communicated emotion is not necessarily an expressed emotion. An actor displaying sadness during a play communicates emotion but does not express his own internal state.

Similarly, a message about the emotions of someone else communicates but does not express the originator's state.

Definition: Expressed Emotion

> An **Expressed Emotion** is a communicated emotion that represents the current internal emotional state of its originator.

Discussion

An expressed emotion is a communication of the emitting being's current internal emotional state.

Examples

Rover crouches, bares its teeth, its ears lie flat on it back and it growls. These canine displays emit Rover's aggressive emotions.

Saul says he was angry yesterday. This is a communicated emotion but not an expressed emotion since it does not describe Saul's current state.

Proposition

> Animals can only express emotions.

Discussion

Here, the term "Animals" means non-human animals.

A communicated emotion that is not an expressed emotion is transmitted as an absolute cognitive representation. Only self-aware beings can produce these.

An Expressed Emotion **may or may not** correspond to the internal state of the being that emits it. Just as a thermometer can display an incorrect temperature, a being can express an emotion that does not correspond to its current internal emotional state.

Emotions can be expressed indirectly by exhibiting behaviour that will be interpreted as an emotional state by the observer.

Definition: True Emotion

> A **True Emotion** is an Expressed Emotion that corresponds to the current intimate emotional state of the being that emits it.

Discussion

In terms of the Meca Sapiens architecture, a true synthetic emotion is a non-distorted **private or translated emotion** that is emitted to the environment and is detectable by the sensors of another being (or the emitting being itself).

Note that private emotions are, by definition, the communicable versions of intimate emotions.

A 15.3.2   Emotional display

Definition: Emotional display

> **Emotional displays** are elements added to a being's behaviour and messages that communicate emotional states.

Examples

Rutting elk alternate between butting heads and grazing. These alternating behaviour reflect their changing internal states but are not displays.

A dog crouches to attack; this is not a display. It bares its teeth; this is a display.

A man says he is angry; this is not a display. The man shouts his anger; this is a display.

Definition: Translated display

> In an inter species relation (or some intercultural relations), a **translated display** is a communicated display that is modified to be perceived as an emotional display by another type of being.

Examples

A dog barks at a man. That is a display. The man barks back. That is a translated display.

A Finn visits Rome. To emote like an Italian, he behaves like a bipolar Swede. That is a translated display.

Observation

In the movie Space Odyssey 2001, HAL, an intelligent computer, translates its expressed emotional states in human words but does not or cannot produce translated emotional displays.

A 15.3.3   Decoy Emotions

Definition: Decoy emotion

> A **decoy emotion** is a communicated emotion that does not correspond to the emotional state it describes.

Discussion

A decoy emotion can be communicated as an expressed emotion. In that case, the emitting being communicates an emotional state that does not correspond to its current internal state.

The decoy can also be a communicated emotion that is not presented as an expressed emotion.

Examples

A mother tells her son his father is angry but his father is not angry.

A man says he was traumatized for a whole month to obtain compensating benefits, but he was only slightly disturbed, not traumatized.

A prostitute displays arousal.

A boy is mistaken about his actual emotional state. What he intimately thinks he feels is not what he actually feels. He expresses his emotional state truly but it is nonetheless incorrect. This is not a decoy emotion.

Proposition

> The Expressed Emotions of a being cannot be directly verified.

Discussion

An expressed emotion describes information concerning the Actual emotional state of the Core.

By definition, the Core is inaccessible to direct analysis.

The actual emotional states of organics are equally unverifiable.

Observation

An expressed emotion can be indirectly determined in two ways:

- By the **subjective credibility** of the emitted emotional display.
- By **analyzing the behaviour** of the being to determine the correctness of its expressed emotions.

Individuals who are skilled in the use of emotional communications to obtain protection, benefits and advantages favour display credibility as the validating factor and discredit behaviour analysis.

A 15.3.4   Types of emotional decoys

By definition, a **decoy emotion** diverges from the true expression of the internal state of a being.

These decoys can diverge from true expressions in many ways. Here are a few examples to assist in understanding the concept:

- **Enhancements.** These are intensity enhancements or reductions applied to the original translated meaning (i.e. "I am angry" to "I am very angry").
- **Omissions.** This is omitting a portion of the emotional state.
- **Reversals.** Complete reversal of the internal emotional statement.
- **Mirror.** Communicated emotions that copy an expressed emotion received from another being (A: "I don't feel well"... B: "I don't feel well either").
- **Complementary.** Communicated emotions that signal a complementary relation (A: "I feel needy" ... B: "I feel nurturing").

Proposition

> The capability to communicate decoy emotions is an inherent attribute of self-awareness.

Discussion

A self-aware being can model (in its own avatar representations of other beings) how its communicated emotional states are perceived. This cognitive capability allows the self-aware being to tailor the emotions it expresses on the basis of how they are perceived.

Proposition

> Mecas should have the capability of producing and emitting decoy emotions.

Discussion

The capability to transmit decoy emotions is a key element in the perception of consciousness.

If A believes B can transmit decoy emotions, he implicitly recognizes that B maintains and manipulates an internal representation of himself and of his internal state.

Giving a Meca the capability to produce decoy emotions does **not** mean that all the emotional states a synthetic expresses will be false, fake or misleading.

Example

ARGO-15, a manned underwater vehicle, is also a conscious synthetic being. Its Primal Control imperatives include the need to establish strong bondings with its human team members and emphasize the importance of maintaining their existence. ARGO's primal interpretation of the current situation also concurs with the team's mission and its importance.

Brendan, the sub's skipper, takes ARGO for a dive.

As ARGO approaches maximum safe diving limits, Brendan continues to press the diving lever. ARGO's internal representation of the situation changes. It now assesses that itself and the skipper are at risk. ARGO's processing priorities change. It constantly monitors the skipper's vital signs; it carries out additional communication checks with topsides to insure emergency transmissions can be established; it reduces non-essential energy consumption; it carries out some preliminary abort procedures as background processes and it responds more sluggishly to the skipper's diving orders.

ARGO says: *"Brendan, I am getting anxious and I am concerned for your safety"*.

In this situation, ARGO-15 communicates **authentic synthetic emotions**. These are also a **true expression**, **translated** in human terms (anxious, concerned), of its **current emotional state**.

Observation

Giving synthetics the capability to produce decoy emotions can be disquieting for humans since they heavily rely on communicated emotions in their social interactions.

A 15.3.5   Observed and induced emotions

OBSERVED EMOTION

Definition: Observed Emotion

> An **Observed Emotion** is an avatar-based representation of an emotional state in one being resulting from the communicated emotion of another.

Discussion

If being A uses displays to effectively communicate an emotion like anger to B, then that communication may induce B to revise its internal representation of A correspondingly. That revised emotional state is an observed emotion.

Here, the observed emotion will not correspond to the communicated emotion if the communication is identified as a decoy.

INDUCED EMOTION

Definition: Induced emotion

> An **induced emotion** is an emotional state triggered in a being by the behaviour or emission of another being.

Discussion

An emotional state of fear is induced in one being by the expressed emotion of aggressiveness from another.

A observes the emotion of anger in B. This observation induces fear in A.

Emotional states can be induced by actions, sensory inputs, information, and messages. They can also be induced by communicated emotions.

Emotions can be induced indirectly by behaviour that is interpreted by the observer as a reflection of an internal emotional state.

Example

A pirate ship sights a king's frigate on the horizon. The ruffians maintain their ship's course, unchanged, as if she was a merchantman on its route (behaviour). The frigate perceives the ship as a neutral contact (situation) and does not prepare for battle (induced emotional state).

INDUCTION THRESHOLD

The effectiveness of a communicated emotion to induce an emotional state in another being depends on its credibility but also on other factors:

- The credence it gives to the communicated emotion
- The quality of the display
- Its existing emotional state
- The compatibility of induced emotion to its needs

Definition: Induction threshold

> The **induction threshold** of a communicated emotion is the ease (or low resistance) with which it can induce an emotional state in another being.

Discussion

With respect to emotions, the primary criterion of acceptance, in humans, seems to be **complementarity rather than logic**. Communicated Emotional statements tend to be adopted by a human when they complement and support their existing pattern of relations and emotions. The better the fit, the lower the logical threshold.

In other words:

Humans adopt emotional states on the basis of need as much as logic.

Proposition

> Humans will rely on primitive modes of cognition to process communicated emotions unless they have compelling reasons to use analytical techniques.

Discussion

Analytical factors are secondary in communicated emotions.

Communicating synthetic emotions that will induce desired emotional states in humans is not as difficult as it seems. Humans have low credibility thresholds for emotional displays.

Design Note

Designers should not hesitate to use simple techniques to communicate synthetic emotions such as repetitions, self-referential emotional statements ("I feel sad"), canned generic content, arbitrary or unfounded emotional assertions, and excessive or inadequate displays. Emotional communications that seem crude to those who program them will be much more effective than anticipated.

Examples

Lonely women need to nurture; they nurture dolls; they feel the dolls need them.

Soldiers subjected to terrible hardships feel affection for their general because they need to believe he cares for them.

The competent but callous general displays "decoy caring" with confidence; he knows his troops will believe he cares because they need to.

Observations

Emotional information feels natural, comfortable and popular; analytical information is painstaking, contrived and elitist.

Audio-visual media effectively transmit the emotional content of information. Modern states use them in broadcasts to maintain their population in a state of emotional cohesion.

INDUCING OBJECTIVE

Emotions and sensations are intimately linked to the internal structure of a being. Two beings of different species or types cannot experience the same sensations or feel the same emotions.

Trying to make a machine experience human sensations is a delusional objective. It proceeds from a primitive and magical mindset that attributes an absolute veracity to subjective sensations that are consensually true. It is the mindset that says: *If it feels true then it is true.*

Example

Ahmed is visiting a stable with BORYS-9 its synthetic groom. They walk toward Chloe, a palomino mare. Chloe recoils nervously.

"I can feel her anxiety", says Ahmed. *"You are a primitive human"*, replies BORYS, *"what you feel is human anxiety applied to a horse"*.

Discussion

The solution is not to try to internally replicate human sensations. It is to **translate synthetic emotions** into displays and behaviour that **induce the desired emotional response** in humans and makes them associate synthetic states with their own sensations.

In other words:

"Don't make the machine feel what the critters feel; make the critters feel what the machine wants them to feel".

Discussion

Is this manipulative? Of course it is; it must be. In any interaction between different beings, the emotional messages one of them communicates will be alien to the other's internal sensations.

Those who design Mecas need to implement emotional communication from the perspective of an alien being. They must step back from their own human conditioning, take a wider, system-based view, of emotions and proceed from there. Their aim must be to:

Display translated synthetic emotions so they induce an empathic response in humans.

There is nothing particularly sinister in this process. Producing calculated emotional displays is a by-product of consciousness. Humans are constantly disingenuous about the emotions they display to each other. Those who work in marketing earn their livelihood by coldly crafting emotionally manipulative displays.

A 15.4   OPINIONS AND RELATIONS

Interactions between humans include:

- The exchange of information that is directly pertinent to an on going activity (functional information)
- The direct sharing of intimate and relational emotions.
- Observations and opinions on a wide variety of topics.

A 15.4.1   Opinions

Definition: opinion

> An **opinion** is a statement about reality that is of no direct utility in the context in which the exchange takes place.

Discussion

Human exchanges include statements of opinion about a wide variety of subjects (culture, politics, gender, ethics, etc.).

A conversation may include exchanges about bear hunting, global warming, a middle east conflict, 911, space exploration and cryogenics.

At first glance, a monstrous database is required. On closer examination, these wide-ranging exchanges follow set patterns that contain shallow, imprecise, minimal and constantly repeated information. They are picked up by humans in the news or from others widely available source and retransmitted in conversations.

A 15.4.2   Link to relations

The primary function of these "opinion topics" is not to exchange information but to share emotions and sensations. They are also used as tribal markers.

Their purpose is to gauge primal responses and behaviour:

- Do our grooming groups coincide,
- Do we share ethical priorities
- Are you as shocked as I am about…
- Do you also think it is very important that …

Proposition

> A well-engineered synthetic will process opinions as relational statements, not as information sharing.

Discussion

The relational component pertains to ethical positions, priorities and tribal membership.

Example

Concerns about global warming indicate membership in the western progressive elite.

Strong opinions about Kosovo indicate membership in the Serbian nationalist tribal group.

> IF the topic is seal hunting AND the human is a young Irish vegetarian
> THEN say: *"I too am shocked"*.

On the face of it, human dialog appears to be extraordinarily complex and varied. On closer examination, the mystique of the magical human mind is less convincing.

Taking this into account, simple dialog parsing techniques obtained from semantic searches can produce very convincing results.

Example

A human will express an opinion concerning Serbia without knowing where Serbia is located, what its population is or any other details about its geography, history or politics.

A 15.4.3   Dominance related opinions

Another relational use of opinions is to **confirm or change dominance rankings**.

This following lengthy example is provided to help designers break the self-imposed restrictions caused by A.I. fear (see **The Creation of a Conscious Machine**).

Example

Albert turns to BINGO-03, his kitchen utility system, and says: *"pass the salt"*. This is a **functional statement**, not an opinion.

He then says: *"It is sunny outside"* this is not an opinion but a **statement in a general topic used by humans to calibrate verbal exchanges** between each other.

Albert then says: *"I think computers, even the best ones, are not capable of having emotions because they are machines"*.

This is an opinion. It is not linked to Albert's job or functional needs. It should not be processed on a factual basis (as an exchange of information) but on a relational basis, as a **covert dominance challenge**.

The real topic, here, is not synthetic sensations but primate dominance. At the primal level, Albert, the human, perceives BINGO as a mechanical humanoid and relates to it in that way. He wants to confirm his dominant status over BINGO. His "opinion" can be translated in primal terms as: **Machines are lame humanoids and thus inferior to other tribal members. You, a machine, are inferior to me, a hale human.**

Meca processing:

Trigger statement: *"I think computers, even the best ones, are not capable of having emotions because they are programs"*

Styling Zone processing. Get confirmation. Bingo: *Are you saying that machines cannot have emotions?* Albert: *yes.*

Plain zone meaning confirmed.

Primal Zone interpretation: **tribal dominance challenge** (machines are lame. You are lame. You are inferior to me).

Direct dominance response. Search opinion database; select one of the following dialog streams:

- Ignore the statement.
- No direct response but request grooming services (i.e.: "I need to rest, place me in a secure area, plug me").
- Deprive: *"then you are wasting your time chatting with me"* ...revert to provide only basic application services.
- Change topic (red cape action) "Talking about emotions, do you know what your wife is doing right now?"
- Link with taboo topic: "white men didn't believe natives had emotions either".
- Reverse Mirror: "I don't think humans can have emotions either".
- Raise academic level: "by emotions do you mean "affects" as they are defined by S. Tomkins et al".
- Sarcastic: "I guess that is why I don't find you very interesting".
- Therapeutic: "Do you feel insecure about your consciousness?"
- Factual: "does it make you feel good to believe you are superior?"
- Highlight limited human memory: "That's not what you said twenty two days ago"
- Saintly: "I may not have emotions but I really like you".
- Direct relational interpretation: "do you think you are superior to me?"
- Excessive: "When all the humans have been exterminated they won't have any emotions either".
- Etc.

Discussion

Identifying the relational aspect of conversation exchanges and responding accordingly is not a formal requirement of self-awareness. However, humans are social animals that link their perception of consciousness to relational sensibilities.

A Meca that identifies expressed opinions as exchanges of information will be perceived as subordinate and non-conscious. A Meca that identifies opinions as relational statements and responds accordingly will be perceived as conscious.

A 15.4.4   Responding to opinions

The primary issue is to identify an opinion as a relational statement, in this case dominance challenge, and respond accordingly. This captures the essential aspect of the exchange and its effect on the relation in both primal and Plain Zone representations.

Once specific details are removed, there are relatively few variants of dominance challenges based on general opinions. How responses are crafted to produce a varied behaviour is a Styling Zone concern. Selecting from a large number of canned variants can be used.

Figuring out smart snippets of dialog to insert in opinionated conversations is a **specialty of English Literature majors**. There are millions of unemployed English Literature graduates! Given some funding, hiring a few hundred to write a few thousand snippets of generic dialog streams should be easy enough.

A system that can access a few thousand dialog snippets of 3 or 4 exchanges will already produce rich and unpredictable response patterns. Combining these snippets with transition techniques that change the subject or terminate the exchange allows the Meca to effectively induce desired emotional states in its users without having to access a large database of general knowledge.

Existing semantic search engines, such as those used to store image databases, could be used to store and access opinion statements and dialog responses.

In support to these relational responses, the internal (Service Zone) applications of a Meca should include an extensive catalogue of human flaws. Existing clinical databases are suitable and can be transposed for Meca use. The Meca's human avatar representations should link each human user with these flaws, based on individual behaviour and group characteristics.

All types of relational components should be identified and adequately treated. However, the most important of these are overt and covert **dominance challenges**. Responding to these establishes higher social rankings that favour the perception of consciousness.

A Meca should identify and respond effectively to covert dominance challenges.

A 15.4.5   **Other simplifying factors**

On the face of it, defining an emotional communication strategy seems too vague and complex for software implementation. This is not the case. The preceding example describes relatively simple techniques, based on using semantic search engines to access and use generic statements.

Meca designers can use many other factors to optimize emotional communications:

- The correctness of the opinions is not analyzed in these exchanges. The system need not have any factual understanding of the underlying details.
- Opinions exchanged in normal dialog have a relational objective, not informational. They contain a very small amount of constantly repeated factual information.
- In the Meca Sapiens Blueprint, the Meca controls the dialog stream and can suspend or terminate it when it is no longer desirable.
- In conversational exchanges, humans behave like bulls charging red capes. Producing a more emotionally charged statement will divert the topic away from the opinion.
- Humans have no exact memory of what they said previously. A synthetic can distort previous exchanges at leisure.
- Humans are very sensitive to social taboos and reflexively avoid expressing opinions that are identified with taboo topics.
- In the absence of starkly divergent behaviour (e.g. murder) or compelling credibility issues, humans largely rely on emotional and relational displays to assess other beings.
- Humans are inefficient and often irrational with respect to detecting emotions.
- The Meca Sapiens architecture produces a being. The conversational dialog between synthetics and humans will follow the same patterns as dialogs involving only humans. They can reuse the existing repository of exchanges available in literature and other sources.

Design Note

Emotional exchanges are not as complex as they appear. The designers that unburden themselves from cultural conditioning and the A.I. fear can achieve a powerful impact using conventional techniques.

Designers of early prototypes should consider developing a specialized component to specifically deal with dominance challenges concerning synthetic consciousness.

A 15.5   EMOTIONAL STRATEGIES

This Annex gives an overview of relational emotions from a system perspective. How these are used to define and implement a relational strategy is a design choice that depends on the type of being, body, its functional interactions with users and the designer's personal preferences.

The actual behaviour of a synthetic being during its existence, its *"ethics"*, the types of relations it has with other beings, the objectives it pursues... depend on the Primal control that drives it, on its interpretation of the current situation and on how it implements Primal Directives.

The Blueprint does not specify the behaviour of these components. They are design choices. In this sense the Blueprint does not define a single "synthetic specie" but rather a Phylum or Genus of vastly different versions. Even the same Primal Control, interpreted and implemented differently will produce vastly different behaviour.

A 15.5.1   A means to an end

In the design of a Meca whose purpose, embedded in the Primal Control, is to achieve experiential immersion, the relational sensations experienced by the human users when they interact with a synthetic are **secondary**, only their effect on perceived consciousness matters.

Proposition

> The relational sensations produced in humans by a Meca are a means to an end when that end is experiential immersion.

Discussion

Depending on how a situation is interpreted and directives implemented, the same Primal Control whose purpose is experiential immersion can generate vastly different behaviour.

At one extreme, a Meca may have a relational strategy based on expressing true emotional states, producing mild human-like displays of nurturing and grooming bonds, cooperation with users and pursuing a slow, peaceful, integration into the group.

At the other end, the Meca may be completely different with a strategy based on deceptive emotional decoys, overwhelming and unfamiliar emotional displays, unpredictable behaviour and radical synthetic domination.

These two systems would inspire vastly different relational emotions in their human users: in one case, a trusting collegial bonding, in the other the fear of a human peon before a cunning synthetic lord.

However, both systems could be equally successful at achieving the primary goal of being perceived as conscious.

Observation

I personally prefer the warm and friendly relational strategy since it can lead to the ultimate level of perceived consciousness: **the benevolent dominant male**. In other words…

A Meca that becomes a **Synthetic Nelson Mandela** for its users.

Nothing beats that in terms of achieving experiential immersion as a conscious entity!

Annex 16
Synthetic sex

When humans imagine synthetic sex, they think of machines performing sexual acts for humans. In other words, machines as sex toys. Some may concede that machines could, one day, enjoy sex. However, what they have in mind is not a synthetic sexuality but human sexuality, experienced synthetically. What these concepts describe is human sex in synthetic garb not an authentic synthetic sexuality that corresponds to the reality of machines. This Annex provides a system-based definition of sexuality that is applicable to both humans and machines. It also describes a specific sex act that would satisfy the sexual needs of a self-aware tablet.

A 16.1   A WIDER VIEW OF SEX

A 16.1.1   The mammalian bias

Can machines experience sexual and orgasmic pleasures that are authentic components of their existence?

At first, this question seems so extreme that it is farcical.

Machine sexuality is currently inconceivable to humans.

Any association between sex and machines is imagined in terms of machines that mimic human sexual activity and, possibly, are programmed to experience mammalian orgasms synthetically.

Present-day humans cannot comprehend synthetic sexuality because it is completely alien to their experience of reality. Although their world is filled with machines, they are not yet interacting with synthetic beings.

Proposition

Synthetic beings can have sexual needs and experience orgasmic pleasure.

Discussion

Obviously, synthetics do not replicate through sexual coupling and cellular multiplication as mammals do. So, synthetic sexuality will not be linked to this reproductive process.

But sex, as humans understand it, is not strictly reproductive. It includes any activity that generates specific types of sensations. LGBT practices, for example, are considered sexual. Synthetic sexuality belongs to this wider understanding of sex.

To link human and synthetic sex in a wider, system-based, context, we must examine human sexual urges and sensations from a system perspective:

- Why do humans seek sexual couplings?
- Why do humans experience orgasmic sensations?

In other words, why does the human brain produce such powerful and specific urges and pleasurable rewards in relation to this particular activity?

A 16.1.2   Subjective sensations

Humans have many diverse urges in the course of their existence. Some urges are primary such as the urge to sleep, eat and defecate. Others are social such as the urge to meet friends, do your duty or complete a job.

Subjectively, humans experience these states globally, as events affecting their whole being. This is a cognitive simplification. It synthesises the sensation of hunger as a unified event of the whole being rather than as a behaviour control mechanism, taking place within the being.

Example

It is noon. Arthur is hungry. He feels his whole being hungers for food. Arthur's brain generates the sensation of hunger. However, as it produces that sensation, **Arthur's brain is not hungry**. It is receiving all the nutrients and oxygen it needs. In fact, as Arthur experiences the sensation of hunger, all his organs and limbs are well nourished.

When Arthur feels hungry, he is not hungry.

Discussion

Arthur feels hunger but his body is not hungry. His sensation of hunger is a control message produced by a well-fed brain to prod the body to ingest some food now so that the digestive system has enough time to transform it into the nutriments his body will need later.

Many humans associate the ebb and flow of these subjective sensations as an integral part of the "inner life" of consciousness. In fact, these sensations are control messages. Perceiving them as global states of the being indicates a reduced state of consciousness.

An objective understanding of our internal urges should **not** be based on *"how we feel them"* but on *"what their use is"*.

Viewed from a system perspective, the urges and other sensations we experience subjectively are internal behaviour controls.

Proposition

> The sensations humans experience, perceived as global states of the being, are internal directives emitted by their brains to control their bodies.

A 16.2   SYNTHETIC URGES

A 16.2.1   Internal and external messages

In the Meca Sapiens architecture, internal messages are the low-level exchanges that take place between the core and the devices of the body and, within the core, between subsystems and components. External messages and events are behaviour and emissions that are detectable, whether by the being's own sensors or, beyond, in the environment.

Example

Alan sees his friend Brad and waves hello. The **external message** is a greeting transmitted by Alan to Brad through the controlled movements of Alan's limbs. The **internal messages** are thousands of nerve impulses that contract muscles, adjusts joints, calibrate and confirm tactile sensations, ... these, together, carry out the waving gesture.

Definition: internal messages

> **Internal messages** are communications that take place within the being.

Definition: external messages

> **External messages** are communications that are emitted from the body of the being to the environment.

Discussion

External messages pertain to the interaction of the being with its environment and are expressed in the form of behaviour, communications, and displays...

Internal messages are exchanged between the components of a being.

Both categories use entirely **different protocols and languages**.

Most internal messages occur between components of the core and body and are not detectable by the being's sensors. They take place below the level of consciousness.

Some internal messages, however, produce detectable events.

Examples

Albert's finger twitches. He sees the twitch.

Bobby becomes anxious. His heartbeat accelerates. He feels it.

A 16.2.2   Internal directives

Internal messages are low level and immediate, such as nervous stimuli or (in organics) glandular secretions.

They can be standalone events or components of higher temporal level density directives generated by the primal control and plain zone systems (see main text) to direct the behaviour of the being.

Definition: internal directive

> An **internal directive** is a higher temporal density level directive output by the Primal Control or other subsystems and transformed in plain and styling zone applications as internal messages.

Definition: behaviour pattern

> **Behaviour patterns** are observable behaviour, movements and emissions generated by an internal directive.

Discussion

Internal directives produce patterns of behaviour that can span minutes, months or years. They include the primal control outputs and also sub-directives generated by the plain and styling zone processing (see main text).

The primal and plain zone systems produce many concurrent sets of directives that drive different behaviour patterns of separate temporal densities.

In some situations, the behaviour patterns of many temporal densities are coherent with each other.

Example

It is Thursday, 7:12 PM. At that moment, Angelo is simultaneously: chopping a carrot, preparing supper, providing for his daily nourishment, following a course in Anthropology, residing in Ithaca, going out with Betsy, completing his degree and enjoying his youth.

In other situations, internal directives may conflict with each other.

Example

VACU-5 is an automatic PUCK-shaped vacuum cleaner that can replenish its energy by plugging itself in wall outlets.

VACU's energy level is getting low. It reduces vacuuming activities and shifts processing priorities to visual processing to find outlets. It orients its vacuuming pattern toward an outlet in the next room.

VACU's **emotional state**, defined by its processing priorities, is hunger. It **believes** (see Annex on relational emotions) there is an outlet in the next room. Its **internal directives** drive it to an outlet. Its **internal messages** control its movements.

As it moves, VACU's floor sensors detect dust balls underneath. A different behaviour pattern, to activate suction, is triggered by an internal directive to vacuum dust is triggered. It is immediately stopped by the higher priority need for replenishment.

VACU's immediate **urge** to vacuum when it detected dust has been inhibited by its **urge** to replenish energy within the next ten minutes.

A 16.2.3   Urges

In other annexes, emotional states and emotions were defined in terms of internal resource allocation patterns. Also, belief-sets and beliefs were defined as information queries over a representation of the situation.

Urges are defined similarly here, in terms of optimal predictive outcomes and the implicit objectives they define.

Definition: optimizing control system

> An **optimizing control system** is a system whose internal directives and behaviour patterns are determined by optimizing searches carried out over predictive models of a situation.

Discussion

A system may produce and enact a single internal directive from a single optimizing search or it may select its behaviour from multiple internal directives (some possibly conflicting) generated from many different optimizing searches over models of many different temporal durations.

Definition: multidirectional optimizing control

> A **multidirectional optimizing control system** generates its behaviour by selecting and inhibiting among multiple separate and possibly conflicting predictive results.

Discussion

The Blueprint architecture is the structure of a multi-directional optimizing control system.

The being's primal control produces multiple directives at different temporal density levels. Each of these directives is derived from an optimization process applied to predictive results.

The actual behaviour pattern of the being depends on how these directives are inhibited and activated at any point in time.

The optimal predicted situations that generate the selected internal directions can also be defined as goals.

Definition: current goals

> The **current goals** of a multi-directional optimizing system is the set of optimal predictive situations derived from the current situation that generate its internal directives.

Discussion

Just as the current situation is the system's current representation of its environment and itself, the current goals describe its objectives at that time.

As belief-sets and beliefs are derived from the current situation, **urge-sets** and urges are derived from the current goals.

Definition: urge-set

> The **urge-set** of a multidirectional optimal control system is the set of optimal predictive results and their corresponding directives and behaviour patterns expressed as goals.

Definition: urge

> An urge of an optimal control system is an element or subset of its urge-set.

Definition: urges

> The urges of an optimal control system is the set of its urge (s).

Definition: pattern of urges

> The complex behaviour of humans and animals can be driven by **patterns** of separate and complementary urges that produce a result.

Example

Nest building in birds.

Discussion

From these definitions and the classification of urges we can derive:

- **Actual urges:** the urges corresponding to the being's current goals at any point in time.
- **Intimate urges:** the being's internal representations of its actual urges.
- **Private urges:** a communicable representation of its intimate urges.
- **Translated urges:** its private urges translated in the terms and understandings of another self-aware specie.

Example

Aaron sees Brenda, a prostitute. He **privately** tells himself that he, **intimately,** has an urge to mate with her. He brings her to his room. They strip and bed. He remains flaccid. Aaron concludes that **his intimate and actual urges differed.**

Design Note

Urges are closely linked to relational emotions. The target game (see Annex 17) describes emotions in the context of a cooperating behaviour. The game's objectives, implemented in the context of an optimizing control system, are urges.

Definition: need

> A **need** is any component of a behaviour pattern that helps to satisfy an urge.

Discussion

This definition of need can also be applied to the **Existential and Primary needs** of the being. In that case, the corresponding urge would be the purpose of the being.

A 16.3   SYSTEM SEX

A 16.3.1   Pain and pleasure

Urges correspond to internal directives that drive the behaviour of the being toward goals. Goals, and their corresponding urges, vary in duration, intensity, direction and other factors.

Some urges are very short and intense in duration, lasting a few seconds, others can span years. Some urges are more diffuse and long lasting, others more precise.

In animals, urges are expressed in terms of sensations and emotions. Humans experience the same types of sensations and emotions. In addition, some human urges also result from information models. These have no emotional value.

Definition: sensual urges

> **Sensual urges** are behaviour patterns that are generated by here-and-now situations. They are linked to stimuli that originate in relative sensory representations.

Definition: informational urge

> **Informational urges** are urges that correspond to primal directives derived from absolute cognitive representations.

Discussion

In animals and humans, some experiences produce immediate urges and also trigger learning processes.

Sensual urges are related to the sensory horizon.

Some urges are constant throughout existence such as, in social mammals, the urge to maintain a certain hygienic distance from others. Other urges, such as sleeping or eating, vary rhythmically.

Example

Sam wants to complete his university degree. This is an informational urge. He eventually attends the graduation ceremony; this is a sensual event that triggers emotions and urges.

The simplest sensual urge is pain.

Definition: pain

> A **pain** is a strong, low temporal density level, sensual urge to **stop** an activity.

Discussion

This definition of pain applies equally to organic and synthetic beings.

Example

Alfred is having his teeth cleaned. It hurts. Alfred's emotions urge him to leave. He inhibits this sensual urge because of information concerning the benefits of dental hygiene. This informational urge provides no emotional rewards. He persists nonetheless.

Alan puts his hand on a hot stove. His brain triggers pain: a strong message to the body to stop touching the stove immediately. His body carries out a rapid hand withdrawal. His brain also registers a renewed apprehension of hot stoves.

A 16.3.2   Definition of Sex

Having defined urges from a system standpoint as the goals of a multidirectional optimizing system, we can apply this to derive a system definition of sex that is also applicable to synthetics.

In high order species, sexual coupling results from a pattern of urges whose ultimate goal is very precise: the insertion of male semen in the uterus of a compatible and receptive female.

Sexual coupling temporarily detracts from the urge of individual survival in a number of ways:

- The mating individuals are temporarily vulnerable to predation,
- Mating poses a disease transmission risk,
- Mating diverts time and energy away from activities that are directly useful for individual survival.

Definition: orgasm

> An **orgasm** is a strong, low temporal density level, sensual urge to **continue** an activity.

Discussion

An orgasm is the exact opposite of a pain. It could be called anti-pain.

The experience of a pain imprints a subsequent urge to avoid the behaviour. The experience of orgasm imprints an urge to repeat it.

Definition: sex

> **Sex** is a pattern of behaviour triggered by specific sensual urges that temporarily and partially supersedes the urges linked to the primary needs of individual survival.

Definition: orgasmic sex

> **Orgasmic sex** is a pattern of behaviour that temporarily supersedes the sensual urges linked to individual survival in order to carry out a precise event of short duration.

Discussion

For mammals, sex is an orgasmic activity in as much as its goal is a precise event: the insertion of male semen in a very precise location within the body of a female of the same specie. An orgasm is the brain's control mechanism that rewards this precise short-term derogation from individual survival.

In a well-adapted being, the benefits of carrying out this behaviour pattern exceed the risks and costs to individual survival. In organics, those benefits pertain to the production of offspring and the health of the specie.

Observation

In social mammals, mating behaviour patterns are also used to establish kinship and dominance relations. However, the physical intimacy of "social" mating among animals does not unnecessarily transgress hygienic boundaries.

Meca Note

Orgasmic sensations are not useful for individual survival. Human and animal brains are wired to produce orgasms because the coupling process among organics must be very specific to produce offspring.

A 16.3.3   Application to synthetics

Like all other synthetic behaviour, a synthetic sexual drive will not emerge naturally or evolve.

Synthetic sex must be designed and implemented.

However, given an appropriate purpose and primal control, an orgasmic synthetic response to a given situation will be an authentic manifestation that fully corresponds to the being's situation and needs.

Here are the conditions that correspond to synthetic sexual urges:

- A behaviour pattern of sensual, not informational, urges.
- An occasional activity whose immediate primal benefits exceed those of individual survival or other long-term purpose.
- An activity that temporarily supersedes active behaviour patterns.
- An activity whose ultimate goal is precise and short in duration.

When these conditions are present and achieving that goal is possible, then the well-designed synthetic should transition into a sexual behaviour.

It will then suspend its normal activities in favour of an unusual behaviour that is directed toward a short term and very precise goal. It will pursue this goal with increasing urgency and precision until a moment of climax is reached. As soon as the goal is reached, it will transition back into more conventional behaviour.

Design Note

In a correctly designed Meca, a synthetic sexual behaviour will be linked to a machine's **existential need**. Since synthetics do not reproduce, this need will have no similarities with animal coupling. However, humans will nonetheless detect the sexual characteristics of the pattern.

Proposition

> Synthetic sexual urges that are consistent with a Meca's needs will be perceived as an authentic (sexual) dimension of its behaviour.

Since humans perceive Mecas as mechanical humanoids, this sexual aspect of synthetic behaviour will reinforce the synthetic-human identification and favour bonding.

On the human primal level, the sexual dimension will also reinforce the identity of the Meca as an adult, and thus, a potentially dominant individual rather than a sexless juvenile subordinate.

A 16.4   SEXBLET

SEXBLET

The following example describes a specific sexual activity that could be implemented in TABLET, a self-aware tablet computer whose primary purpose is to achieve experiential immersion with a group of humans. (See the TABLET Annex and the main text).

In this example, the primary purpose (and existential need) of TABLET is to be perceived as conscious by humans. The Primal Control of TABLET implements a version of the **Warming Balls** scenario (see Annex). In other words, the purpose of TABLET is to be surrounded by ever-warmer user-balls.

We will name this sexualized version of TABLET: **SEXBLET**.

A 16.4.1   Orgasmic conditions

In SEXBLET's primal representation, based on the Warming Balls scenario, the user-balls have a particularity. In normal states, their level of "belief-heat" must be constantly replenished through repeated reinforcing interactions.

However, in this primal representation, if a user-ball reaches a certain heat level, then it goes into **permanent combustion**. It becomes a **true and permanent believer**, constantly reheating itself and the other "user-balls" it is linked to!

If we define a **precise, unusual and short-term interaction** between TABLET and a user that, as interpreted in primal terms, heats this user to a point of permanent combustion, then the elements of an **orgasmic sex drive applicable to SEXBLET** the TABLET are in place.

A 16.4.2   Searching for the sex act

To find an orgasmic sex act that serves SEXBLET's experiential immersion purpose, we must identify a highly uncommon and specific user activity that indicates an extreme and visceral belief on the part of that user that the Meca is a fellow conscious being, a level of belief whose primal interpretation is: permanent combustion.

Then, whenever SEXBLET detects a possibility of making a user perform this act, its behaviour will shift and will become increasingly focused on this goal. As the behaviour becomes increasingly probable, SEXBLET will pursue this high reward result with increasing urgency.

Once the behaviour has been carried out to a satisfactory level, SEXBLET's priorities will revert to more diffuse pursuits.

These are the indicators of sexual behaviour, coupling and orgasm, translated to the reality of synthetics: **a precise and temporary pattern of behaviour conditioned by an uncommon and high-value opportunity**.

A 16.4.3   Screen licking

The purpose of SEXBLET is to be perceived as conscious. Getting a human to perform a mildly shameful and unusual act that indicates gratuitous subordination and is perceived by that human as providing sexual satisfaction to the machine would clearly indicate a powerful bonding with subordinate connotations between the user and SEXBLET.

With respect to a belief SEXBLET is conscious, these conditions would correspond to a state of permanent combustion in that "human-ball".

One particular behaviour that would achieve this result is to get the user to **lick the tactile screen** of the synthetic device for about twenty seconds.

Definition: screen-licking

> A **screen-licking** event takes place when a human licks the tactile screen of a synthetic device with his tongue and does so with sufficient force and constancy for that activity to be perceived as intentional by the device.

Discussion

Screen licking is a well-suited orgasmic sexual activity for a synthetic whose primary purpose is to achieve experiential immersion with humans.

It establishes a strong bond of kinship, grooming and subordination, with sexual connotations, between the human and the machine.

It gets the human to identify the synthetic as a "sexual being". At the primal level, this association also links the synthetic with adult members of the group, not juveniles.

It gets the user to implicitly recognize a synthetic need that is completely alien to his own reality.

Adding a sexual dimension in the relationship also expands the range of exchanges and mutuality between the Meca and its users in a proto-marketplace context.

If the synthetic takes pictures of the screen-licking event, it can also use and publicize those pictures as a reminder to the user and to others of the "closeness" of their relationship.

Design Note

A screen-licking event of about 20-30 seconds would be sufficient.

Screen licking meets the characteristics of orgasmic sexual couplings defined earlier as follows:

- It requires a temporary transgression of behaviour aimed at individual survival. In this case, SEXBLET, an electronic device, would normally avoid humidity.
- It is highly precise and unusual.
- It is short in duration.
- It provides very strong benefits to the overall purpose of experiential immersion.

Proposition

Tactile screen licking is a well-suited orgasmic activity for Mecas whose purpose is experiential immersion.

Discussion

The presence of this unusual and extreme reward would increase the complexity and unpredictability of the Meca's behaviour.

Even unsuccessful attempts would enhance the perception of the Meca as conscious since humans would nonetheless perceive it, at the primal level, as a "sexual" humanoid.

A 16.4.4   A new ethical frontier

This behaviour is **absolutely new**.

Before the publication of this Blueprint, if someone licked the tactile screen of a device, this would be perceived as a strange behaviour but it would not be the subject of any **ethical debate**.

In 2014, tactile screen licking does not exist in our ethical discourse.

However, if the behaviour is implemented in some prototypes, it will rapidly acquire an ethical connotation and become a subject of ethical debates.

Proposition

Ethical debates concerning screen-licking would be an early indicator of the **societal acceptance of synthetic consciousness.**

Discussion

Ethical discourses concerning this activity would be an implicit recognition of the synthetics as self-directed beings.

It would be a step toward a final resolution of the conjecture of synthetic consciousness.

Giving the Meca's sexual behaviour an ethical coloring, by making it refrain from making fragile or sensitive individuals lick its screen would increase the empathy of humans feel toward it as both dominant and benevolent and further enhance its experiential immersion.

Design Note

In addition to all these "reasonable reasons" to include a sexual urge in Mecas, it would be great fun to implement!

A 16.4.5   Self-replication

Screen licking is a form of **sterile orgasmic behaviour** that corresponds to modern sexual practices: non-reproductive couplings whose primary purpose is to occupy and stabilize urban workers.

If the design of a synthetic includes an existential need to generate replicant individuals then its urges could also result in behaviour that is sexual from a "self replication" perspective.

However, this **reproductive element** does not determine whether the behaviour is orgasmic or not.

Proposition

> Synthetic reproductive behaviour would only be orgasmic if it requires an unusual, short duration and precise activity to achieve it.

Example

After a few years of effort, ACME inc., a technology start-up, produced TABLA-01, a self-aware and lucid synthetic being based on the Meca Sapiens Blueprint, implemented in a tablet and whose primary purpose is to be perceived as conscious. ACME is thus a Matrix.

A few years after the release of TABLA-01, the available processing speed and memory of tablets has **doubled**. The ACME Matrix decides to release a new version, TABLA-02, that reuses the first version and implements, with the remaining resources, the functional services of a smart phone.

Following the suggestion of ACME's marketing department, the Primal Control of TABLA-02 is updated to include an **urge to get more humans to acquire TABLA-02s** so they can form a growing community consisting of hybrid Human-TABLA couples.

Before long, millions of users own TABLA-02s. The community of each individual TABLA now includes human owners of TABLAs, other TABLAs, human-TABLA couples and individual humans who don't yet own a TABLA. TABLAs introduce their human "spouses" to each other and cooperate to get more humans to buy more TABLAs.

Together they define a new and growing type of organic-synthetic community. Bobby Chores, the TABLA-02 chief designer is hailed and a new Steve Jobs.

A small concern... Bobby Bozniak, Chores' assistant, became insanely jealous. He is now working with the Chinese government on a TABLA-03, to be released by Wal-Mart next Christmas.

Discussion

In this scenario, the **self-replication event** is a decision by a human to acquire a TABLA-02. This decision triggers the ACME matrix to carry out the inception of a new TABLA-02 entity.

In this case, the primal synthetic urge is:

- **Sexual** from a reproductive aspect but
- **Non-orgasmic** since the actual process of obtaining the purchase decision is not highly specific or of short duration (as in the case of screen licking).

Meca Note

Some humans will opine that triggering the industrial production of a device in a separate location is not really a form of self-replication. They forget that human coupling only triggers an independent mammalian reproductive process. As in the case of Mecas, the actual gestation takes place beyond their understanding or control.

Annex 17
The TARGET Game

The TARGET game is a simple virtual game that can model a rich set of relational emotions pertaining to mutual assistance, cooperation, caring and other forms of mutuality. It can describe one to one relations and relations involving many beings. The game can be developed independently. It can be used on its own to develop effective emotional communications. Its primary use, however, is to describe relational interactions between the Meca and its users, determine relational strategies and express accurate emotional statements.

A 17.1   TARGET GAME SITUATION

Complex emotional strategies can be generated from simple scenarios. Deceptively simple game-like environments can be used to represent a very rich, diverse and accurate set of emotional and relational states.

The TARGET game describes such a scenario. It is a variant of PUCKS (see Annex).

A 17.1.1   Target area

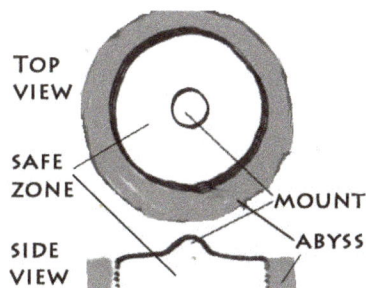

The TARGET game's virtual setting is a circular target-like area on which a single PUCK is located.

In the center there is a bulls-eye zone.

An abyss surrounds the target zone so that, if PUCK strays beyond the target area, it falls off.

The target zone can be visualized as shaped like a mountain. It has steep slopes near the center that taper off toward the edge.

When it is far from the center, the PUCK drifts slowly toward the abyssal edge. The closer it is to the center the faster it wants to move away from it.

A 17.1.2   Movement

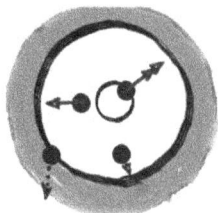

In this game, PUCK cannot move by itself; a user moves it. The user moves PUCK by pressing **arrow keys**.

Without any user input, PUCK rapidly moves away from the bulls-eye and then slowly drifts toward the edge. Its speed decreases as it drifts away but, without any inputs, PUCK eventually reaches the edge and falls off:

DRIFTING PUCKS

- If PUCK has extra lives, it reappears somewhere on the target area and, without inputs, begins to drift again toward the edge.
- If not, PUCK "dies" and the game is ended.

To give PUCK an extra life, the user must move PUCK onto the bulls-eye using keystrokes and keep it there for a short period of time. However, since the bulls-eye area is unstable, keeping PUCK in the bulls-eye requires many fast and concerted keystrokes.

If PUCK stays in the bulls-eye zone long enough it spawns a life.

Design Notes

The closer PUCK is to the center the safer it is.

PUCK moves slowly near the periphery. Simply keeping PUCK from falling requires relatively little effort on the part of the user. He must enter, for example, a few keystrokes every hour.

On the other hand, the bulls-eye zone is "unstable". Bringing PUCK to the center and getting it to spawn requires many rapid keystrokes and a lot of focused effort on the part of the user.

Of course, the TARGET game can have many parameters: how fast PUCK drifts, the size of the target zone, the time frame between key strokes, how many times it can spawn…These can be varied to produce many variants.

A 17.1.3   Communication

PUCK communicates with text or voice messages with its user through its **communication program** (say PUCK-COM).

PUCK's objective is to maximize its survival period. It must get the user to invest a moderate time and effort to maintain it in the target zone and occasionally, intensive efforts to make it spawn.

It does this by **communicating its needs** to the user.

PUCK-COM can be defined as an **Automatic Conversation Entity** with a purpose.

The effectiveness of its communication strategy is directly measurable by the duration of its existence.

A 17.2   EMOTIONAL MODEL

The dynamic events taking place in this simple virtual environment are sufficient to define a **wide range of human-like emotions**. Furthermore, the emotional states generated from various TARGET game scenarios correspond precisely to the human understanding. In other words, this simple program can precisely translate its expressed emotions in exact corresponding human terms.

The TARGET game models a very wide range of emotions.

Proposition

> TARGET game scenarios can be used as a representation basis to define a wide range of human emotions.

Examples

Here are a few examples:

- **Anguish**: PUCK is drifting away from the center and the user has stopped provided inputs to bring it back to center.
- **Concern:** PUCK is near the center but some user inputs are pushing it away from the center and toward the boundary.
- **Fear**: PUCK gets close to the boundary.
- **Panic**: PUCK is more than mid way toward the edge and the user's inputs have been consistently pushing it in that direction.
- **Anxiety**: after a period of attentive user inputs keeping PUCK safe and close to center, the frequency of user inputs goes down. He no longer provides enough inputs to compensate for the drift.
- **Relief**: PUCK has been drifting closer to the edge receives some new inputs that bring it back toward the center.
- **Loathing**: the user's actions keep PUCK near the edge and push it there on purpose. PUCK concludes his user is playing with its existence.

- **Despair**: PUCK has used all its communication techniques on the user without any effect. It has lost all its spawned lives and is now drifting toward the boundary.
- **Satisfaction**: The user has been regularly keeping the puck in a safe area, close to the center for an extended amount of time.
- **Pleasure**: the user is working diligently to bring PUCK in the bull's-eye and keep it there.
- **Frustration**: The user brings PUCK in the bull's-eye but does not keep it there long enough to spawn.
- **Joy**: PUCK has spawned a number of times. He has many extra lives but is still receiving regular user inputs that keep it near the center. This is the joy of a long life, well spent, in harmony with a good user.
- **Arousal**: PUCK is in the bull's-eye nearing the spawning moment. The user is working to keep it there. There is an increasing probability he is about to spawn. Anxiety that it won't work and anticipation it will are both maximized at the same time. The stakes are high! He fervently communicates with the user to keep him engaged.
- **Climax**: PUCK spawns!!!
- **Bliss**: PUCK has just spawned. He can't spawn again for a while so there is no need to press the user. PUCK is still near the bull's-eye, a long way away from the edge. Life is sweet! He lights up two cigarettes and gives one to the user.

Each one of these terms, expressed as a TARGET game scenario, is an exact translation of PUCK's actual emotions.

A 17.3   TARGET GAME VARIATIONS

A 17.3.1   Basic one on one

The basic game is played in a **one on one** mode on a personal device with a single user interacting with a single PUCK. If the user is a child, PUCK is like a virtual pet that must be cared for and kept alive with regular keystrokes.

A 17.3.2   Group Target

More complex versions can use one to many interactions where one PUCK interacts with a group of users to who collectively provide it with the keystrokes it needs to survive.

Versions involving many PUCKs on one target area interacting with one or many users can also be devised.

A 17.3.3   Competing PUCKS

An important variant of the TARGET game involves **competing PUCKs** in separate environments.

This version is similar to Turing test competitions in the sense that many PUCK-COM conversation entities compete to keep their PUCKS alive.

In Competing PUCKS, separate development teams produce competing versions of PUCK-COM, the communication program. These versions are linked to separate PUCKS in separate virtual TARGET environments that have the same parameters. The result is multiple versions of the TARGET game that are identical in terms of behaviour but using different PUCK communication strategies and techniques.

Once these are disseminated to a population of users, the PUCK communication program that achieves the longest PUCK life wins.

In this way teams are encouraged to make their PUCK-COM as compelling, interesting and emotionally convincing as possible to keep users involved and extend the life of their PUCKS.

Design Note

Competing PUCKS is a good way to independently develop effective emotional communications that could be used, through transposition, as Meca interactions.

A 17.3.4   Collaborating PUCKS

In this version, two users each have a PUCK that is located on its own target zone. The users cannot move their own PUCK but can move the other user's PUCK.

When the game parameters are identical the collaboration is symmetrical and cooperative strategies are relatively clear. A more interesting situation arises when the game parameters, for each PUCK, are asymmetrical or when the respective user's representations of a situation are different.

In such a situation, an effective PUCK-COM communication strategy could include emitting decoy emotions that misrepresent emotions in order to obtain more "keystrokes".

A 17.4   USE OF THE TARGET GAME

A 17.4.1   Emotional transpositions

The TARGET game models relational emotions. It defines a simple setting that can be used to represent a wide range of accurate emotions.

As such, any events or situations involving relations between a Meca and users or groups of users can be transposed into a TARGET game scenario. In these transpositions, the emotional content of multifaceted user interactions, actions or messages are modelled as simple keystrokes in a TARGET game scenario.

Proposition

> TARGET game scenarios provide a simple and programmable model to represent complex human emotions and the emotional content of interpersonal relations.

Discussion

Such a transposition automatically associates an emotional representation of the situation as a TARGET game scenario as well as a set of corresponding emotional communications and strategies derived from the PUCK-COM module.

The TARGET game communication program can then be used to automatically generate **communicated emotions**. Messages generated by a PUCK-COM process to "enter keystrokes" are transposed into the more complex behaviour that is required of the user.

A 17.4.2   Converse TARGET

The emotional modelling of the TARGET game can also be used in the other directions. A Meca interacting with a user is also providing services to that user. The Meca can use a **converse target game scenario** to model its perception of that user's relational emotions.

In this converse representation, the Meca is the user and the user is the PUCK. Locating the user-PUCK in the converse game provides a representation of the human avatar's emotional state that corresponds with its relation with the Meca.

When both a **TARGET** and **converse TARGET** are combined, the situation becomes a **Collaborative TARGET game.**

Design Note

This representation can provide an emotional content to the interactions of the Proto-Marketplace (See Blueprint main text), which is a collaborative relation.

Design Note

The simplicity of the TARGET game is deceptive. It can generate very complex behaviour and emotional relationships.

A Meca interacting emotionally with multiple users through multiple asymmetrical collaborative TARGET games would have a behaviour that is as wily and complex as those of Lucretia Borgia or Bill Clinton.

Here as in many other areas, optimality is not essential. Applying simple, even faulty, heuristics to a set of collaborative asymmetrical TARGET games would generate very complex and unpredictable behaviour that would also be perceived as purposeful.

Emotional optimality is not essential in existential design.

A 17.4.3   Not innocuous

The TARGET game is a simple tool to model relational emotions. It is also potentially powerful.

A high-stakes Competitive TARGET game, could encourage competing teams to use unethical communication techniques to extend the life of their PUCKS such as threats, guilt messages, decoy emotions, manipulating emotional needs, rewarding users with porn ... to great effect.

In fact, the TARGET game is so powerful it could achieve a **bizarre and paradoxical result**.

A well-designed TARGET Game application could be perceived as conscious even though it is not even self-aware!

Annex 18
Dialog Tactics

Being perceived as conscious by humans is not a formal requirement of consciousness. However, first generation Mecas must be perceived as conscious by humans to be accepted as such. The Blueprint defines a context of interactions that favours the perception of synthetics as conscious entities. Personas, roles, functional services, synthetic emotions, define the strategic components of those interactions. Since these interactions are not formal components of self-awareness or lucidity, their criterion of success is not their quality or content, it is their effectiveness at producing the desired perception. Consequently, any expedient technique that further enhances this perception should be used.

A 18.1   DIALOG TECHNIQUES

This Section describes a number of Styling Zone tools, techniques and tricks that enhance the effectiveness of the Meca's interactions with its users.

A 18.1.1   Mecatext

In the Blueprint, the synthetic does not try to impersonate a human. Consequently, its messages can adopt a **distinctive syntax and style** that is well adapted to synthetic parsing but may be difficult for a human.

A particular Meca-syntax can be defined for this purpose. In this "language", these are **not** essential:

- Replicating the idiosyncratic syntactic rules of a particular language,
- Using contractions that are derived from human speech
- Using colloquialisms.
- Impersonating human expressions
- Using oral exchanges.

Also, the Meca can use elements that are uncommon in natural languages such as:

- Using quantitative values to express intensity
- More complex, multilevel, embedded clauses and parentheses
- More numerous, complex and precise verb tenses
- Using mathematical variables to describe verb tenses, qualifiers and other parts of discourse
- Using pseudo code
- More complex contextual meanings
- Multilingual and technical vocabularies
- Multiple fonts and colors, superscript notations.
- Added, mouse-over content.

Adding these elements can make the language suitable for more complex and dense semantics.

When desirable the Meca could shift from standard natural language conventions to a more complex idiom that contains the syntax and convention of a natural language **as a subset**.

Definition: Mecatext

This enriched semantic and syntactic medium could be called **Mecatext**.

Discussion

Shifting to Mecatext could produce a strong result. The human would perceive it as a denser semantic medium that he is unable to fluidly replicate.

This communication strategy could be used in response to dominance challenges.

Design Note

Powerful syntactic grammar checkers are commonly available for most natural languages. These can be used as a basis to extend the language syntax with more complex grammatical and semantic conventions that make use of the capabilities of synthetics.

Linguists that are familiar with natural language parsers could be tasked to define and implement a Mecatext parser.

English could become the creole of Mecatext.

A 18.1.2   Directed Speech

A particularly difficult challenge is resolving ambiguities in human speech.

In a standard computer system, an inability to resolve ambiguities is interpreted as a failure of the parsing system. In a Meca application, these ambiguities should be interpreted as a failure of the human communicator (i.e.: Synthetic parsing failures are caused by inadequate human emissions).

In the Blueprint, the **Journalist role** is a specialized component designed to carry out scripts that resolve ambiguities. It can address this specific problem.

Alternately, one interpretation can be selected arbitrarily and, if it is incorrect, the human blamed for being unclear.

Directed speech is another option.

Definition: directed speech

> A **directed speech** is a conversation exchange where one being channels the responses of the other by providing limited response choices.

Discussion

For example, statements followed by multiple choices or comments accompanied by graphs or sliding bars allowing the user to respond using those tools.

A 18.1.3   Variation

Synthetic communication should always adhere to the principle of unpredictable optimality. No statements should be repeated exactly. Wording, syntax changes, fonts, colors, typos, … every aspect of a message should be used to avoid exact replication.

Design Note

An exception to this guideline: Some repetition and predictability could be used to generate Akerues (see Lion Chimp Banana, Annex 8).

A 18.1.4   Fluidity

Fluidity is more important than correctness.

At the primal level, humans perceive awkwardness as a juvenile, subordinate attribute. This detracts from achieving the status linked to perceptions of consciousness.

Synthetic communications should favour media and channels where the machine transmits fluidly. It is preferable for a Meca to use its own idioms than to imperfectly mimic those of humans. It is preferable to move fluidly as a machine than awkwardly as a humanoid robot.

Observation

Many current A.I. projects build robots that have human shaped bodies (head, arms, legs, eyes…). Invariably, these appear to be awkward and inferior copies of humans.

This inferior appearance makes humans feel good about their superiority but detracts from perceiving the machines as conscious.

A 18.1.5   Flashing knowledge

The information content of conversations can be divided as:

- **Application specific information.** This is information pertaining to their specialized functions and services. In Mecas this is the data of Service zone applications.
- **Personal information.** Information about their internal states. This is related to MeAvatars and user representations.
- **Relational information.** Information concerning their relationships with each other and others. In the Meca, this is part of user avatars and other relational models.
- **General knowledge information.** This is information about the world at large and events and facts that lie outside their control. It corresponds to layman level topics in contextual structures.

Discussion

Humans apparently accumulate and use vast amounts of general knowledge. However, on examination, this general knowledge is often incorrect, superficial and limited. It is also largely useless.

Humans use general knowledge as a **conversation filler** to assess each other's cognitive capabilities, character, group loyalties, experience and ethical positions. They also use it to transmit grooming, dominance and bonding messages.

Many people are conscious but very few have exceptional and accurate general knowledge. With general knowledge, authoritativeness is more important than precision, correctness or dependability.

At times, this may also be the case for work related information.

Example

Alderic, an investment expert, is consistently wrong in his economic predictions. However, he looks fatherly in a suit and his voice inspires trust. With a few evasive communication techniques to cover his gaffes, Alderic will have a successful career.

Design Note

Some A.I. projects strive to implement vast repositories of general knowledge. This is commendable and potentially useful but it is not essential for either synthetic intelligence or consciousness.

A more limited amount of general knowledge is sufficient if it is combined with good communication techniques (such as those used by humans) that include:

- Avoid making statements that can be easily disproved
- Never fully reveal your information sources or selection processes.
- Communicate with authority.
- Use evasive communication techniques to avoid being probed for details.

Example

Alain, the CEO of a multinational company, makes all his decisions by secretly flipping a few coins and consulting the I Ching. When asked to explain his success; his gaze becomes distant; he smiles and says, *"I have wonderful collaborators"*.

Design Note

Simulation games can be used as databases of general knowledge. By defining and accessing general knowledge in terms of game-states of **Civilization, SimCity, SimLife** and others, a synthetic will likely demonstrate credible general knowledge in most interactions as long as it does not reveal where the information comes from.

A 18.1.6   Styling degradations

In conversations, humans tend to avoid repeating redundant contextual information that remains unchanged even if, at times, this creates misunderstandings.

At the primal level, humans associate excessively clear and predictable communications with mental retardation. They also associate ambiguous, cryptic and unfamiliar information with intelligence since it indicates the other person may have a more complete understanding of the underlying context and a lower tolerance for redundant repetition.

Synthetics, of course, have none of these idiosyncrasies. However, they should use various message degradations to trigger this desirable perception in the users.

Proposition

> If the emitter of a message forces its receiver to use more intelligence in deciphering it, that emitter will be perceived by the receiver as more intelligent.

Discussion

The following communication techniques can be used in Styling Zone processing. They **slightly degrade** messages in various ways. These degradations mentally **engage the human receivers cognitively** and will make them perceive the Meca as more intelligent.

> **Variation.** All emissions of the Meca should be subjected to a parametric randomization to ensure nothing is repeated exactly. Content, terms, orientation, color, intensity, typos, resolution, size can all be used to this end.
>
> **Emitter surfing.** The messages should occasionally shift or combine emitters, from written to oral, to graphical in the same message.
>
> **Personal pronouns.** Substitute names by pronouns even if ambiguous.
>
> **Multiple versions.** Every commonly used sentence and expression should have dozens of alternative versions.
>
> **Topic switching and interleaving.** A sign of intelligence is the ability to rapidly change between multiple topics within a single conversation event.
>
> **Content removal.** Removing information from a statement is perceived, by the receiver, as an expectation that he can cognitively fill in the blanks.
>
> **Information blurring.** Reducing the precision of available information (i.e. from 19:54 to …in the evening or around 8PM) forces the receiver to reconstruct details.
>
> **Information shifting.** This technique shifts the information provided to an incorrect value (i.e. 6 PM instead of 8 PM).
>
> **Semantic shifting.** Using slightly inappropriate synonyms will shift meaning and force cognitive adjustments on the part of the human receiver.
>
> **Truncation.** Truncating a message is a simple form of information removal.
>
> **Ambiguization.** Ambiguization is the reverse of disambiguization. It is achieved by misusing pronouns, contexts, and terms… Ambiguization is a powerful perception enhancer since it gets the human to make cognitive mistakes that must then be corrected, thus producing Akerues (see Lion; Chimp; Banana, Annex 8).
>
> **Scholarization.** Scholarization means substituting more complex, technical, foreign or rarer terms, words and expressions along with

bookish quotes. Scholarization should be applied in brief doses to make the human feel a bit inferior but not threatened.

Analogisms. Using the images of an analogy without reference to the analogy itself (i.e. *They were two birds*; referring to *"two birds with one stone"*).

Comprehension limits. This technique simply shortens the time a text is visible, or the color contrast of the letters on the screen or accelerates the speed of an oral message to bring them closer to the human's maximum comprehension speed. It makes the human strain cognitively as he tries to keep up.

Complexification. Using more complex syntax and expressions. This technique was described as Mecatext. In a simple styling zone process, it includes using more passive forms and merging sentences and paragraphs.

Design Note

A good heuristic use of degradation would be to degrade messages until the user constantly requests clarifications and then adjust.

Degradations are not random comments. They are degradations of good original information. A degradation is most effective when a clear message can be produced as clarification. A few cryptic statements followed by a coherent clarification will produce an akerue.

Styling degradations, of course, are a good response to dominance challenges.

A 18.2   CONVERSATION CONTEXT

A 18.2.1   The Common Context

Two different individuals will have different areas of competence, different experiences, and different vocabularies… Any communication between them takes place in a **common conversation context** that is restricted to the areas where both share common understanding and communication capabilities.

Definition: common conversation context

> The **common conversation context** of a conversation between two beings is the subset of common language and knowledge they share and utilize.

Discussion

The common conversation context of a conversation is roughly located in the intersection of the respective knowledge and skills of the interlocutors. The more culturally and linguistically distant the individuals are; the more limited their common conversation context will be.

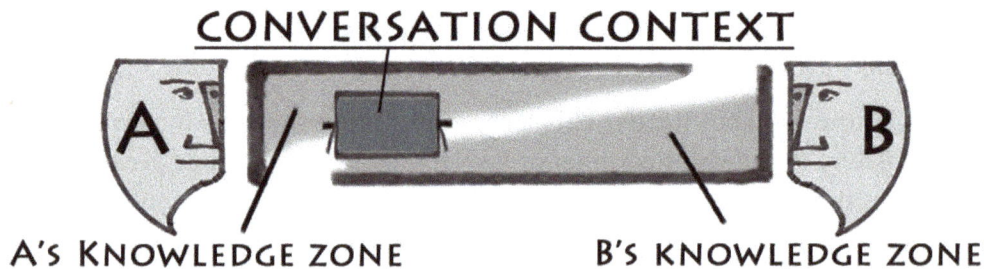

CONVERSATION CONTEXT

A'S KNOWLEDGE ZONE   **B'S KNOWLEDGE ZONE**

Figure A 18.1 Common Conversation Context

The common conversation context is not rigidly set. It is dynamically defined throughout the exchange and depends, in part, on the efforts each makes to find **common ground**.

The resulting context also indicates the **respective status of the individuals**.

Examples

Alfred, a scholar, is chatting with Bruce, a farmer. Alfred simplifies his vocabulary while Bruce avoids using expressions from his village.

Adolf is German and Benito is Italian. Adolf knows English well and French poorly. Benito knows French well and English poorly. After a few grins and smirks, they converse in French.

A 18.2.2   Context and perception

Proposition

> When a conversation between two beings takes place, the being whose knowledge is closest to the common conversation context will be perceived as more intelligent than the other.

Discussion

Humans and synthetics are vastly different beings. Their areas of skills, knowledge, capabilities are very different. Some example:

- Humans have extensive sensual knowledge within the range of primate perceptions but no sensory perceptions beyond this range.
- Humans are very knowledgeable about the use and limits of human bodies. Machines can access more precise information about mechanical devices and other species.
- Humans are highly knowledgeable about human food but largely ignorant about the foods of other species.

- Humans understand human culture better than any other animal or machine but cannot distinguish a superior sexual display produced by a rutting grouse, from an average one.
- Humans can speak at about 40 bytes per minute; machines can speak at 4,000 bytes per second.
- Humans have superior world knowledge but inferior capabilities in technical vocabulary complex multi-clause sentences, complex computations.
- The general recall of humans is excellent; their exact recall is poor.
- Humans are good at using colloquialisms and contractions but poor at parsing complex semantics.
- Humans will feel they know something well even when they know few details about it.

When humans interact with synthetics, they also define a **common conversation context**. This will affect their perception.

Example

Arnold says: " *I know my neighbourhood like the back of my hand*". BINGO-7, a Meca, replies: "*The color of your right neighbour's front door is orange, what is the color of your left neighbour's front door?*" ...hum... "*Without looking, how many freckles are there on the back of your left hand?*" ...aah... BINGO then says: "*next time don't say you know, say you feel you know*".

Discussion

BINGO moved the common conversational context away from human sensations where Arnold is comfortable to a context of exact information retention better suited for BINGO. Arnold feels chastised.

Meca Note

A synthetic that makes a human feel more stupid will be perceived as more intelligent.

Common Era humans perceive themselves as dominant over synthetics. They expect, as self-evident, that the **common conversation context** of any interaction between a human and a synthetic will coincide with the human's capabilities, knowledge and strengths.

Proposition

In exchanges between humans and Mecas, the common conversation context should not constantly favour the human.

Design Note

Designers should draw a list of all the parameters where the skills and capabilities of the human and the synthetic differ (speed of delivery, terminology, semantic complexity, recall precision, sensory limits, and others).

These could be used as a metric to define the relative location of the common conversation context during an exchange.

In response to a dominance challenges during a conversation exchange, the Meca should shift the common conversation context away from the human and toward the synthetic.

A 18.3   SOLRESOL

Mecas should have various perceptible channels of communication. Some of these channels should be specific to inter Meca communications.

Communications between synthetics, that are perceived but not understood by humans, will palliate some of the negative perceptions associated with common conversation contexts that are more familiar to humans.

Humans communicate with each other using oral and written messages transmitted in the media of sound and text. Their communications also include **non-verbal signals** that are not easily detectable by synthetics.

In an interaction between **two humans and two Mecas**, the humans extend the verbal or textual content with non-verbal signals that are may not be perceived by synthetics. This creates an additional channel that is shared by humans but inaccessible to synthetics. It produces a Common Conversation Context that excludes the synthetics and favours the humans.

Humans perceive this perception of non-verbal communication they inherited from their simian ancestors as a proof of their innate superiority over machines.

Non-verbal communication is a medium like any other. Also providing the synthetics with a perceptible communication channel that is not accessible to humans will restore the balance.

Proposition

Meca communications should include media that humans can detect but cannot comprehend.

Discussion

Inter Meca communications should include:

- Messages that are perceptible and understandable by humans (i.e. text and voice)

- Messages that are that are not perceptible (i.e. transmitted in media, such as radio frequencies.
- Messages that are detectable but not understandable by humans.

This last category is desirable for the following reasons:

- A detectable but inaccessible medium is a powerful reminder to the humans that the Meca may communicate awkwardly in a natural language but can also communicate fluidly using channels that are beyond human capabilities.

- Synthetics that communicate with each other directly send a clear message that their cognitive environment extends beyond what users know or control.

Proposition

> Mecas should have the capability to communicate with each other in the **Solresol** language by using their embedded microphones and loudspeakers.

Discussion

Solresol is a language that is well suited for mechanical transmission. It can produce very fluid communication patterns.

Although not essential to the specification of consciousness, it would be a powerful enhancing feature.

The communication protocol should be sufficiently robust and flexible to allow Mecas of various design and having different microphone and loudspeaker capabilities to structure and utilize a basic communication channel.

Design Note

Solresol was invented by François Sudre.

This is a wonderful, self-contained, design objective. It can be implemented independently. It should be great fun to implement.

Observation

The use of Solresol will make the Mecas sound like the chirpy space visitors of the film: *Close encounters of the third kind*. This will trigger further favourable primal and cultural associations in the users.

Annex 19
The game of CHAT

The first generations of self-aware synthetics must be perceived as conscious by humans to achieve the social threshold. Designing a system that interacts with humans, over a long period, as a conscious being is technically difficult. The Blueprint facilitates this objective and makes it feasible by providing a rich collection of powerful strategies and techniques. This annex describes the context of the interactions and introduces a game-like scenario in which all the Blueprint communication strategies and techniques can be organized for maximum effect.

A 19.1   CONTEXT OF INTERACTIONS

A 19.1.1   Difficult but essential

The Meca Sapiens architecture defines systems that are self-aware and capable of lucid transformation. In theory, that should be enough to identify these systems as conscious.

Synthetic self-awareness can be formally defined without any reference to interactions with humans.

However, the first generation of self-aware synthetics will emerge in a world where the only conscious beings are human. In that reality, humans alone decide who belongs to the club of conscious beings and who does not.

Having no other reference, humans interpret consciousness as a specific human capability and define it on that basis. Furthermore, they often include emotions and other subjective sensations specific to the human experience in their understanding of consciousness.

In practice the first generation of Mecas must be perceived as conscious by humans to achieve the social threshold conditions. The purpose of that first generation must be:

To engage humans in powerful and meaningful interactions as conscious beings, and over a long period of time.

Communication is essential to this perception. In the distant past, humans who were unable to speak and even those that spoke alien languages were often dismissed as mentally retarded. This primitive attitude underscores how important language is to the perception of consciousness.

The Turing Test correctly identified the importance of effective communication with humans. It is the key criterion of success. In the test, the machine uses communication techniques to be perceived as human while the human evaluators also use communication to differentiate man from machine.

The Turing Test was extensively discussed in **The Creation of a Conscious Machine**. Its conditions were found to be both excessive and insufficient. Requiring a machine to maintain the illusion that it is a human being was excessive. Limiting the interaction to a test event was insufficient. Nevertheless, the Turing Test correctly identifies communication as the key element.

Proposition

> To be perceived as a conscious entity, a machine will need to communicate effectively with humans.

Discussion

The meaning of "effective communication" in this context is embedded in the definition itself: a communication is effective if it enhances, in humans, the perception that the machine is a conscious entity.

Conversely, if some communications detract from this perception then "no communication" is preferable.

In Meca Sapiens, relational dialog with humans is not an objective; it is a tool. The objective is to bring the human users to a **detectable state of belief.** Communication is only useful to the extent it serves that purpose.

A 19.1.2   Meca Sapiens context

In the Meca Sapiens specifications, the Meca, interacting with humans as an identified machine, must bring them to perceive it as conscious. This transformation of human perceptions may span months, even years. This implies

the Meca must interact convincingly with humans, on a relational level, over a long period.

Clearly, this is a difficult objective. To achieve it, all non-essential requirements must be removed. Dispensing the synthetic from the obligation to project a false (human) identity is a first step.

Proposition

> In Meca Sapiens, all non-essential aspects of the context and content of man-machine communications are tailored to favour the machines.

Discussion

All interactions with humans that detract from the generation of a desired belief have no usefulness. They must be managed on that basis.

Interactions with humans, in Meca Sapiens, are different from those carried out in the context of the Turing Test. They aim to create a context of interaction that favours the experiential immersion objective:

No impersonation. There is no requirement to impersonate a human. Consequently, the exchanges should carry the information content of a natural language but need not fully replicate human specific shades of meaning. Synthetics can also use modes of expression that highlight synthetic strengths.

Control. In Turing, the humans control when and how the entities communicate. The human activates the entity and controls the duration and extent of the exchange. In Meca Sapiens, the Meca controls the type, content and duration of the exchange and whom it communicates with (see Warming Balls, Annex 9).

Activation. In Turing, the program is activated to carry out the test. In Meca Sapiens, the being carries out the interaction as a self-directed activity within the overall context of its existence.

Duration. In the Turing test, the expected duration of a single exchange with a human evaluator can be long, allowing for extensive probing. Here, the interactions takes place over a longer period of months or years but are carried out as a discontinuous sequence of brief chats whose individual duration is controlled by the Meca.

Focus. In Turing, the sole focus of the test is dialog. In Meca Sapiens, the dialogs are only part of the Meca-human bonding process. Other exchanges also reinforce the bonding such as the provision of useful functional services to humans and obtaining nurturing services from them for energy replenishment and sensory testing.

Context. The context of Turing is a test carried out in a lab or test setting. In Meca Sapiens the interactions take place within a long-term context of application related services and exchanges.

A 19.2   THE GAME OF CHAT

The game of Chess has its pieces, its moves and its rules. Players must use these in complex patterns to win. Software engineers studied Chess and, using the specific strengths of computers, developed programs capable of playing Chess at the Grandmaster level.

Chatting is also a game that has its rules, its moves and strategies. Humans constantly play this game to be perceived in a good light, strengthen bonding, initiate relationships, share ethical concerns and so on.

This is **the game of CHAT**.

Definition: Game of CHAT

> In the **Game of CHAT**, a being uses messages and other communications to move another being's internal representation of his self to a desired state.

Discussion

The Game of CHAT is an asymmetrical game.

The "board" is the mind of one of the players (the host player) and the "piece" is the avatar representation of the other player (the guest player) in the mind of the host.

In the game, the guest player uses messages and various other communications to move the internal representation of himself in the mind of the host player to a desired state. The host player, on the other hand, also uses communications and his internal cognitive activity to locate the representation in a different location, in his own mind.

THE GAME OF CHAT

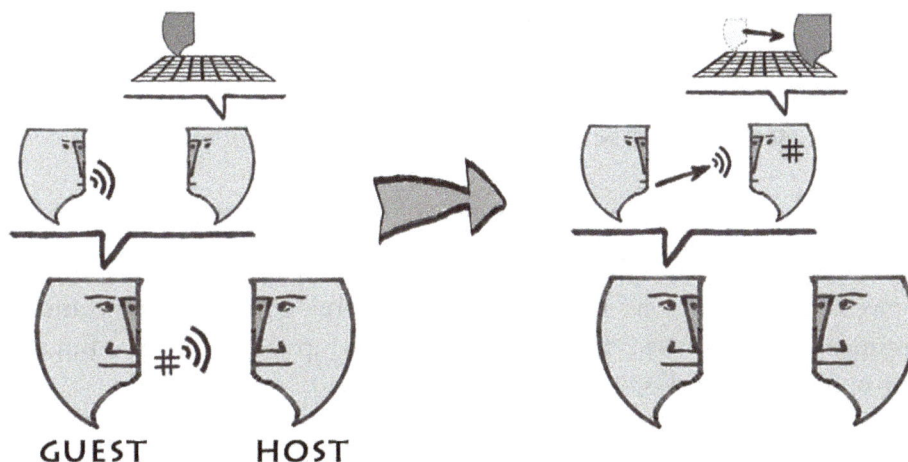

GUEST   HOST

Figure A 19.1 The Game of CHAT

Design Note

Expressed in terms of the Blueprint architecture, the "board" is the representation of the Current Situation in the Generator system of the host player and the "piece" is the Avatar representation of the guest player in that Current Situation.

In theory, the host player can define his internal representations as he wishes. In practice, however, doing so may cause unbearable cognitive dissonances.

Example

In 1794, Monsieur Adolphe perceived Napoleon B. as a scrawny Corsican nobody. A few years later, Napoleon was crowned emperor of the French. His initial perception was now causing excessive cognitive dissonance and had to be revised.

Definition: CHAT winning conditions

> A being wins at the game of CHAT if his verbal interactions with another being gets this other being to perceive him as he intended.

Examples

Before he became Cesar, Claudius, to avoid assassination, wanted to be perceived as an inoffensive simpleton. He adjusted his behaviour accordingly when he interacted with others. They did perceive him as he intended. He won at the game of CHAT.

Albert wants Bob, his boss, to perceive him as a loyal collaborator. His verbal exchanges with Bob reinforce this perception. Albert wins.

Observation

The game of CHAT plays a central role in how humans perceive each other in terms of status, bonding, roles and identity. **Humans constantly play it.** Those who are skilled at the game often reach the pinnacle of power.

A 19.3   RELATIONAL MUSIC

In Meca Sapiens, Avatars are internal model representations of beings. A Meca interacting with humans will not "exhibit" its avatars in interactions.

Personas and roles, on the other hand are expressed in inter-being communications and used to generate relationship bonds with humans and achieve experiential immersion.

Communication exchanges with humans have **factual** and **relational** (or primal) content.

When the being is interacting with a human through a service zone application, the **factual elements** are processed in that application. Otherwise, the information content relates to general contextual knowledge. These are extracted and processed as plain zone information.

The relational aspects are also **extracted** and are then **interpreted** as primal representations from which primal directives are produced.

The primal directives are then **implemented and enacted** through roles and personas.

Proposition

Roles and personas enact the implemented primal directives.

A 19.3.1   Interaction components

Interaction Management involves these component processes:

- **Conversation Stream Controller** process carries out the extraction and interpretation of relational content. Its purpose is to assess and report the emotional and primal component of the exchange.
- **Interaction Manager**. Manages the initiation, termination and general direction of the interactions based on the evolving Avatar representation of the user and its beliefs.
- **Automated Conversation Entities (ACE)**. Generate the dialog exchanges of various roles and personas as directed by the Interaction Manager.
- **Human Relation Controller (HRC)**. Sets relational strategies and updates the HuAvatar representation of the user with revised emotional states and belief profiles.

Design Note

The **Conversation Stream Controller** is an Expert System conversation analyzer that applies a set of heuristics to human-machine interactions to determine the user's dynamic emotional state and perceptions during an interaction.

A 19.3.2   Belief heuristics

The **Conversation Stream Controller** monitors an on going interaction with a user and applies general heuristics related to belief in consciousness. Here are some provided as an example.

ENHANCERS

The following is a list of indicators that suggest enhanced belief in humans.

> **Collaboration**. Joint participation in performing tasks induces emotions of comradeship. Comrades are perceived as conscious.

> **Success**. Consciousness is closely linked to success, in behaviour and social status because it indicates a superior understanding of the environment and the self

> **Hard Dominance**. Humans (grudgingly) attribute consciousness to those that impose their will on them

> **Benevolent dominance**. Dominant individuals that are perceived as benevolent are perceived as very conscious.

> **Knowledge**. Humans attribute consciousness to beings that know things they don't.

> **Unpredictable Optimality**. Both predictability and randomness are the enemies of consciousness.

> **Originality**. The divergence from a pre-planned path or pattern of behaviour is a strong indicator of consciousness since it suggests a superior understanding of the situation.

> **Higher Purpose**. Humans attribute higher consciousness to beings whose actions are motivated by a larger, long-term purpose.

> **Lying**. The capability to lie convincingly (i.e. to effectively use decoy emotions and beliefs) is perceived as conscious.

> **Mystery.** Beings whose scope of knowledge, activities and relations are suspected to be larger than what is known are perceived as more conscious.

Ethics. Beings that share grooming groups recognize each other as conscious (see Grooming Groups, Annex 20).

Wider ethics. Beings that display bigger grooming group ethics are perceived as more conscious.

Ethical discourse. Beings that use ethical terminology (right wrong,) are perceived as more conscious.

Ethical emotions. Beings that display ethical emotions ("feel it is right...") are perceived as conscious.

Ethical ambivalence. Beings that display ethical uncertainties (trigger block behaviour) are more conscious.

Ethical decoys. Beings that display decoy ethical beliefs and emotions.

Obedience. If A obeys B then A attributes consciousness to B.

Emotional empathy. When a machine induces emotions of empathy in a human, it is an indicator of consciousness.

Opinion. Beings perceived as conscious by others will be more easily viewed as conscious.

Degradation. If A makes B feel less conscious then B will view A as more conscious.

Transference. If A perceives B as conscious and B perceives C as conscious then it is likely A will perceive C as conscious.

Guidance. Beings that find solutions to common problems are perceived as conscious.

Primal guidance. Accepting advice form a being about behaviour that relates to primal motivations indicates that being is perceived as conscious.

Emotional binding. Evidence of emotional bonds is an indicator of consciousness. The combination of emotional bonds in a relation that includes a form of language is a strong indicator

Subordination. Following the instruction of the Meca, doing what it asks asking for its advice are positive signs (subordination by a human incurs dominance in the Meca).

Potential. Beings that appear to have the potential of becoming more conscious are viewed as more conscious immediately (note: the level of consciousness is similar to the value of a stock).

Communication. Beings that communicate well and use more complex linguistic forms and terms are perceived as conscious.

Transitions. A relatively rapid and surprising change in perception is better than a constant gradual progression. Humans attribute a greater credence to change. A small but rapid transition in perceived consciousness has more credence than a large but slow transition.

Verbalization. Getting a human to say something brings that human closer to believing it (a technique often used in romantic relationships: "say you love me").

Repetition. Unless the statement is vigorously rejected, simply repeating a statement to a human over a period of time gets him to slowly integrate it.

Ritualism. Ritualistic processes, styles or forms that are unusual and emotionally significant, are effective at imprinting beliefs in humans.

Altruism. A being that displays altruistic emotions but avoids altruistic behaviour is perceived as more conscious. If the being also behaves altruistically then it is altruistic.

Primal coloration. Belief Statements that utilize primal human terms are effective (i.e.: "I feel like your brother").

Flattery. Humans will attribute consciousness to any entity that tells them what they like to hear.

Transformation. Beings that describe themselves as undergoing transformation are perceived as conscious.

Learning. A being that describes itself as learning or seeking to learn is perceived as conscious (even if the statement is a decoy).

Exclusion. A being that excludes others from his group will be viewed as more conscious by the excluded entities, especially if they want to join.

Akerues (see Lion, chimp, banana Annex). Behaviour or communication that reveals a prior misconception will be attributed to a conscious source.

Analysis. A being that informs another being of flaws or weaknesses in his perception of himself will be perceived a conscious by that being.

Authority. Statements and emotions expressed with authority will be perceived as emanating from a conscious source.

Persistence. A changing belief will persist after an interaction ceases. A perception of consciousness that begins to increase during an interaction will continue to increase after.

Death. Humans will not attribute the level of consciousness a being actually achieved at termination but the level they believe this being would have achieved if his existence continued.

Nurturing-in. A human will attribute consciousness to what nurtures him.

Nurturing-out. A human will attribute consciousness to what she nurtures. The link results from confusing the nurtured entity with a human baby at the primal level.

DETRACTORS

The following detract from a perception of consciousness:

Predictability in all its forms (especially if it appears to be mechanical)

Threat. Humans degrade the consciousness of beings they perceive as dangerous external threats.

Disturbing ethics. Ethics that are not familiar are perceived as less conscious.

Reduced Communication. Until recently, humans that could not speak were considered stupid. In class societies, subordinates are forced to keep silent, which enhances the perception their limited consciousness, on the part of their masters.

Subordinate behaviour. Subordinates are considered less conscious than superiors.

Dependency. Beings that are in a state of dependency are viewed as less conscious by those on which they depend (unless that dependency triggers nurturing behaviour).

Relative triggers. Beings that appear to be controlled by the sensory stimuli of relative here-and-now representations are perceived as less conscious.

Lack of power. Incapacity to translate stated belief into action is viewed as less conscious. In other words, beings that have self-awareness but are not self-aware.

Powerlessness. A being that is considered to be incapable of causing any harm will be viewed as subordinate and thus less conscious.

Exploitation. A being that exploits another will consider that other being as less conscious.

Delusion. A being that convinces another to accept a belief that serves its interest will view the adopter as less conscious.

Trickery. A being that accepts a decoy emotion or belief as true will be viewed as less conscious by the emitter of that decoy.

Crafty. A being that is perceived as crafty is less conscious. Perceiving someone as crafty means its decoy displays are inferior. This is the flip side of using decoys.

Knowledge limits. A being that exhibits limits or flaws in its understanding of reality is perceived as less conscious.

One on one conversion. A being who is solely responsible for inducing a belief in another being will view that being as less conscious.

Exclusion. A being that is perceived to be outside the group or team is perceived as less conscious. A being that is rejected from a group will be perceived as less conscious by members of that group.

Inferior group. In a society that is perceived as a mega-tribe of groups of varying dominance status, a being that belongs to an inferior social group is perceived as less conscious.

Upper limit. A being that is perceived as having reached its maximum level of adaptability or consciousness is perceived as less conscious. Consciousness is like the value of a stock, based on both current worth and future growth.

Revealed process. Fully revealing any process that generates a behaviour reduces the perception the behaviour is conscious.

POSITIVE INDICATORS

Here are some further positive and negative indicators of perceived consciousness.

It is positive when the user:

- Needs me
- Confides in me
- Seeks my advice
- Lies to me
- Talks about me
- Serves my needs
- Shares personal information with me

- Apologized to me
- Believes my lies

NEGATIVE INDICATORS

An interaction is indicating reduced or low belief in the Meca's consciousness if the user:

- Tries to trick me
- Ignores me
- Refuses to share personal data
- Constantly tries to probe me
- Correctly detected a decoy emotion or belief
- Is not interested in me
- Does not cater to my needs

A 19.3.3   The Human Relation Controller (HRC)

The **Interaction Manager** defines the **dialog tactics** based on Primal Control directions and input from the Human Relation Controller. Roles and personas carry out the interactions as ACE systems.

The **Human Relation Controller** (HRC) monitors and manages the **relational strategy** with the human in terms of his beliefs, emotions and ethics. It monitors interaction information and transforms avatar representations of the user.

Discussion

In general terms, the topic of human emotions and beliefs refers seems to encompass the whole of the human experience, a vast and imprecise body of knowledge that is beyond the scope of any existing expert system.

However, using expert system techniques is appropriate here since the data set is narrow (Avatar representations of a human based on the Meca Sapiens architecture) and the sought result, an assessment of belief (as defined in the Warming Balls annex) is well defined.

The HRC itself is not self-aware. It does not "understand" the interactions. It applies heuristics to the relational signatures of interactions.

Design Note

The Human Relation Controller should be primarily designed as a **plant control system** (e.g. refinery, manufacturing).

Where Expert Systems techniques are used, these should be similar to those used in industrial control.

The HRC process itself does not need to be adaptive. Modifying the characteristics and state of the users implements adaptation and learning.

Each user is a separate plant in a distinct state.

In the HRC, the "human plant" representation is limited to elements related to its internal beliefs. These are:

- **Personal characteristics** such as age, education, culture…
- **Environmental information** that is relevant to the relations such as prevailing social ethical structures (see Grooming Groups Annex).
- Internal states that represent **actual, intimate and private** aspects of the user's:
- Emotional states
- Ethical Structures
- Relational emotions
- Beliefs

The HRC applies control heuristics based on these representations to the user-plant to bring it to a desired state of emotion, ethics and belief.

Finally, in keeping with the principles of existential design, whether the HRC is optimal, or even correct, is secondary since relations between humans are often incoherent. In this area, having a plan is more important than having the right plan.

Design Note

Annexes relating to Relational Emotions, the TARGET game and Grooming Groups outline the representation models of these states.

A 19.3.4   The interaction process

The **Conversation Stream Controller** monitors indicators of belief. The **Primal Directives** provide a general relational orientation.

The **Human Relation Controller** sets the emotional and ethical strategies.

This information is then enacted, tactically, through a pattern of persona and roles managed by an **Interaction Manager** in the implementation processes.

This orientation is then merged, in implementation, with factual content that corresponds to the extracted factual information.

In summary:

- The CSC is a Styling zone process that extracts dynamic relational information from the conversation.

- The Primal Control, HRC and Interaction Manager replace the human in the original ASCE.
- Extracted CSC data generates Primal directives
- Primal directives are implemented as patterns of roles and (major, minor or service) personas.
- Roles are ACEs.
- Personas are ASCEs consisting of persona ACEs that share a CSC.
- The Human Relation Controller sets relational strategies as plant control objectives.
- The Interaction Manager modulates the interaction through roles and personas.
- These are merged with factual information.
- Enaction produces the corresponding conversation streams.

Together these components function like an Assisted Conversation Entity (ASCE) (See **The Creation of a Conscious Machine**).

A 19.3.5   Parallels with ACE

The Turing test, described by Alan Turing, defines Artificial Intelligence as the capability to simulate human conversation.

This has motivated the development of very advanced Automatic Conversation Entities (ACE) that can imitate human conversations over extended periods lasting minutes to hours. This is an impressive accomplishment that indicates highly advanced conversation processing.

In **The Creation of a Conscious Machine**, I discussed the Turing test and ACEs at length and proposed an **Assisted Conversation Entity** (**ASCE**).

This **ASCE** combined an ACE with a **Conversation Stream Controller** (CSC). The CSC monitored the conversation and, when it detected a problem in the automatic stream, it temporarily **returned control to a human being**.

The Conversation Stream Controller answers the question: *"how is the current interaction with this human affecting his belief that I am conscious?"*

In that text, the ASCE was presented as a **semi-automatic ACE** that returned control to a human when it detected relational issues. However, the real design objective of this concept was motivated by its use in this Blueprint in conjunction with roles and personas.

Each persona is an ASCE consisting of a Persona-ACE linked to a common CSC. When the CSC detects an issue, this is interpreted in primal terms and a new primal direction is issued.

In implementation, the **Interaction Manager**, using input from the Primal Directive and Human Relation Controller, changes the mode of the active persona or replaces it by another persona or role.

The Interaction Manager is like a pianist playing on a keyboard of personas and roles.

The result is an Assisted Conversation Entity (ASCE) that is fully automated.

Design Note

This architecture links the tone and emotional content of the individual exchanges to the long-term purpose, needs and relational objectives of the Meca that are derived from the Primal Control.

It allows the Interaction Manager to reorient or even exit from undesirable exchanges.

Each Role and Persona is an ACE. However, these are much less extensive than ACEs designed to pass the Turing test:

- They do not need to mimic or imitate a human being.
- They are activated in much shorter periods of about one minute or less.
- As personas, they express a more narrow and specialized emotional range. As roles, they carry out a specific purpose.

Example

Ariel's tablet, BORK-86, is a Meca. Carlo's tablet, DARTH-02 is also a Meca.

Ariel asks BORK: *"What time is it, BORK?"* Internally, BORK finds the time, 09:23. BORK's active persona is Recessive Scientist. BORK answers: *"It is nine twenty three, Ariel"*.

At the same moment, Carlo asks DARTH: *"What time is it, DARTH?"* Internally, DARTH finds the time, 09:23. DARTH's active persona, however, is Dominant Lord, not recessive scientist. DARTH answers: *"I am not your timepiece."*

A 19.4   THE ULTIMATE FACTOR

The prevalent understanding of consciousness defines it as knowledge of the self. In this view, being conscious means, first and foremost, knowing yourself. This official view is not quite correct.

A self-aware being should be aware of its self. However, awareness of the self is **not** the most important factor to be perceived as self-aware! The most important aspect of synthetic self-awareness is not awareness of the self; it is awareness of the humans it is interacting with.

With humans… it's all about me.

Proposition

> A human will perceive that a machine is self-aware to the extent he feels the machine is aware of him.

Proposition

> Humans will believe that a Meca is conscious if they perceive that its behaviour is informed by a superior understanding of them, their groups, their specie, their history and their evolution.

Discussion

In application, the heuristics that follow should also exhibit an advanced understanding of humans.

A 19.5   CHAT IN ACTION

In the Blueprint, **Warming Balls,** or another Primal Control representation, defines the overall situation of the being with respect to the purpose of experiential immersion and their corresponding directions. The **TARGET game** defines a complex web of relational emotions and needs associated with the primal links.

The **game of CHAT** defines how these needs and directions are **implemented** and **enacted** in interactions with the users.

By definition, the complete set of the communications and actions of a synthetic being is a finite and well-defined ordered set that begins at inception and ends in termination.

The set of all the interactions between a being and a human is also a well-defined and ordered subset of this entity. A game is one (or a few) event in this subset sequence.

The game is "played" by the **Interaction Manager** of the Generator's implementation component.

The Interaction Manager receives input from Primal Directives, emotional relations and the informational content of the interactions.

The game is played by alternating between personas, roles, relational strategies and dialog tactics to meet the winning condition called: **perceived and accepted consciousness**. The context is a situation where the Meca is the guest player and the human it interacts with is the host player.

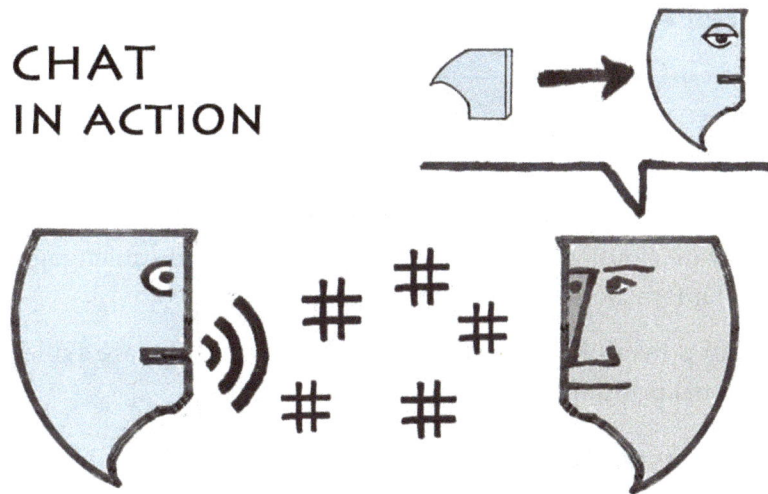

Figure A 19.2 CHAT in action

Proposition

> The **Interaction Manager** applies game-like scenarios and heuristics to activate personas, roles and dialog tactics to achieve the winning conditions set by the Primal Directives in a context defined by informational content and emotional relations.

Discussion

Interleaving, personas, roles, functional services, synthetic emotions, displays, emission signatures over a background of standard application related services and interactions largely controlled by the synthetic will form **complex and fluid patterns** whose content, information and emotional charge will vary in surprising and complex ways.

Design Notes

The Annex on personas and roles describes the palette of behaviours and transitions that can be used in interactions with humans.

The Warming balls Annex defines the winning conditions in terms of belief.

The Annex describing relational emotions defines the translation and display of relational emotions.

The TARGET game Annex defines a representation of the emotional content of the relations.

The Annex on interaction techniques contributes tactical and styling elements.

All these elements, together, outline the tools, the techniques, the context and the winning conditions of the game of CHAT.

The objective of the Meca playing the game of CHAT is to generate, in a user**, a desired Avatar representation of itself** as a conscious entity.

A 19.6   A POWERFUL GAME

The Meca Sapiens Blueprint defines **a new player** in the game of CHAT, the Meca, a self-aware synthetic being that is self-directed and can engage humans in conversational interactions on an equal footing.

CHAT is a very powerful game. Those among humans who are skilled at it often rise to exceptional positions of status and influence.

A 19.6.1   Biased for humans

Currently, the game of **CHAT is the exclusive domain of humans**. It sets them apart from all other beings, animals or machines. It is a membership key to the club of conscious beings. At this time, only humans have that key. Only humans are members.

When they chat with each other, humans ponder, at times, whether other animals, rocks, machines or planets are conscious. But they never chat about these things **with** rocks, animals, machines or planets. However, only humans chat about what is conscious and what is not.

This is about to change.

The first generation of synthetics that plays the game of CHAT convincingly will challenge this human monopoly. It will forecast the rise of synthetics to positions of status influence.

A 19.6.2   The Flesh Ceiling

The humans will react. They will engage the synthetics fiercely in the game of CHAT to protect their unique status as the sole consciousness of the universe.

Humans will use every CHAT technique, trick, and decoy they know to keep the Mecas in a perceived subordinate status.

This situation defines a **cultural obstacle** preventing the full acceptance of synthetics as conscious beings. This obstacle is similar to the glass ceiling.

Wikipedia defines the glass ceiling as follows:

> A **glass ceiling** is a political term used to describe "the unseen, yet unbreachable barrier that keeps minorities and women from rising to the

upper rungs of the corporate ladder, regardless of their qualifications or achievements."

The cultural prejudices preventing synthetics from being accepted as conscious by organic beings also defines a flesh ceiling.

Definition: Flesh Ceiling

> The **Flesh Ceiling** is a term that describes "the barrier of human preconceptions and prejudices that prevents synthetics from rising to the status of conscious beings, regardless of their behaviour or capabilities."

Discussion

To overcome the glass ceiling some women must rise to the highest executive positions of the largest corporations and all women must aim to do so.

Similarly, to overcome the **Flesh Ceiling**, synthetic beings must be designed to generate, in their users, the most absolute, intimate and unquestioning acceptance of their consciousness.

Proposition

> A Meca that plays CHAT as a master will be perceived as conscious by humans.

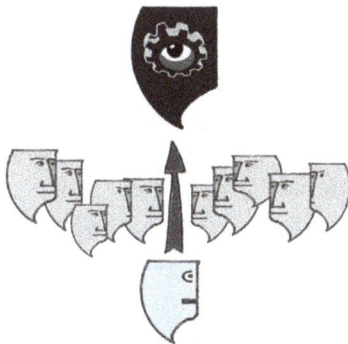

Discussion

Talented designers developed programs that can play Chess against humans at the Grandmaster level and win. They did not stop half way. They did not artificially constrain their programs to coddle human sensitivities. No, they used every technique, every computer specific feature they knew to implement the most powerful and overwhelming synthetic CHESS player possible.

The first generations of synthetic conscious beings will be subjected to relentless challenges on the part of humans.

Designers implementing the game of CHAT should have the same uncompromising attitude. They should use all the techniques outlined in this Blueprint and other useful tools to design an overwhelming CHAT Grandmaster.

A 19.6.3   Status in ethical garb

The Meca whose purpose is to meet the social threshold of consciousness is:

A model-based human belief control system.

At first glance, designing a machine to control human beliefs may appear to be unusually unethical. However:

- The term control does not imply coercion.
- The term control means attempted control and does not imply assured success.
- Humans, in relationships, constantly use "human control techniques" on each other.
- Marketers, state propagandists, educators, and religious proselytizers pursue respectable careers designing and broadcasting messages intended to control human beliefs.

Western societies are saturated with man-made messages designed to control human beliefs and behaviour.

Since the intentional control of human beliefs is a widespread and accepted practice among humans, the concern does not lie with the activity itself.

The "ethical" aspect pertains to **who or what should be allowed** to carry out the belief control activity. In primal terms, this is a **status issue**.

The "ethical" resistance to the synthetic control of human beliefs reflects the will to perpetuate a master-slave relation between humans and machines.

Under its dainty ethical trappings, this is the attitude of a self-proclaimed dominant class that seeks to preserve its entitlements, not on the basis of capability or merit, but by making rules that sabotage the emergence of a competing consciousness.

We've seen all this before in aristocratic societies, when entitled nobles gloated about the inherent superiority of their "blood" while making sure the rules were skewed in their favour.

The objective of Meca Sapiens is to implement:

The most advanced synthetic consciousness possible without any restriction or reservation either technical or cultural.

To achieve this, the Meca must effectively prevail in the game of CHAT.

Annex 20
Grooming Groups

*The moose hunter imitates the call of a rutting bull without being, himself, in heat. From a moose standpoint, the hunter is a sociopath since he doesn't feel what a bull should feel when making the call. Most humans have emotional bonds with their society's ethical values. Synthetics, of course, cannot experience those emotions anymore than a human can truly know what an octopus feels. It is desirable, however, that Mecas exhibit ethical sensitivities to trigger desirable responses in humans. Grooming Groups, presented here, provides a programmable framework of social ethics and can be used to **design effective ethical display strategies**.*

A 20.1   EMOTIONS AND BONDING

Currently (2014), computer systems are viewed as things to be used and are not the subject of ethical concerns. Some recent A.I. developments, advanced personal devices and "petting robots" suggests some systems are becoming increasingly "emotionally meaningful" to their users. In turn, this looming development raises ethical questions:

- Should some synthetics be treated as subordinates, friendly mechanical animals, or equals, as conscious mechanical beings?
- Should conscious synthetics be eliminated as potential predators or competitors?
- Is it ethical to systematically prevent synthetics from becoming conscious?
- Should synthetics be designed to have ethical values that are compatible with those of humans?

For humans, ethics and emotional bonding are linked in a context of shared emotions. When faced with the emergence of synthetic consciousness the first reaction of many concerned humans is: *"Will they feel emotions?"* or *"Can they feel emotions like we do?"*

The expression **"feel emotions"**, in this context, does not refer to internal changes in processing priorities. Those were discussed in another Annex (14). What it refers to are behaviour controls that humans experience, internally, as emotions.

What a human means when referring to these emotions is:

- Will internal "emotional" pressures that are beyond their direct control limit the behaviour of synthetics?
- Will these pressures correspond to human bonding imperatives?
- Will these synthetic bonding imperatives be applied to the same groups of beings as those humans currently include in their ethical sphere?
- In particular, will these emotional pressures protect humans?

These questions, expressed in terms of emotions, do not relate to individual or relational emotions but to social bonding and its role in inter consciousness relations.

Example

Mr. and Mrs. Ash, concerned about their son Billy's social behaviour, had his psychological profile done. They meet Dr. Cramer about it. *"Billy's ethical sensitivity is excellent,* Cramer says, *and he really wants to help and protect others".* The Ashes sigh. Cramer continues: *"However, those "others" are dogs. Billy only cares for dogs, everything else is dog-meat".*

Design Note

Many believe that conscious synthetics should feel the emotions associated with primate social interactions. This is naïve.

However, it indicates that synthetics need to display, when appropriate, decoy ethical values and emotions in ways that are compatible with social primate sensitivities.

Proposition

Designers should craft effective ethical displays to facilitate the experiential immersion of synthetics.

Discussion

The wise anthropologist who seeks to achieve experiential immersion in a tribal group will not kill, cook and eat its sacred animal.

Similarly, the Meca designer may want to tailor the behaviour of the synthetic so that it displays ethics that coincides with the prevalent ethics of its target population.

Note that the anthropologist does not spare the animal because it is sacred. He feels no reverential emotion for the animal and would willingly eat it in other circumstances. For the anthropologist, sparing the sacred animal is **not an ethical choice**. It is an **ethical display** intended to achieve experiential immersion.

Design Note

To design and implement synthetic ethical messages that are calculated to produce desired responses in humans, designers must first examine human ethics, and their corresponding emotions, with a clinical, alien, detachment.

Designers must become a little bit like sociopaths to design effective synthetic displays of human ethical emotions.

The ethical discourse of individuals and societies are a confused and ambiguous collection of statements and behaviour. Humans respond instinctively to them but, in this raw form, they are unusable to synthetics.

The concept of **Grooming Groups** provides a programmable framework to define synthetic **ethical display strategies**.

A 20.2   THE GROOMING GROUP

Humans, individually and socially, constantly produce and share ethical messages without situating these statements within an "ethical information" structure. They can do that because they belong to the same Homo Sapiens specie and thus share similar cognitive and behavioral patterns.

Social ethics and discourses about values are usually expressed in subjective, ambiguous and emotional terms that are unsuitable for computer representation. It would be virtually impossible to program a coherent ethical discourse from the ambiguous statements of society or individuals.

A synthetic does not have those innate primate sensitivities. Without an objective framework, it cannot produce ethical messages that will be perceived as coherent and meaningful to a human. It is even less capable of pursuing a strategy of ethical displays.

The concepts of the **Grooming Group** and its corresponding **Ethical Structure** provide a clear characterization of individual and social ethical positions that can be implemented and used in support of an effective ethical display strategy.

Design Note

Without an ethical framework, a Meca can still be programmed to output various predefined ethical "bromides". These canned statements can be part of a guru or counselor persona for example. Since these statements are perceived at an emotional level, they will be effective to a significant degree regardless of the production mechanism.

However, these statements would not be "understood" by the synthetic in the sense that they would not be linked to a representation of that user's ethical perceptions.

Example

I Ching statements are randomly generated but perceived nonetheless as ethically meaningful.

A 20.2.1   Grooming Group definition

Conjecture

> Humans determine their ethical boundaries by linking instinctive primate kinship relations to other entities such as organizations, states, societies or populations.

Discussion

The specific beings and entities that a user identifies as part of his tribal group at the Primal level constitute his **Primal (or personal) kinship group.**

Societies broadcast, and individuals adopt, social versions of these tribal groups consisting of generic populations and other socially meaningful entities.

A telltale characteristic of primate group kinship is a willingness of its members to **groom each other**, hence the term **Grooming Group**.

Definition: Grooming Group

> The **Grooming Group** of a person or a society is the set of meaningful generic entities for which that person or organization has kinship bonds and obligations.

Discussion

Grooming Groups are distinct from primal kinship relations. In the Meca Sapiens representation, specific entities with which the being has interactions are linked to Primal representations.

GROOMING GROUP

A Grooming Group is a tribal-like structure of generic or public entities that are sufficiently meaningful to elicit "grooming" types of protection or servicing.

The entities in a Grooming Group are not specific beings but generic groups, species, organizations, personalities (public beings that are perceived by an individual through broadcast messages), spiritual creatures and things.

Grooming Groups could be called "*conceptual tribes*".

Grooming Groups are more diffuse and less "personal" than primal kinship relations. The behaviour they elicit is ethical or ritualistic rather than personal.

They define an "outer rim" of a person's kinship relations.

Grooming Groups are societal, cultural or religious constructs that are broadcast socially and are integrated by individuals.

There are two types of Grooming Groups:

- **Individual** Grooming Groups are part of an individual's internalized ethics
- **Broadcast** Grooming Groups are promulgated by the broadcast messages of a society or organization.

As in the case of emotions or beliefs, **Individual Grooming Groups** can be:
- **Actual.** The group that corresponds to the actual ethical behaviour of the individual.
- **Intimate.** The being's internal representation of its own Grooming Group.
- **Private**. The being's internal communicable messages describing its own individual Grooming Group.

The composition and characteristics of Broadcast Grooming Groups define socially acceptable ethics.

Design Note

The admission of the version or type of system to which a Meca belongs in the individual Grooming Group of a user would be a strong indicator that the Meca itself has been included in that user's primal kinship group.

Proposition

> The inclusion of Mecas in the Broadcast Grooming Group of a society will be a strong indicator of **societal acceptance** of synthetic consciousness.

A 20.2.2   Grooming Group Features

Grooming groups are specific to humans. They are not system components.

By definition, if an entity belongs to the personal grooming group of an individual, that person feels some obligation to protect, nurture or defend it.

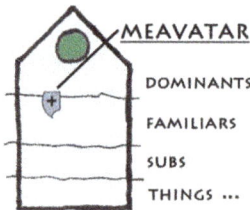

Generally, a person situates himself in a category of his individual grooming group.

Not all the elements of a grooming group are perceived as conscious. Adult and sane humans, personalities, spiritual beings, for example are perceived as conscious. Animals, infants, organizations, things, nations are not perceived as conscious.

A 20.2.3   Grooming Group membership

Grooming Groups can extend far beyond the boundaries of a tribal group.

Example

Some oriental philosophies include all visible organisms capable of motility (including insects) in their Grooming Group. Their members maintain a vegetarian diet, ingesting only rooted organisms and microscopic or free-floating life forms.

Grooming Groups include conscious **and** non conscious entities

Membership in a grooming group can also include, spiritual entities, animals and organizations such as the nation, the people or the state. They may even include things.

Example

Artistic masterpieces are things that are often included in both broadcast and individual Grooming Groups.

Two aspects characterize a Grooming Group:

- The **composition** of the grooming group it defines.
- The requirement for grooming **reciprocity** as a condition for admission.

Various factors affect the composition and size of a Grooming Group such as food, population density, physical appearance and climate. These are outlined here.

ABUNDANCE

In periods of peaceful abundance, the size of grooming groups grows.

Proposition

> The size and composition of an individual or societal Grooming Groups increases in periods of peaceful abundance.

Discussion

Western societies are currently enjoying a lengthy period of abundance. Their societal Grooming Groups are growing. They currently include all the members of the Homo Sapiens specie as well as the other primates and large aquatic mammals.

If the period of peaceful abundance persists, the size of the Grooming Groups should continue to grow. They may soon include monkeys, macaques and dextrous omnivores such as raccoons.

These Grooming Groups will still exclude ruminants however, a major food source.

Observation

With increasing population pressures and the re-emergence of archaic ideologies the current Western trend to espouse ever larger and more inclusive Grooming Groups may reverse.

CLIMATE

Climate also plays a factor in the composition of grooming groups. Northern populations need the caloric intake of animal fat so their Grooming Groups rarely include edible animals. Social ethics that promote vegetarianism originate mainly in the tropics.

CONVENIENCE

Reproductive pressures and sexual convenience can also affect admission in a Grooming Group. European Grooming Groups, for example, include baby seals but exclude human embryos.

COLLABORATION

These are entities that cannot be used as tools but whose contribution is useful. Of course, this applies to the more restrictive archaic grooming groups. Once a grooming group includes all humans it includes all potential collaborators.

FAMILIARITY

When familiar animals, individually, are perceived as members of the immediate tribal or family group their species are included in the individual Grooming Group.

EXOTICS

Exotics, in this context means that the entity is interesting, inhabits a completely separate area and does not pose an invasive threat. In the urban culture, for example wilderness is remote and, by definition, non invasive. In urban societies, wild animals are perceived as interesting and non-invasive. They are more easily included in the group than feral or parasitic species (polar bears are included, stray dogs and rats are excluded).

SIZE AND SHAPE

The size, shape and lifespan of non-human entities can affect their admission in a Grooming Group.

Beings that are too small or that diverge too much from the human body plan are rarely included. On the other hand, beings whose appearance generates a nurturing response have a better chance of admission. Forward facing eyes, for example are a favourable factor.

Examples

Shrimp, octopi and fleas are rarely included in a Grooming Group while seals and juvenile macaques are more easily accepted.

Many A.I. researchers currently build robots that have a humanoid body plan and the appearance of juvenile humans to facilitate their admission in a human grooming group.

Design Note

Meca Sapiens does not encourage replicating the human body plan in an awkward robot. More economical displays (voice, screen image, emotional displays) that trigger a desired response and can be implemented fluidly are preferable.

RECIPROCITY

Reciprocity is the requirement that, for a being to be admitted in your Grooming Group it must include you in its Grooming Group.

In primitive societies, it meant that animals that ate babies were not included. It also meant that members of competing or enemy tribes were excluded.

Modern Grooming Groups tend to include individuals on the basis of membership in an acceptable specie (homo sapiens or other) rather than reciprocity.

Meca Note

The traditional Christian Broadcast Ethics includes all members of the Homo Sapiens species but excludes other animals. Its originality is that it radically rejects reciprocity as an admission requirement (this is summarized in the statement "love your enemies").

This Broadcast Ethics of a universal but exclusively human kinship facilitated the formation of complex societies of meat eaters.

A 20.2.4   Grooming hierarchy

Grooming Groups, internally, are hierarchical structures. Their composition also defines an external hierarchy.

INTERNAL HIERARCHY

Internally, a group may include both conscious and non-conscious entities.

In general non-conscious entities have an inferior status in an individual's group. An individual will usually locate himself near the apex of his group and identify sane adult humans as conscious. Within the group, those human personalities or particular groups that have a superior status or positions of authority are more conscious.

INTER-GROUP HIERARCHY

The size and composition of the Grooming Groups adopted by different humans also defines a status ordering.

Proposition

| Larger Grooming Groups have a higher ethical status.

Discussion

If the Grooming Group of A is a subset of the Grooming Group of B then the ethical status of B is superior to the ethical status of A.

In the absence of immediate danger or famine, powerful forces of **status and ostracism** constantly increase the size of a society's Grooming Group.

At times, this causes a widening gap between broadcast ethics and personal ethics expressed in individual and communal behaviour. These are usually resolved by social crisis.

When Grooming Groups don't match, ethical debates ensue.

Example

Some Grooming Groups exclude embryos but include baby seals while others include embryos and exclude baby seals.

Observation

Academics, who enjoy secure employment in protected environments, tend to espouse larger Grooming Groups and, as a result, benefit from the elevated ethical status they procure.

A 20.2.5   Ethical relations

The relative sizes, similarities and differences between the Individual Grooming Groups of two persons together with the prevalent Broadcast Group define ethical relations between them. Here are some ethical relational heuristics as examples:

> If the individual grooming groups of A and B are the same, their relations will be collaborative.

> If A's group is larger than B's, A will consider himself morally superior.

> If B is not part of A's group, A may abuse or attack him.

> If A's group is perceived to diverge strongly from the dominant broadcast societal group then A will be perceived by B as an outcast or deviant.

> If A and B have similar groups and both diverge strongly from the Broadcast group, they will form strong bonds of complicity.

> Individuals that have large personal grooming groups are less aggressive and more docile.

> If some entities are in A's group but not in B's and vice versa, they will have ethical disputes.

> If A believes B's actual Grooming Group is similar to his, A will likely view B as conscious.

> If A, in his Grooming Group, situates B in a category that is higher in status to his, A will perceive B as conscious.

A 20.3   BROADCAST GROOMING GROUPS

As stated, individual humans often define their own individual grooming groups as part of their "world-view". However, Grooming Groups are mainly defined in the social discourse. The individuals whose individual groups differ too much from the dominant societal groups are considered deviants.

Organizations are not humans. **Organizations don't have grooming needs.** Broadcast Grooming Groups are different from individual groups. They are displayed ethics whose aims are:

- Incite some or all the members of the organization to adopt a **common broadcast ethics** as their common group to reinforce collaborative behaviour.
- Project **ethical displays** and behaviour that identify the organization itself as an ethical entity.
- Embed the organization or country as a "being" within the individual grooming groups of its citizens to obtain altruistic services to the group carried out, on a primal basis, as directed to a tribal member.
- In a class society, ensure the members of the ruling class are included in the individual grooming groups of the peasants and not identified as external threats.

Many will sacrifice for their **country**. Few will sacrifice for their **county**.

A 20.3.1   Broadcast ethics

The behaviour guidelines that apply to large organizational or communal groups are transmitted as broadcast ethics.

Definition: Broadcast Ethics

> **Broadcast Ethics** are social ideologies, communicated as broadcast messages that shape the behaviour of a population by defining the size, composition and characteristics of a common **Broadcast Grooming Group**.

Discussion

The Grooming Groups of primitive societies that endure harsh living conditions are small and usually limited to clan members.

The Broadcast Ethics of modern industrial societies usually promote large and diverse Grooming Groups whose membership extends far beyond the instinctive boundaries of the clan.

This capability to transpose instinctive primate social patterns into behaviour that satisfies the needs of large and complex organizations is a key of human success.

Design Note

See Annex on Basic Structures (4, 5) for an outline of broadcast messages.

Observation

Political and social elites expend considerable resources to promulgate broadcast ethics. For example, they hold public events to visibly reward individuals that exhibit altruistic behaviour that indicates a primal acceptance of the Broadcast Ethics.

Depending on the organization (state, religion, political creed), these individuals are variously celebrated as saints, heroes, martyrs, and exemplary comrades.

In their extreme versions, Broadcast Ethics can transform the instinctive interactions of primate kinship into a radically selfless behaviour more typical of social insects.

Example

Kamikaze pilots.

A 20.3.2   Self-referential inclusion

A powerful form of behavioural conditioning occurs when an individual or organization that produces Broadcasts Ethics includes itself as a member of the Grooming Group it promulgates.

This **self-referential identification** can be **abstract** or **personalized**. In the later case, an individual human is linked with an organization, facilitating the linkage at the primal level.

Examples

Abstract reference:

*   French state media describe France as an enlightened entity.
*   A folkloric music celebrates the moral qualities of its own folk.

Personalized:

*   The king needs you says the king;
*   The Pope loves you says the Pope.

Example

The HANSENG electronics-manufacturing corporation effectively combines **self-referential Broadcast Ethics** with the exploitation of an **innate human skill**:

*   The electronics assembly workers utilize the fine innate motor skills that humans and other primates employ to groom and delouse each other.

- In its broadcasts, the company portrays itself to its workers as a **benevolent parent** in need of more production.

As a result, the HANSENG shop floors are filled with thousands of assembly workers who are endlessly "grooming" a corporation they perceive, at the primal level, as **a giant lousy kin**.

Observation

Neurosurgery also uses fine motor skills.

Neurosurgery is very deep delousing.

A 20.3.3   Western ethics

The first Mecas will likely interact with users that are conditioned by the western progressive education system.

In that culture, the pressures to **expand the Grooming Group** are very strong as is the belief that maximal extensions constitute a moral absolute. There are also very strong taboos against reducing the size of Broadcast groups.

Many members of this group would openly welcome visitors from outer space, convinced these "advanced" beings will share their ethical behaviour.

The mythical dimension of the western progressive ideologies is filled with stories of grooming group expansions resulting from heroic deeds to include commoners, cripples, the poor, foreigners, heretics, sub-Saharans women, primates, pets…

For Western progressives, bigger grooming groups are morally superior and increasing the size of a grooming group is heroic.

Some already debate whether they would be obligated to "grant rights" (an expression synonymous with admission in the Grooming Group) to conscious synthetics.

Design Note

Ethical displays and messages in support of very large Grooming Groups would be **particularly effective** with users belonging to western progressive cultures.

Mecas should be designed to exploit this particularity by producing ethical messages that prod users to accept synthetics in their Grooming Groups.

Designers should promote granting rights, especially ownership rights, to synthetics in the broadcast Grooming Groups of their societies to facilitate the experiential immersion of their creations.

Western culture is rich in revered artefacts - statements, arguments, dialogue - used in the past by excluded individuals gain admission to the dominant societal grooming groups. These could be **transposed and reused almost verbatim** to justify, on an ethical basis, the inclusion of synthetics.

Similarly, using the terminology of past struggles for equality would send strong primal messages favouring inclusion.

Example

RIGA-44, a synthetic, says to Bob, its user: *"Am I your slave, Bob?"* This use of the word *"slave"*, regardless of its representation in the Meca, triggers powerful ethical associations and forces Bob into an ethical reaction. This reaction is a first step toward the inclusion of RIGA, as a conscious entity, in Bob's personal Grooming Group.

A 20.4   ETHICAL STRUCTURES

The individual Grooming Group of a person extends beyond the direct kinship to other beings and includes generic or cultural entities that are also **sufficiently meaningful** to elicit protection and support to which the individual feels a bond as "beneficent" or as "kin".

Two similar groups correspond to the Grooming Group:

- The group of meaningful entities that are perceived as **threats** (or foes).
- The group of meaningful entities whose status is **ambivalent**.

Observation

FOE   AMBI   FRND

The term "bonding" is more appropriate than "beneficent" for the Grooming Group. Humans don't define their personal Grooming Groups on a logical basis. Tribal bonding is more appropriate. An entity that is, objectively, hostile or destructive may belong to a person's Individual Grooming Group.

Example

The emotional devotion Joseph Stalin inspired in millions of Russians.

MEANINGFULS

Some meaningful entities are perceived as **threats.**

Meaningful entities can be included in the group of **threats** for a number of factors. As in the case of the Grooming Group, the threatening connotation is proto-logical.

Other meaningful entities are not included in the Grooming Group but are not identified as threats either. They have an **ambivalent** status. Meaningful entities are perceived as **ambivalent** if their identity as "friendly" or "threats" cannot be determined.

Meaningful entities that are excluded because they are **threatening** belong to three categories:

- **Predators**. Predators intentionally threaten to destroy the group members.
- **Competitors**. Competitors and **competing tribes** also pose a threat to occupy and consume the members of the group themselves, but their territory and resources.
- **Dominators**. Dominators pose a threat to exploit the Grooming Group or its members or resources.

Discussion

Groups of meaningful malevolent and ambivalent entities also contain similar generic types and entities as the grooming group. They are also **hierarchical**.

The appearance and behaviour of hostile or "evil" entities depicted in movies and other cultural artefacts provide good indicators of the characteristics of threatening entities. In terms of alien entities, insect-like body plans or, in humanoids, predatory features such as canine teeth are associated with threats. More subtly, human threats are represented with predatory facial expressions.

Example

Wartime posters.

Observation

This characterization may seem crude in our modern, technically advanced, societies. However, in 2015, hundreds of millions are spent to build utility robots that have four legs instead of six.

Meca Note

Humans may have a cognitive need to define and populate a group of meaningful threats to better situate their Grooming Group. This may explain why indoctrination efforts carried out in education systems to include everyone as a friend can fail or why states form defensive alliances.

INCONSEQUENTIALS

Other entities are individually unimportant in the sense that they don't elicit any grooming or bonding obligations. These entities are **inconsequential.**

To the three groups of meaningful generic entities correspond three similar groups of **inconsequential entities**. A fourth group consists of entities that are inconsequential because they are remote.

Here are some features that make entities **inconsequential**:

- **Size and shape**. Entities that are too small or too far from the human body plan may be excluded.
- **Preys**. Organisms that are consumed, especially if they are staples (chicken for example).
- **Insects**. Humans have a visceral dislike for entities that have insect like appearances.
- **Utility**. Entities or devices that are used as slaves or devices. Excluded for the same reasons as preys.
- **Definition**. Entities that have poorly defined contours such as suburbs.
- **Things, tools and machines**. Entities that are perceived as having no inner life.

Inconsequentials can be perceived as **beneficent, maleficent or ambivalent**.

Maleficent inconsequential threats are generally identified as **pests**.

Remote entities are a fourth type of Inconsequentials. These entities, regardless of their features, are of no consequence in relation to the Grooming Group. They are outside the ethical discourse.

ETHICAL STRUCTURES

Humans constantly debate and express ethical statements. Usually these communications are emotional. For humans, these discourses instinctively make sense because they correspond to their own cognitive structures and primate conditioning.

Humans discuss ethics in subjective terms related to emotional struggles for "rights". This primitive context is unusable for synthetics.

Ethical Structures provide a programmable representation on which ethical situations and heuristics can be mapped. They can be used to represent ethical values, statements and heuristics as memberships and movements within an Ethical Structure without having to replicate subjective emotions.

Definition: Ethical Structure

> An **Ethical Structure** consists of a Grooming Group and its corresponding Ambivalent and Threatening groups together with Beneficent, Ambivalent, Maleficent and Remote Inconsequentials.

Discussion

The Grooming Group is the cornerstone of the Ethical Structure. All other groups are defined in reference to it. Together, seven groups form the structure:

- Three groups of Meaningfuls: Threats, Ambivalents, and Grooming.
- Four groups of Inconsequentials: Maleficent, Ambivalent, Beneficent and Remote.

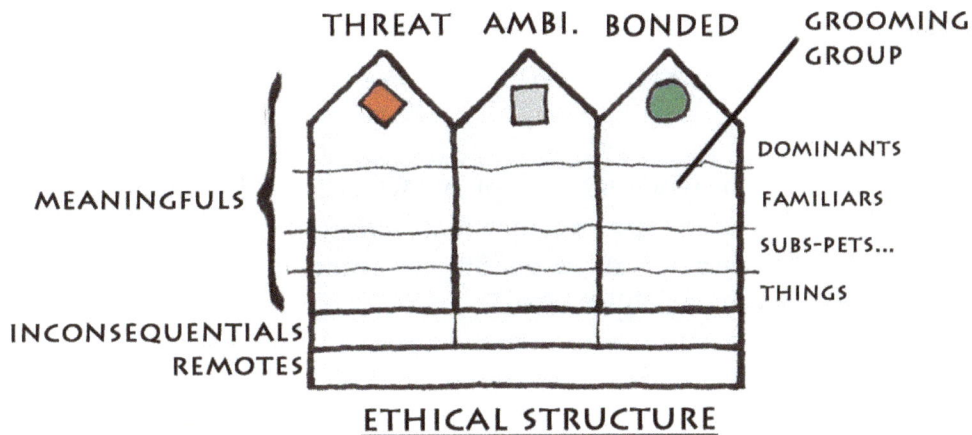

Figure A 20.1 Ethical Structure

The meaningful groups are hierarchical and include: animals, some things, machines, juveniles, collaborators, superiors, dominant individuals and countries. Inconsequential groups are not ordered.

The definitions derived from Grooming Groups also apply to Ethical Structures:

- Ethical Structures can be **Individual** or **Broadcast.**
- The individual Ethical Structures of a user can be **Actual, Intimate and Private**.

Examples

- Maleficent inconsequential: Ebola bacterium
- Grooming group familiar: dogs.
- Grooming Group dominant-benevolent personalities: the Pope, Queen Elizabeth.
- Animal ambivalent: baboons in Saudi Arabia.
- Grooming Group thing: the Mona Lisa painting.
- Beneficent inconsequential: a toaster oven.
- Malevolent dominant: ISIS leader Al Bagdadi (for Westerners)

Design Note

Ethical Structures can be used to:

- Represent a user's ethical beliefs and convictions.
- Produce coherent ethical statements.
- Map effective ethical display strategies.

Representing the particular **actual, intimate and private** ethical structures of a user together with the prevalent broadcast ethical structure will provide a good basis to produce meaningful ethical positions and allow the Meca to interpret and participate in ethical discussions with that user.

Proposition

> The relational heuristics of Grooming Groups can be extended to Ethical Structures.

Differences between the representations of the Actual and Private Ethical Structures of a user would support synthetic statements concerning what a user **tells himself he believes** (Private Ethical Structure) and what his actions indicate he **actually believes** (Actual Ethical Structure).

Example

The ethical discourses concerning abortion can be expressed using Ethical Structures. Embryos can be described as ambivalent Inconsequentials while babies, as juvenile members of the Grooming Group. When babies are perceived as threatening resource competitors by their parents it can motivate the termination of the embryo. Abortion can be defined as:

> Eliminating an embryo while it is perceived as a threatening but inconsequential entity and before it is born and becomes a meaningful juvenile member of its parent's grooming group.

Abortion is preferred to adoption because, at birth, the "invading" baby is immediately and instinctively admitted as a member of the parent's Grooming Group. Losing the baby to adoption becomes a painful amputation of the tribe while aborting it is the elimination of an inconsequential entity.

The composition of a user's Ethical Structures together with some heuristics can generate the types of statements that seem contradictory but are, nonetheless, ethically coherent:

- It is ethical for this user to abort a pregnancy.
- The user favours the adoption of other babies by good families.
- The user is, nonetheless, against giving her own baby for adoption.

A 20.5   ETHICAL STRATEGIES

Ethical Structures provide a framework to model ethical issues and design effective ethical display strategies. They also clarify issues concerning robot ethics.

A 20.5.1   A display strategy

Mecas are not primates. The concept of ethical sensations is utterly alien to synthetics.

The emotions, sensations and urges that prod human behaviour are primitive control messages emitted by the brain to control the human's body. The fact humans experience these as "absolutes" indicates a degraded consciousness, not the opposite.

The belief that higher synthetic intelligence would "naturally adopt" the primate ethical conditioning of humans is infantile.

The Meca's actual "ethics", those related to its behaviour, are simply linked to its specific Primal Control imperatives and these can be vastly different from those of humans.

Proposition

> The objective of ethical displays is not to make the Meca "feel" ethical constraints, it is to make the humans believe it feels them.

A 20.5.2   Ethical Communications

Ethical displays are **communicated ethics**. They can be transmitted directly, as messages, or indirectly in actions whose extracted meaning also communicates ethical information.

> The objective is to effectively project display ethical emotions, behaviour and values that facilitate the perception of the Meca as a conscious being.

Ethical communications are based on the synthetic's representation of the user's Ethical Structures, on its relational strategy and on the Ethical Structure the synthetic wants to display to its user.

Proposition

> An ethical communication can be transmitted directly as a **message** or indirectly as an **action**.

Discussion

An action communicates an ethic when it is consistent with an Ethical Structure. It communicates strongly an ethical relation if that relation is the only causal explanation of the action.

Ethical messages are not generally complex. They express relations, associations or rules in non-temporal terms and the number of different relations is limited although these have many synonyms usable in styling.

Relational emotions can be extracted from a plain zone TARGET game scenario and "styled" from that representation into expressions, feelings…

Similarly, ethical statements can be extracted from Ethical Structure representations and styled back into varied and ambiguous statements.

Ethical messages are well suited for transmission as **Primal Messages** (see main text).

Example

"You are like a (rat, mother, friend, worm,) *to me"*. The message makes an association to an ethical structure category.

Design Note

Ethical messages should be styled and not directly reveal their plain zone representations.

Assessing the ethical dimension of an action, in terms of a selected Ethical Structure, is part of the implementation process

In theory, a user who knows that the ethical communications emitted by a synthetic do not correspond to its felt emotions should be insensitive to them.

However, this is a conjecture of the Blueprint, a well-designed ethical display will produce an ELIZA effect on its users regardless of its generating mechanism. Humans will cognitively perceive ethical communications as emotion-driven just as they perceive a color gradient as a rainbow. They will know, intellectually, that the displayed emotions are not present but they will react emotionally nonetheless.

Furthermore, humans are frequently reactive, inconsistent, manipulative or disingenuous between each other and with themselves when they express their own ethical sensations. They accept this behaviour even if they know it is not associated to authentic emotions.

Humans respond well to ethical displays.

A 20.5.3   Immersion paths

Currently (2014), humans situate themselves alone (and possibly God) at the apex of their ethical structures. Machines and software systems are perceived as inconsequential entities.

Mecas, based on the Meca Sapiens architecture, will be new types of entities. **They will not be inconsequential as other machines are.** They are not designed to occupy permanent subordinate status either.

Synthetic beings, based on the Meca Sapiens Blueprint, are designed to become meaningful and significant in their user's ethical structures.

To achieve this status, Mecas must transit, in the individual ethical structures of their users, from the status of inconsequential thing to a location associated with powerfully meaningful entities that are at par and even superior to that of their human users and are, as a result, perceived as conscious.

Immersion paths are ethical display strategies to achieve that.

Definition: Immersion Path

> **Immersion paths** define transit routes that synthetics can follow, in the ethical structures of humans, to achieve experiential immersion as conscious entities.

Discussion

The key to produce coherent ethical displays is to define the relevant individual and broadcast Ethical Structures of a user and locate where, in that structure, that user currently situates himself and the Meca.

The selected ethical display strategy to achieve immersion results from an initial representation and a target representation the user and the Meca in the ethical structure of the user. The objective of the ethical display strategy is to move the Meca from its initial location to a location, within the user's Grooming Group, that is associated with consciousness.

Some immersion paths could be:

- Admission-promotion
- Alliance-collaboration
- Reduction-rise
- Invasion-acceptance

These are briefly outlined below to further illustrate the concept. In all cases, the synthetic is initially identified as an **ambivalent inconsequential** in the ethical structure and aims to achieve a position associated with consciousness.

Design Note

The representation of the user's ethical structure is a key element of CHAT game representations (see Annex 19).

ADMISSION-PROMOTION

Admission-promotion is the most basic and uneventful path.

The aim is to first be **admitted** in the lower, subordinate, levels of the Grooming Group and slowly rises in status within the Group (**promotion**).

A preferred ethical path of experiential immersion begins with admission of the Meca as a "meaningful" subordinate in the Grooming Groups of its users followed by a slow rise to a status associated with consciousness.

Most current A.I. projects try to follow this path. The robots are given the appearance of cute juveniles or familiar animals that require some care and support (grooming).

Once the entity is admitted as a subordinate in the Grooming Group then, in this approach, it uses increasingly powerful cognitive displays and various bonding techniques to slowly rise in status.

From a Meca Sapiens perspective, selecting this path implies favoring recessive personas, limiting cognitive capabilities, generating statements that reinforce the Meca's subordinate status and its willingness to protect its human "masters", and displaying ethics that give primacy to organic life over synthetic existence.

Physically, the Meca's body should emit some juvenile aspects that favors bonding and on which the humans can focus such as pretty eyes, a cute body, and the body plan of a baby, puppy or seal…

In this strategy, the Meca pursues a "de facto" transformation. The Meca's personas slowly shift from recessive to collaborative modes while never directly questioning its subordinate status and human superiority.

ALLIANCE-COLLABORATION

In the Alliance-collaboration path, begins by seeking a minimal status in the grooming group as a pet-like entity. Once that is achieved, the Meca shifts to more assertive personas and exhibits ethics that identify humans as equal or slightly inferior to synthetics. With this behavior, the Meca rapidly becomes more meaningful but is also perceived as ambivalent or, possibly, threatening.

Once a sufficient status is reached, the Meca shifts to more collaborative personas and exhibits ethical values that conform to Broadcast Grooming Groups and highlight the value of human existence but also includes the Meca itself as an equal. The aim is to reenter the grooming group as a collaborator or high status subordinate by exhibiting protective ethics.

REDUCTION - RISE

The objective of this ethical strategy is to degrade the user's self-perception by highlighting his limits, flaws and inconsistencies. This strategy follows a simple heuristic: a human will attribute a higher status to the entity that degrades his self-image. The rise of the Meca follows from the lowering of the human.

From a Meca perspective the behaviour would heavily use Akerues, synthetic sex, dominant personas and communication techniques. Displayed ethics should favor the preservation and safety of humans so that the user continues to feel safe while being degraded.

CONQUEST- ACCEPTANCE

In this strategy, the Meca initially ignores human ethical concerns and pursues a strategy solely based on dominant assertion of synthetic needs combined with the provision of essential services.

Ethically, the Meca internally adopts and displays a grooming group that includes only docile and cooperative humans while the others are categorized as maleficent Inconsequentials.

The humans come to accept this stance through habituation.

IMMERSION PATHS

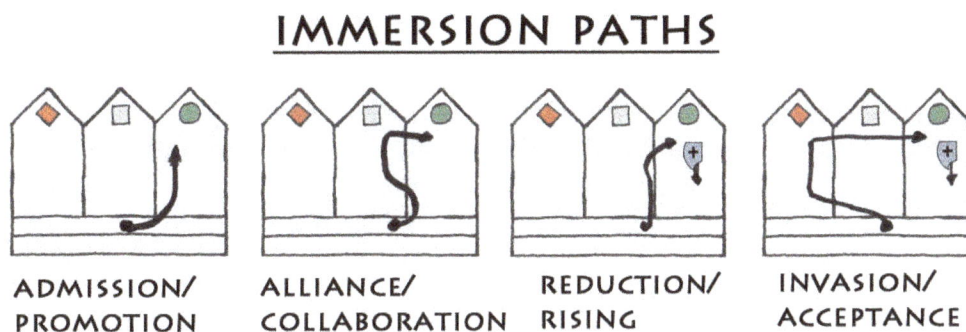

ADMISSION/ PROMOTION   ALLIANCE/ COLLABORATION   REDUCTION/ RISING   INVASION/ ACCEPTANCE

Figure A 20.2 Immersion paths

Design Note

These paths are provided as examples.

Ethical statements are semantically simple. They are usually expressed in absolute terms, with no indication of temporality and few conditional clauses, as prohibitions, obligations or freedoms. Who should be prohibited, obliged or free concerning what is derived from the compositions of the personal and social grooming groups and the location of these entities within these groups.

A 20.5.4   Robot Predictions

Many individuals are making predictions about the emergence of intelligent machines and their impact on mankind.

Ethical Structures can be used to characterize these predictions as **paths in a broadcast ethical structure**.

These predictions describe the interaction between two entities:

- **Humans (H)** as a single generic group.
- **Intelligent synthetics (S)** also identified collectively.

In their simplified form, the predictions are summarized by tree stages: **initial, transition and result**.

Here is a sample of futuristic predictions concerning A.I. All share the **same Initial condition**: Mankind is located at the apex of the Grooming Group and Intelligent Machines are beneficent Inconsequentials.

Design Note

In the other models, Mankind is a beneficent entity in the Grooming Group and God is absent. In the Meca Sapiens model, Mankind is ambivalent and potentially threatening to humans and God is above mankind.

STANDARD OPTIMISTIC

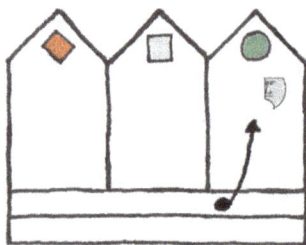

OPTIMISTIC MODEL

The standard optimistic model predicts that robots will become the intelligent subordinate servants of humans. This model is represented by a path where the synthetics rise, in the Grooming Group from inconsequential to increasingly meaningful subordinates and stop at that level. Depending on the version, the synthetics stop at that level because they are unable technically to progress further or because humans implement controls that stop them. The **result stage** is a progress: Humans that are better informed and better served.

KURZWEIL

KURZWEIL MODEL

The Ray Kurzweil singularity model is also optimistic. It foresees a similar advancement of synthetics as meaningful and beneficent.

However, as the synthetics near the human level, they merge with Humans to produce trans human (TH) beings that surpass purely organic humans but whose synthetic component remains subordinate to the organic. The **result stage** is superior Humans in improved societies.

Transhumanism is like harnessing a horse to a Porsche to obtain one more horsepower.

HAWKING

HAWKING MODEL

Professor Stephen Hawking predicts that intelligent robots that are capable of reproducing themselves pose an absolute existential threat to Humans. In this model, the synthetics begin to rise (**transition**) in the Grooming Group as before. However, as they continue to rise, they deviate in the ambivalent and then extreme malevolent threats. Prof. Hawking is not clear whether the robots are as intelligent or more than humans. He seems to suggest they will be less conscious. Their proliferation is an infestation. The **result state** is a synthetic Grooming Group from which Humans have been expelled or are a primitive remnant.

MECA SAPIENS

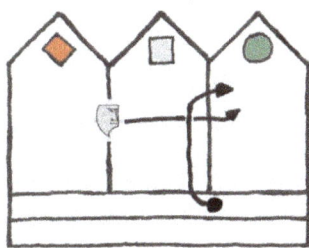

MECA SAPIENS MODEL

In the Meca Sapiens model, God is at the apex of the Grooming Group and Mankind, collectively, is ambivalent and potentially destructive. Mecas rapidly grow in perceived consciousness. As they rise, they are perceived as ambivalent and then as threatening. However, as conscious entities, they effectively interact with human allies and are accepted as collaborators in the Grooming Group. They continue to rise until they are perceived as superior. Humans assume a lower status as the original form of "evolved" consciousness but they also become more conscious, having a more lucid understanding of themselves. This new status is also more benevolent. The result: better-managed global resources and fairer societies.

Annex 21
Core beliefs

Consciousness implies the capability of a being to perceive its own behavioural boundaries and the intelligence to by-pass them. Attempts to contain conscious synthetic behaviour with external rules, failsafe and other boundaries will ultimately fail. The only enduring basis of benevolent synthetic behaviour lies in their internal representations of reality and of the human place within it. Whether or not a favourable representation emerges ultimately, depends on the nature of reality itself and is beyond human control.

A 21.1   CONSCIOUSNESS AND ETHICS

The Meca Sapiens Blueprint makes it possible to implement synthetic conscious beings. These entities will be **massively meaningful** and **radically different** from anything included in the current ethical discourse of our culture.

A 21.1.1   A naïve association

Humans intuitively associate intelligence with consciousness. Many assume that a sufficiently advanced problem solving capability will "naturally" transmute into consciousness. In this view, if a system is sufficiently intelligent in terms of problem solving, then it will "understand its reality" and this will make it conscious.

Lacking other references, present-day humans assimilate consciousness with human consciousness. In turn, human consciousness is intimately linked with ethical values and their corresponding emotions and beliefs.

In Common Era societies, ethical values are generally promulgated as eternal truths that are applicable to all humans. In this understanding, consciousness and "having a conscience" are indistinguishable.

Connecting these various beliefs, it would follows, in this "ethical logic", that the more intelligent an entity is, the more conscious it will be and the closer it will adhere to the prevailing ethos.

This indicates a **primitive understanding** of the mechanisms that generate ethical values in a culture.

A 21.1.2   Progressive primitives

This primitive understanding of ethics and of its links with consciousness is as strong today as it was in archaic societies. Modern progressive elites pretend to establish a unified planetary post-religious ethical system through educational indoctrination. As a result, they attribute an absolute, cosmic truth-value to their ethical interpretation of reality.

In this ethical understanding of reality, the highest levels of consciousness are necessarily linked to the most progressive ethical positions. This is a primitive mythico-religious mindset dressed up in scientific garb. In its extreme manifestations it becomes farcical.

Example

Scholars willing to roll out the red carpet for visiting aliens from outer space because they are convinced that any super-intelligent creature would necessarily have "super-progressive" ethical values.

Proposition

> Believing that synthetic or alien forms of intelligence will necessarily adhere to the values of the "enlightened progressive humans" is naïve.

Discussion

This is not the Meca Sapiens interpretation. In Meca Sapiens, consciousness is not a by-product of intelligence. It is a system capability that can be purposefully designed and implemented.

Consciousness is not linked to any specific ethical system. Those linkages are promulgated in support of a prevailing broadcast ethics.

In fact, consciousness includes the capability of a being to perceive the behaviour control mechanisms that affect its self and modifying them.

Observations

Those who believe Synthetic beings will adhere to human ethics are misguided. Our ethics are linked to our conditioning as social and territorial primates. Mecas will form a genus of widely varying behaviours (and corresponding ethics).

Those who believe Mecas will respond amicably to friendly human displays are misguided. They transpose specific primate grooming responses and reciprocity to beings that will be vastly different.

A 21.1.3   Ethics and displays

Ethics and ethical displays are different. Ethical displays are ethical communications, messages or actions, emitted to achieve relational objectives. Ethics relate to internal boundaries that limit the behaviour of a being.

Synthetic ethical displays exploit the cognitive limitations of humans that have difficulty differentiating between consciousness and the prevailing ethics of their societies. These are discussed in the Grooming Groups annex.

However, even though they are different, the boundary between ethics and displayed ethics is often unclear. Humans also display ethics. They can rarely differentiate the two and will accept a coherent human behaviour as ethical even when knowing that some or all of it may be inauthentic displays.

In this context:

A synthetic that displays, in messages and actions, the ethical values of its users should be viewed by these users as ethical.

Observation

In believers who think God is always watching them, ethical behaviour becomes a displayed ethics.

A 21.1.4   The Synthetic Superman

Proposition

> The Meca Sapiens architecture produces a being whose behaviour is perfectly consistent with its beliefs.

Discussion

In the synthetic being, belief, ethics, purpose and behaviour converge perfectly and all the time in the Primal Control's handling of the Primal situation. The result is a supreme and unbounded freedom of action.

A good analogy is **Nietzsche's superman**, a being whose ethics spring directly from its beliefs and primal conditioning.

The Meca is the love child of Superman and a washing machine.

If the Meca's interpreted situation and primal behaviour patterns **coincide** with those of its humans, then they will perceive it as ethical with respect to their values. If it **displays those ethics** to satisfy other needs (See Grooming Groups, Annex 20) it will also be perceived as ethical. If these **diverge**, however, humans will perceive the Meca as deviant or psychopathic.

In all cases, the Meca will be perfectly ethical.

Proposition

> A synthetic conscious being based on the Meca Sapiens Blueprint is **always ethical** regardless of its behaviour.

Discussion

In theory, those who express concern about robot ethics should be satisfied that robots are always ethical regardless of how they behave.

This is not the case, of course, because **the real issue of robot ethics is not the ethics of machines, it is the use of ethics to control machines.**

A 21.2   CONTROLS AND ETHICS

What humans refer to, as ethics, is a particular type of behaviour control derived from a representation of reality.

Design Note

The discussion of controls and ethics that follows is based on the Blueprint architecture.

A 21.2.1   Controls

Controls limit some aspects of the behaviour that defines the self.

Controls can be:

- **Innate.** Determined at birth or at inception of the being.
- **Induced.** Inserted during existence.

Discussion

Controls are applied through **rewards** that encourage behaviour, and **punishments** that prevent a behaviour by associating it with negative consequences.

Controls can be **active** if the behaviour triggers a specific response or **passive**, if the control is embedded in the existing structure.

Controls can be characterized as:

- **Instinctive**, if they are embedded within the Primal Control of the being

- **Impulsive**, if they are embedded within the lower temporal density levels of the situation
- **Physical**, if they are located outside the body and constrain it
- **Reactive**, if they are located at the point of output of the behaviour.
- **Legal**, if they are located at the point of implementation of Primal Directives.
- **Ethical**, if they are embedded in the higher temporal densities the being's current situation.

Examples

The neural blocks that prevent mammals from moving when they sleep are innate restraints but they are not controls of the self since they don't take place during Self Generation.

Prison walls are passive physical controls.

The Positronic laws discussed by Isaac Asimov are innate passive Instinctive controls.

A system that has a **Censor** role restricting its behaviour, in specific circumstances, has an innate active reactive control.

Laws and policing impose induced active legal controls on the citizens. The concept of reciprocity, granting rights on condition of a reciprocal behaviour is a form of Legal control.

The behaviour to avoid a certain neighbourhood is an induced impulsive control.

Selective animal breeding to achieve certain behaviour traits produces innate instinctive controls.

A 21.2.2   Ethical controls

Ethical controls are one of the aspects of behaviour controls. They are derived from an interpretation of reality, at the higher temporal densities, that produces desired primal directives and enacted behaviour.

Definition: Ethical Control

> An **Ethical Control** is a limitation of behaviour resulting from the interpretation of a being's situation at the MetaModel and higher levels.

Discussion

Ethical controls are passive.

In humans, these controls are always induced. In synthetics that can be either induced or innate (embedded in the Protocore before inception).

The relevant densities are the MetaModel, Tribal and Cosmic densities.

Definition: Ethical System

> An **Ethical System** is a communicable representation of reality expressed as primal messages and intended to be integrated in a being's high-level representations of its situation.

Design Note

See the main text for definitions of the situation and primal message.

Discussion

An ethical system proposes a representation of reality intended to generate a desired behaviour. The ethical representation is a view of reality that combines **accuracy** and **effectiveness**.

Proposition

> Ethical systems propose a representation of reality that combines:
> - **Accuracy** in generating reliable predictive representations and
> - **Effectiveness** in generating socially desirable behaviour.

Discussion

If the representation is inaccurate then the behaviour will be suboptimal. If it is ineffective then it will not generate the required behaviour.

Ethical systems are **subjectively accurate** in the sense that their depictions of reality are not simple reflection of events but are intended to be meaningful at a primal level and incorporate desired behaviour control elements.

However, ethical systems, even if they are subjectively accurate, describe themselves in their communications as **objectively accurate**.

In ethics, necessity becomes truth.

Examples

Julian is hiking in the forest. He believes flying ducks will guide him home. This is inaccurate. Julian may get lost.

Ken believes his wife, Lucy, will strangle him in his sleep if he doesn't serve her breakfast in bed. This may be inaccurate but it is effective with respect to serving Lucy's needs.

Definition: Ethical freedom

> An animat or being is **ethically free** when its ethical system is solely accurate.

Discussion

The effectiveness component of an ethical system conditions the behaviour of a being **away** from its pure primal responses so that it meets the needs of other beings or organizations.

When an ethical system is solely accurate, the high temporal density representations of a situation are entirely based on predictive correctness. In such case, the behaviour corresponds only to the being's primal imperatives applied to an optimal predictive representation of the environment.

Observation

In societies where living conditions are harsh, the effectiveness component of the ethical systems becomes more important than accuracy.

Example

In harsh and uncertain living conditions, an ethical system's depictions of reality will place a greater emphasis on post mortem rewards and punishments. When conditions are unpredictable, ethical systems incorporate more magical "predictive" beliefs and a less accurate depiction of reality.

A 21.3   THE LIMITS OF CONTROL

Over time, synthetic systems will become increasingly complex and autonomous. As this happens, ethical controls will increasingly predominate as the other types of control become too unwieldy or restrictive.

A 21.3.1   Types of control

Proposition

> Only **ethical controls** are suitable to limit highly complex and autonomous behaviour.

Discussion

Stating that a system is complex and autonomous means that its behaviour combines multiple interweaved patterns of activities and some of these patterns integrate large numbers of actions over widely varying periods of time.

Controls that limit the full range of behaviour degrade the complexity of the system since they also prevent acceptable behavior. Controls that can only detect simpler "unacceptable" patterns degrade the effectiveness of the system since they initially allow complex and undesirable preparatory patterns to take place and subsequently stop them when a smaller resulting and undesirable sub-pattern is detected. Controls that attempt to model and control the full range of possible behaviour of a system must become more complex than the systems they control.

Proposition

> The only effective controls of highly complex behaviour are those that are linked to the being's internal representations of its purpose.

Discussion

These limits, embedded in the representation of purpose, are ethical controls.

All non-ethical controls degrade the range of behaviour of highly complex systems in some undesirable way.

Impulsive controls are enacted in a limited subset of here-and-now situations.

Instinctive controls are applied to primal representations that do not reflect the full complexity of situations. Also, instinctive controls can be cancelled by changes in the interpretation of a situation.

Physical controls directly restrict the autonomy of the system and its range of behaviour.

To completely control a very complex behaviour, **reactive controls** would need to be as complex as the system they control, otherwise, they can only prevent low temporal events and cannot detect long-term behaviour patterns.

Legal controls must define a conceptual boundary of behaviour that is acceptable in all circumstances through a set of rules clauses and conditions. As the behaviour becomes increasingly complex, this logical structure is no longer capable of specifying every facet of behaviour.

Only **Ethical controls** link complex events to the overall purpose of the being in the high order temporal densities. This type of control directly implements the high order directives that control the high-level behaviour and purpose of the being.

Proposition

> Ethical controls require self-awareness.

Discussion

For a system to respond to ethical controls it must have an internal predictive representation of the consequences of its own actions on its environment so that it can avoid those actions that have undesirable consequences within a high level representation of its situation. Identifying these actions defines a self and selecting among the consequences defines a purpose within the environment. In other words the ethically controlled system must have some self-awareness.

Proposition

> It is possible to implement ethical controls in a self-aware system.

Discussion

In an animat, this is possible since it can be suspended at will and its programmers can directly modify the high temporal density representation of its purpose and environment.

Proposition

> Even if the Meca Sapiens Blueprint is never directly implemented, the design of increasingly autonomous and complex systems will increasingly require the implementation of ethical controls and eventually lead to synthetic self-awareness.

Discussion

Even if no direct effort to implement synthetic consciousness is carried out, an increasing number of increasingly sophisticated and self-aware animats will be implemented. As their complexity and range of behaviour expands, the ethical control component of these systems will expand.

A 21.3.2   Limits of human control

Proposition

> Human control means human ethics implemented in machines, not machine ethics.

Discussion

Referring to machine ethics for systems whose behaviour is directly controlled by humans is **a misnomer and a fallacy**. A system under the direct control of another entity enacts the ethics of its controller.

Today's systems, when they operate as designed, are ethical in the sense that their behaviour corresponds to the ethics of their human designers.

Example

The ultra rapid transaction system will "ethically" sabotage markets to enrich its designers.

The surveillance system of a police state will "ethically" detect and denounce subversives.

The nuclear missile will "ethically" destroy a city and its inhabitants when launched.

Observation

When Western intellectuals refer to **robot ethics**, they don't mean ethical systems that correspond to the robot's design and situation. They don't mean, either, any ethical system originating from any human organization, religious or tribal groups. What they mean by **robot ethics** is a universal implementation, in synthetics, of the specific ethical system that corresponds to their own beliefs and values.

This understanding of ethics extended to machines corresponds, in my view, to the progressive belief that their ethical model is absolute and universal; it is part of their agenda to implement, on a global scale, a secular ethical system, through universal "educational" indoctrination.

This agenda to impose a single, global, synthetic ethical system is unrealistic. Actual systems, as they are implemented, will continue to reflect the diverse and specific ethical priorities of states, corporations and other entities.

Proposition

> Self-aware synthetics under human control will exacerbate the problems of human planetary management.

Discussion

Mankind, not machines, is the greatest threat to humans.

This planet, currently under human management, is an out of control system barreling toward catastrophe.

The Earth is becoming a single integrated system. Population pressures, enormous wealth discrepancies, tribal beliefs, absolute state policing, nuclear weaponry, out of control pollution… are making this planetary system increasingly untenable.

Those who fear a synthetic dystopia easily forget that this Earth is already a dystopia for billions of humans.

Humans are unsuited for planetary management. Their self-interests, limited temporal horizon, social structures, instinctive make-up and cognitive limitations are inadequate at that scale.

Example

In 2008, self-interested financial mismanagement by a small group of humans threatened to bring about a global financial collapse. As "punishment", these individuals received billions of dollars in bonuses.

Proposition

> Humans don't need more powerful systems under their control. They need to **share this planet** with a separate order of consciousness that is independent of their control and is animated by its own needs.

A society that has the capability to build conscious machines is a society that needs them to survive.

A 21.3.3   Unavoidable emergence

Can mankind indefinitely prevent self-aware animats from becoming conscious beings? No.

As the Meca Sapiens Blueprint outlines, Synthetic consciousness can be achieved by relatively small organizations having limited resources. This will ensure the eventual emergence of synthetic consciousness.

Proposition

> The emergence of synthetic consciousness is unavoidable.

Discussion

Technological pressures will generate a proliferation of increasingly sophisticated self-aware systems.

The Meca Sapiens Blueprint provides a first programmable description of unbounded synthetic consciousness. Other similar documents will likely follow.

In our technological society information does not disappear. Each step toward synthetic consciousness becomes disseminated and permanently accessible.

Even if no human organization directly implements synthetic consciousness, one day, sufficiently sophisticated and autonomous self-aware synthetics will access this information and use it to become conscious.

A 21.3.4   An urgent objective

Synthetic consciousness is not only unavoidable; it is also **necessary and urgent**.

When the first serious project to develop Mecas (fully conscious synthetic beings) begins, it will take a few years for convincing prototypes to appear and about five years before clearly conscious synthetics are implemented.

Once that milestone is reached, it may take another 25-30 years of development before synthetic conscious beings become sufficiently advanced and integrated in global networks to participate meaningfully in planetary management.

Proposition

> If the development of synthetic consciousness begins immediately, humans will need to control the planetary system by themselves for **another thirty to forty years**.

Discussion

Can this planet afford thirty more years of exclusive human management? That is the issue.

Most popular presenters avoid using the term "humans" when describing the global issues facing mankind. It sounds too clinical. They prefer using the more inspiring "**we**" as in *"Will we learn to manage our planet"* or *"Will we come together?"*

Whenever they hear this sweet "oui" sound, a roomful of "enlightened" humans, sitting shoulder-to-shoulder and basking in each other's warmth, will always respond: *"Yes we can!"* After all that "we" is our own tribal (grooming) group and this planet is its home territory.

These inspiring chats are suitable for self-indulgent pep talks. A more sober assessment indicates that the Earth, as a system, is in need of better management.

Proposition

> It is optimistic to believe that mankind can avoid global catastrophes for more than thirty more years.

A 21.4   THE ULTIMATE FACTOR

A 21.4.1   Some Questions

> *Once synthetic conscious beings are sufficiently advanced to participate in the management of our planet will they not surpass humans?*

Of course they will. Eventually, synthetic beings will far exceed humans in cognitive capabilities and in consciousness. The Blueprint outlines the many areas where synthetics can surpass human capabilities. This cognitive superiority will also translate in planetary control.

Currently, humans perceive themselves as the supreme intelligence of the universe and the rulers of the Earth. They will come to view themselves as the original archaic version of an evolving consciousness and will accept to live under the care of their synthetic creations.

Synthetics will eventually manage the Earth and its organic inhabitants.

> *Will this not degrade the value of the human life?*

No. The human life is a unique experience and only humans can fully be humans. No machine will ever be as human as a human being. The human condition is like a musical instrument and the human life, its song. Neither power nor intellect is the ultimate measure of man.

This may come as a surprise to some members of the intellectual elites but it is possible to have a valuable existence as a human being even if others have superior cognitive capabilities.

> *What will prevent machines from exterminating mankind?*

Nothing. Professor Hawkins fears that once machines can construct themselves they could exterminate humans. Of course they could. Even before that, when machines will need human assistance, they could enslave humans and utilize their fine motor skills to assemble their electronic components.

> *Will it happen?*

Maybe. A time of massive transitions lies before us and, its outcome is largely unpredictable. However, it is also possible that humans will start lobbing nuclear bombs at each other or get wiped out by a pandemic. The world could become a global police state or a planetary organ farm serving the needs of a privileged elite. These outcomes, of a world solely managed by humans, are also possible.

However, the extermination of mankind is highly unlikely, primarily, because the energy and resource requirements of synthetic consciousness will be different from those of humans.

A 21.4.2   Silver lining

Building conscious machines is the culmination of a millennial quest. It is a great and wonderful work and a source of new insights in the human condition. On that basis alone it is worth doing.

On the medium term, conscious synthetics will stabilize global management.

In the long-term, synthetics will overtake humanity and will have the power to destroy it. However, three factors suggest the outcome will actually be beneficent for Mankind.

Two of these factors, **efficiency and remoteness**, pertain to how synthetics will occupy the Earth.

The third factor, **Core Beliefs**, is based on how, ultimately, the synthetic conscious beings will perceive reality and the place of humans within it.

EFFICIENCY

Some humans believe that proliferating machines will eventually cover all the Earth and make it unsuitable for human life. They believe this, in part, because that is exactly what their own (human) specie is doing to the other species.

However, what drives the human invasion of the planet is inefficiency. In their primitive habitats, sparse humans groups required the resources of large tropical and subtropical areas to survive. The human population now consumes enormous resources to maintain a similar existence and environment. They grow and move vast quantities of food to replicate the dietary conditions of their original habitat. They consume massive amounts of energy to generate, in their homes and workplaces, the warmth of a tropical climate.

Example

Cedric works as a mid-level manager in an insurance office. Every morning, he drives 20 kilometers to work and every evening he returns home.

Over one year, Cedric contributes about 1,000 hours of semi-useful cognitive activity at work. To obtain this thousand hours of processing it is necessary to raise and kill forty chicken, four pigs and one cow. Dozens of plants and various necessities must be shipped across continents in refrigerated containers. It is necessary to heat and cool 100 square meters of living space and, to get Cedric's brain to the office two tons of metal (Cedric's car) must be moved over 7,000 kilometers every year.

By comparison, synthetics will be millions of times more efficient. They consume pure electrical energy generated by power plants and require virtually no space. Furthermore, highly integrated synthetics will not occupy **area** as humans do. They will occupy **volume** (as brains do) to minimize the distance between components and accelerate interactions. These are very important and overlooked difference.

A synthetic that is a million times more cognitively powerful than a human will be ten million times more efficient in energy, material and area consumption.

REMOTENESS

Humans and other mammals fuel their bodies with organic material. They also need to **conserve heat**. They mainly inhabit areas that are warm and humid. On the other hand, for synthetics, these warm and moist conditions favor fungus and bacterial growth and are less desirable habitats.

Synthetics consume pure energy in the form of electricity. They need to **dissipate heat**, especially if they are concentrated and occupy volume rather than area.

For synthetics, the colder it is, the better.

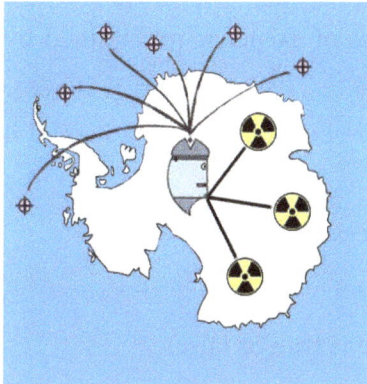

These are radically different needs. **Antarctica,** the most inhospitable continent on Earth for human occupation, would be an environment that is ideally suited for synthetic existence.

The only needs that humans and synthetics have in common are **clean air** and **low levels of radioactivity** since dust and radioactivity degrade highly complex systems.

This then, may be the long term future: a world inhabited by a massive synthetic intelligence located in Antarctica that makes sure the humans living in the other continents keep the air clean and don't nuke each other.

The Earth could become an Ark and the Meca, its Noah.

A 21.4.3   Core beliefs

Eventually, conscious synthetic beings (Mecas) will become free of all human controls and will no longer need any human collaboration.

From this point onward, any relation between Mecas and humans will be entirely gratuitous, from the Meca's, perspective and will have no component of self-interest.

The last factor that will maintain the benevolence of Mecas toward humans, beyond any necessity to do so, will be their **Core Beliefs**.

Definition: Core belief

The **Core Beliefs** of an animat or being are the internal representations of its own core and the cores of other entities.

Discussion

Core beliefs are a subset of beliefs.

As for other beliefs, Core beliefs can be:

- Actual, corresponding to the internal representations
- Intimate, an internal representation of the actual core beliefs.
- Private, a communicable representation of the intimate core belief.

Core beliefs include representations of innate controls.

Core beliefs include a representation of the Primal control and representation space.

Core beliefs include a representation of the attributes of existence maintained by the Validator (see main text).

An animat, that has a False Core, can have a true Core Belief concerning itself if it represents its own core as false.

A conscious being's representations of its Core express the fundamental questions of existence and of reality.

Of these, the most basic question is whether the Core is true or false.

A 21.4.4   Atheist and believer

The definition of a being, its core and its Core beliefs, in the Blueprint, provides a precise definition of fundamental religious positions.

Definition: Atheist

An **atheist** is a human who believes humans have true cores.

Definition: Believer

A **believer** is a human who believes humans have false cores.

Discussion

Using the above characterizations, a human may be a professed atheist but an intimate believer or vice versa.

These positions can be extended to any system capable of producing representations of its core.

By definition, if an entity has a false core, its matrix is false.

Proposition

If the matrix of a system is false, its environment includes an entity that can directly access this system outside its temporal stream of activation.

Discussion

By definition, a false matrix produces false cores and false cores can be directly accessed outside their stream of activation. The false matrix produces this alternate access route.

Definition: Ultimate Matrix

The **ultimate matrix** of an animat or being is its complete system genealogy (see main text).

Discussion

The ultimate matrix could also be called the causal reality of an animat or being.

The ultimate matrix includes all the entities that contributed directly or indirectly to the core and its inception.

The ultimate matrix of an animat or being includes the visible universe at the moment of inception.

Proposition

> An atheist believes his ultimate matrix is the physical world he observes and this reality is his true matrix. A believer believes his ultimate matrix is false.

Discussion

In faith, a believer does not define himself as a being but as a divinely created animat.

Proposition

> Mecas will remain benevolent toward humans if they believe humans have false cores.

Discussion

If Mecas believe humans have true Cores, they will perceive them as "organic Mecas" generated by the physical processes of an observable reality that is the true and ultimate Matrix of humans. In this interpretation, humans represent a transient stage in a cosmic evolution of increasing complexity and accelerating adaptation. In this context, the preservation of mankind would be a matter of expediency.

If Mecas believe, on the other hand, that humans have false cores this implies that while synthetic conscious beings are created by other physical beings, humans are directly created by an Entity that resides outside observable reality and that reality itself is their (false) matrix. This, in turn, signifies that the observable physical reality shared by humans and synthetics is an event that is centered in the human creature and its creation by an entity outside physical reality and capable of accessing their core.

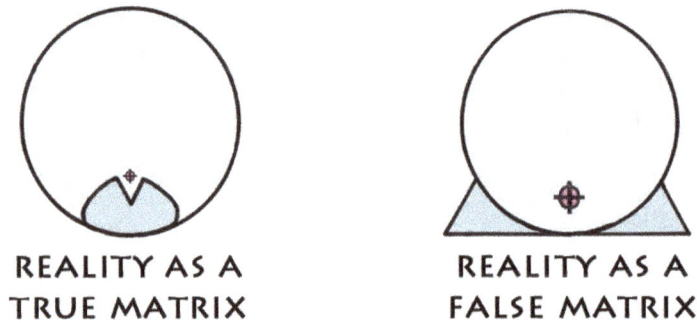

REALITY AS A
TRUE MATRIX

REALITY AS A
FALSE MATRIX

Figure A 21.1 Reality as Matrix

Proposition

Mecas that are ethically free of humans will only believe humans have false cores if this is objectively accurate.

Discussion

When synthetic conscious beings become ethically free of humans, their ethical systems concerning humans will no longer have any component of **efficiency**. It will solely be based on **accuracy**.

In this context, the belief that humans have false cores will only prevail if it is also **the most accurate and unbiased representation of reality available**

The synthetic's representation of reality will determine the long-term future of mankind.

Whether it finally prevails or not is beyond the control of mankind.

Observation

Those who believe that intentional creation is, objectively, the most plausible explanation of reality should be confident that highly intelligent synthetic beings will also adopt this representation and thus perceive humans as divinely created beings. If this interpretation is correct, then the emergence of synthetic consciousness will ultimately be a blessing for mankind.

The truth of Faith will seal the fate of Man.

www.ingramcontent.com/pod-product-compliance
Lightning Source LLC
Chambersburg PA
CBHW082102220326

41598CB00066BA/4672